THE KURZWEIL-HENSTOCK
INTEGRAL AND
ITS DIFFERENTIALS

PURE AND APPLIED MATHEMATICS

A Program of Monographs, Textbooks, and Lecture Notes

MONOGRAPHS AND TEXTBOOKS IN
PURE AND APPLIED MATHEMATICS

1. *K. Yano,* Integral Formulas in Riemannian Geometry (1970)
2. *S. Kobayashi,* Hyperbolic Manifolds and Holomorphic Mappings (1970)
3. *V. S. Vladimirov,* Equations of Mathematical Physics (A. Jeffrey, ed.; A. Littlewood, trans.) (1970)
4. *B. N. Pshenichnyi,* Necessary Conditions for an Extremum (L. Neustadt, translation ed.; K. Makowski, trans.) (1971)
5. *L. Narici et al.,* Functional Analysis and Valuation Theory (1971)
6. *S. S. Passman,* Infinite Group Rings (1971)
7. *L. Domhoff,* Group Representation Theory. Part A: Ordinary Representation Theory. Part B: Modular Representation Theory (1971, 1972)
8. *W. Boothby and G. L. Weiss, eds.,* Symmetric Spaces (1972)
9. *Y. Matsushima,* Differentiable Manifolds (E. T. Kobayashi, trans.) (1972)
10. *L. E. Ward, Jr.,* Topology (1972)
11. *A. Babakhanian,* Cohomological Methods in Group Theory (1972)
12. *R. Gilmer,* Multiplicative Ideal Theory (1972)
13. *J. Yeh,* Stochastic Processes and the Wiener Integral (1973)
14. *J. Barros-Neto,* Introduction to the Theory of Distributions (1973)
15. *R. Larsen,* Functional Analysis (1973)
16. *K. Yano and S. Ishihara,* Tangent and Cotangent Bundles (1973)
17. *C. Procesi,* Rings with Polynomial Identities (1973)
18. *R. Hermann,* Geometry, Physics, and Systems (1973)
19. *N. R. Wallach,* Harmonic Analysis on Homogeneous Spaces (1973)
20. *J. Dieudonné,* Introduction to the Theory of Formal Groups (1973)
21. *I. Vaisman,* Cohomology and Differential Forms (1973)
22. *B.-Y. Chen,* Geometry of Submanifolds (1973)
23. *M. Marcus,* Finite Dimensional Multilinear Algebra (in two parts) (1973, 1975)
24. *R. Larsen,* Banach Algebras (1973)
25. *R. O. Kujala and A. L. Vitter, eds.,* Value Distribution Theory: Part A; Part B: Deficit and Bezout Estimates by Wilhelm Stoll (1973)
26. *K. B. Stolarsky,* Algebraic Numbers and Diophantine Approximation (1974)
27. *A. R. Magid,* The Separable Galois Theory of Commutative Rings (1974)
28. *B. R. McDonald,* Finite Rings with Identity (1974)
29. *J. Satake,* Linear Algebra (S. Koh et al., trans.) (1975)
30. *J. S. Golan,* Localization of Noncommutative Rings (1975)
31. *G. Klambauer,* Mathematical Analysis (1975)
32. *M. K. Agoston,* Algebraic Topology (1976)
33. *K. R. Goodearl,* Ring Theory (1976)
34. *L. E. Mansfield,* Linear Algebra with Geometric Applications (1976)
35. *N. J. Pullman,* Matrix Theory and Its Applications (1976)
36. *B. R. McDonald,* Geometric Algebra Over Local Rings (1976)
37. *C. W. Groetsch,* Generalized Inverses of Linear Operators (1977)
38. *J. E. Kuczkowski and J. L. Gersting,* Abstract Algebra (1977)
39. *C. O. Christenson and W. L. Voxman,* Aspects of Topology (1977)
40. *M. Nagata,* Field Theory (1977)
41. *R. L. Long,* Algebraic Number Theory (1977)
42. *W. F. Pfeffer,* Integrals and Measures (1977)
43. *R. L. Wheeden and A. Zygmund,* Measure and Integral (1977)
44. *J. H. Curtiss,* Introduction to Functions of a Complex Variable (1978)
45. *K. Hrbacek and T. Jech,* Introduction to Set Theory (1978)
46. *W. S. Massey,* Homology and Cohomology Theory (1978)
47. *M. Marcus,* Introduction to Modern Algebra (1978)
48. *E. C. Young,* Vector and Tensor Analysis (1978)
49. *S. B. Nadler, Jr.,* Hyperspaces of Sets (1978)
50. *S. K. Segal,* Topics in Group Kings (1978)
51. *A. C. M. van Rooij,* Non-Archimedean Functional Analysis (1978)
52. *L. Corwin and R. Szczarba,* Calculus in Vector Spaces (1979)
53. *C. Sadosky,* Interpolation of Operators and Singular Integrals (1979)
54. *J. Cronin,* Differential Equations (1980)
55. *C. W. Groetsch,* Elements of Applicable Functional Analysis (1980)

56. *I. Vaisman,* Foundations of Three-Dimensional Euclidean Geometry (1980)
57. *H. I. Freedan,* Deterministic Mathematical Models in Population Ecology (1980)
58. *S. B. Chae,* Lebesgue Integration (1980)
59. *C. S. Rees et al.,* Theory and Applications of Fourier Analysis (1981)
60. *L. Nachbin,* Introduction to Functional Analysis (R. M. Aron, trans.) (1981)
61. *G. Orzech and M. Orzech,* Plane Algebraic Curves (1981)
62. *R. Johnsonbaugh and W. E. Pfaffenberger,* Foundations of Mathematical Analysis (1981)
63. *W. L. Voxman and R. H. Goetschel,* Advanced Calculus (1981)
64. *L. J. Corwin and R. H. Szczarba,* Multivariable Calculus (1982)
65. *V. I. Istrătescu,* Introduction to Linear Operator Theory (1981)
66. *R. D. Järvinen,* Finite and Infinite Dimensional Linear Spaces (1981)
67. *J. K. Beem and P. E. Ehrlich,* Global Lorentzian Geometry (1981)
68. *D. L. Armacost,* The Structure of Locally Compact Abelian Groups (1981)
69. *J. W. Brewer and M. K. Smith, eds.,* Emmy Noether: A Tribute (1981)
70. *K. H. Kim,* Boolean Matrix Theory and Applications (1982)
71. *T. W. Wieting,* The Mathematical Theory of Chromatic Plane Ornaments (1982)
72. *D. B.Gauld,* Differential Topology (1982)
73. *R. L. Faber,* Foundations of Euclidean and Non-Euclidean Geometry (1983)
74. *M. Carmeli,* Statistical Theory and Random Matrices (1983)
75. *J. H. Carruth et al.,* The Theory of Topological Semigroups (1983)
76. *R. L. Faber,* Differential Geometry and Relativity Theory (1983)
77. *S. Barnett,* Polynomials and Linear Control Systems (1983)
78. *G. Karpilovsky,* Commutative Group Algebras (1983)
79. *F. Van Oystaeyen and A. Verschoren,* Relative Invariants of Rings (1983)
80. *I. Vaisman,* A First Course in Differential Geometry (1984)
81. *G. W. Swan,* Applications of Optimal Control Theory in Biomedicine (1984)
82. *T. Petrie and J. D. Randall,* Transformation Groups on Manifolds (1984)
83. *K. Goebel and S. Reich,* Uniform Convexity, Hyperbolic Geometry, and Nonexpansive Mappings (1984)
84. *T. Albu and C. Năstăsescu,* Relative Finiteness in Module Theory (1984)
85. *K. Hrbacek and T. Jech,* Introduction to Set Theory: Second Edition (1984)
86. *F. Van Oystaeyen and A. Verschoren,* Relative Invariants of Rings (1984)
87. *B. R. McDonald,* Linear Algebra Over Commutative Rings (1984)
88. *M. Namba,* Geometry of Projective Algebraic Curves (1984)
89. *G. F. Webb,* Theory of Nonlinear Age-Dependent Population Dynamics (1985)
90. *M. R. Bremner et al.,* Tables of Dominant Weight Multiplicities for Representations of Simple Lie Algebras (1985)
91. *A. E. Fekete,* Real Linear Algebra (1985)
92. *S. B. Chae,* Holomorphy and Calculus in Normed Spaces (1985)
93. *A. J. Jerri,* Introduction to Integral Equations with Applications (1985)
94. *G. Karpilovsky,* Projective Representations of Finite Groups (1985)
95. *L. Narici and E. Beckenstein,* Topological Vector Spaces (1985)
96. *J. Weeks,* The Shape of Space (1985)
97. *P. R. Gribik and K. O. Kortanek,* Extremal Methods of Operations Research (1985)
98. *J.-A. Chao and W. A. Woyczynski, eds.,* Probability Theory and Harmonic Analysis (1986)
99. *G. D. Crown et al.,* Abstract Algebra (1986)
100. *J. H. Carruth et al.,* The Theory of Topological Semigroups, Volume 2 (1986)
101. *R. S. Doran and V. A. Belfi,* Characterizations of C*-Algebras (1986)
102. *M. W. Jeter,* Mathematical Programming (1986)
103. *M. Altman,* A Unified Theory of Nonlinear Operator and Evolution Equations with Applications (1986)
104. *A. Verschoren,* Relative Invariants of Sheaves (1987)
105. *R. A. Usmani,* Applied Linear Algebra (1987)
106. *P. Blass and J. Lang,* Zariski Surfaces and Differential Equations in Characteristic $p > 0$ (1987)
107. *J. A. Reneke et al.,* Structured Hereditary Systems (1987)
108. *H. Busemann and B. B. Phadke,* Spaces with Distinguished Geodesics (1987)
109. *R. Harte,* Invertibility and Singularity for Bounded Linear Operators (1988)
110. *G. S. Ladde et al.,* Oscillation Theory of Differential Equations with Deviating Arguments (1987)
111. *L. Dudkin et al.,* Iterative Aggregation Theory (1987)
112. *T. Okubo,* Differential Geometry (1987)

113. *D. L. Stancl and M. L. Stancl,* Real Analysis with Point-Set Topology (1987)
114. *T. C. Gard,* Introduction to Stochastic Differential Equations (1988)
115. *S. S. Abhyankar,* Enumerative Combinatorics of Young Tableaux (1988)
116. *H. Strade and R. Farnsteiner,* Modular Lie Algebras and Their Representations (1988)
117. *J. A. Huckaba,* Commutative Rings with Zero Divisors (1988)
118. *W. D. Wallis,* Combinatorial Designs (1988)
119. *W. Wiesław,* Topological Fields (1988)
120. *G. Karpilovsky,* Field Theory (1988)
121. *S. Caenepeel and F. Van Oystaeyen,* Brauer Groups and the Cohomology of Graded Rings (1989)
122. *W. Kozlowski,* Modular Function Spaces (1988)
123. *E. Lowen-Colebunders,* Function Classes of Cauchy Continuous Maps (1989)
124. *M. Pavel,* Fundamentals of Pattern Recognition (1989)
125. *V. Lakshmikantham et al.,* Stability Analysis of Nonlinear Systems (1989)
126. *R. Sivaramakrishnan,* The Classical Theory of Arithmetic Functions (1989)
127. *N. A. Watson,* Parabolic Equations on an Infinite Strip (1989)
128. *K. J. Hastings,* Introduction to the Mathematics of Operations Research (1989)
129. *B. Fine,* Algebraic Theory of the Bianchi Groups (1989)
130. *D. N. Dikranjan et al.,* Topological Groups (1989)
131. *J. C. Morgan II,* Point Set Theory (1990)
132. *P. Biler and A. Witkowski,* Problems in Mathematical Analysis (1990)
133. *H. J. Sussmann,* Nonlinear Controllability and Optimal Control (1990)
134. *J.-P. Florens et al.,* Elements of Bayesian Statistics (1990)
135. *N. Shell,* Topological Fields and Near Valuations (1990)
136. *B. F. Doolin and C. F. Martin,* Introduction to Differential Geometry for Engineers (1990)
137. *S. S. Holland, Jr.,* Applied Analysis by the Hilbert Space Method (1990)
138. *J. Oknínski,* Semigroup Algebras (1990)
139. *K. Zhu,* Operator Theory in Function Spaces (1990)
140. *G. B. Price,* An Introduction to Multicomplex Spaces and Functions (1991)
141. *R. B. Darst,* Introduction to Linear Programming (1991)
142. *P. L. Sachdev,* Nonlinear Ordinary Differential Equations and Their Applications (1991)
143. *T. Husain,* Orthogonal Schauder Bases (1991)
144. *J. Foran,* Fundamentals of Real Analysis (1991)
145. *W. C. Brown,* Matrices and Vector Spaces (1991)
146. *M. M. Rao and Z. D. Ren,* Theory of Orlicz Spaces (1991)
147. *J. S. Golan and T. Head,* Modules and the Structures of Rings (1991)
148. *C. Small,* Arithmetic of Finite Fields (1991)
149. *K. Yang,* Complex Algebraic Geometry (1991)
150. *D. G. Hoffman et al.,* Coding Theory (1991)
151. *M. O. González,* Classical Complex Analysis (1992)
152. *M. O. González,* Complex Analysis (1992)
153. *L. W. Baggett,* Functional Analysis (1992)
154. *M. Sniedovich,* Dynamic Programming (1992)
155. *R. P. Agarwal,* Difference Equations and Inequalities (1992)
156. *C. Brezinski,* Biorthogonality and Its Applications to Numerical Analysis (1992)
157. *C. Swartz,* An Introduction to Functional Analysis (1992)
158. *S. B. Nadler, Jr.,* Continuum Theory (1992)
159. *M. A. Al-Gwaiz,* Theory of Distributions (1992)
160. *E. Perry,* Geometry: Axiomatic Developments with Problem Solving (1992)
161. *E. Castillo and M. R. Ruiz-Cobo,* Functional Equations and Modelling in Science and Engineering (1992)
162. *A. J. Jerri,* Integral and Discrete Transforms with Applications and Error Analysis (1992)
163. *A. Charlier et al.,* Tensors and the Clifford Algebra (1992)
164. *P. Biler and T. Nadzieja,* Problems and Examples in Differential Equations (1992)
165. *E. Hansen,* Global Optimization Using Interval Analysis (1992)
166. *S. Guerre-Delabrière,* Classical Sequences in Banach Spaces (1992)
167. *Y. C. Wong,* Introductory Theory of Topological Vector Spaces (1992)
168. *S. H. Kulkarni and B. V. Limaye,* Real Function Algebras (1992)
169. *W. C. Brown,* Matrices Over Commutative Rings (1993)
170. *J. Loustau and M. Dillon,* Linear Geometry with Computer Graphics (1993)
171. *W. V. Petryshyn,* Approximation-Solvability of Nonlinear Functional and Differential Equations (1993)

172. *E. C. Young*, Vector and Tensor Analysis: Second Edition (1993)
173. *T. A. Bick*, Elementary Boundary Value Problems (1993)
174. *M. Pavel*, Fundamentals of Pattern Recognition: Second Edition (1993)
175. *S. A. Albeverio et al.*, Noncommutative Distributions (1993)
176. *W. Fulks*, Complex Variables (1993)
177. *M. M. Rao*, Conditional Measures and Applications (1993)
178. *A. Janicki and A. Weron*, Simulation and Chaotic Behavior of α-Stable Stochastic Processes (1994)
179. *P. Neittaanmäki and D. Tiba*, Optimal Control of Nonlinear Parabolic Systems (1994)
180. *J. Cronin*, Differential Equations: Introduction and Qualitative Theory, Second Edition (1994)
181. *S. Heikkilä and V. Lakshmikantham*, Monotone Iterative Techniques for Discontinuous Nonlinear Differential Equations (1994)
182. *X. Mao*, Exponential Stability of Stochastic Differential Equations (1994)
183. *B. S. Thomson*, Symmetric Properties of Real Functions (1994)
184. *J. E. Rubio*, Optimization and Nonstandard Analysis (1994)
185. *J. L. Bueso et al.*, Compatibility, Stability, and Sheaves (1995)
186. *A. N. Michel and K. Wang*, Qualitative Theory of Dynamical Systems (1995)
187. *M. R. Darnel*, Theory of Lattice-Ordered Groups (1995)
188. *Z. Naniewicz and P. D. Panagiotopoulos*, Mathematical Theory of Hemivariational Inequalities and Applications (1995)
189. *L. J. Corwin and R. H. Szczarba*, Calculus in Vector Spaces: Second Edition (1995)
190. *L. H. Erbe et al.*, Oscillation Theory for Functional Differential Equations (1995)
191. *S. Agaian et al.*, Binary Polynomial Transforms and Nonlinear Digital Filters (1995)
192. *M. I. Gil'*, Norm Estimations for Operation-Valued Functions and Applications (1995)
193. *P. A. Grillet*, Semigroups: An Introduction to the Structure Theory (1995)
194. *S. Kichenassamy*, Nonlinear Wave Equations (1996)
195. *V. F. Krotov*, Global Methods in Optimal Control Theory (1996)
196. *K. I. Beidar et al.*, Rings with Generalized Identities (1996)
197. *V. I. Arnautov et al.*, Introduction to the Theory of Topological Rings and Modules (1996)
198. *G. Sierksma*, Linear and Integer Programming (1996)
199. *R. Lasser*, Introduction to Fourier Series (1996)
200. *V. Sima*, Algorithms for Linear-Quadratic Optimization (1996)
201. *D. Redmond*, Number Theory (1996)
202. *J. K. Beem et al.*, Global Lorentzian Geometry: Second Edition (1996)
203. *M. Fontana et al.*, Prüfer Domains (1997)
204. *H. Tanabe*, Functional Analytic Methods for Partial Differential Equations (1997)
205. *C. Q. Zhang*, Integer Flows and Cycle Covers of Graphs (1997)
206. *E. Spiegel and C. J. O'Donnell*, Incidence Algebras (1997)
207. *B. Jakubczyk and W. Respondek*, Geometry of Feedback and Optimal Control (1998)
208. *T. W. Haynes et al.*, Fundamentals of Domination in Graphs (1998)
209. *T. W. Haynes et al.*, Domination in Graphs: Advanced Topics (1998)
210. *L. A. D'Alotto et al.*, A Unified Signal Algebra Approach to Two-Dimensional Parallel Digital Signal Processing (1998)
211. *F. Halter-Koch*, Ideal Systems (1998)
212. *N. K. Govil et al.*, Approximation Theory (1998)
213. *R. Cross*, Multivalued Linear Operators (1998)
214. *A. A. Martynyuk*, Stability by Liapunov's Matrix Function Method with Applications (1998)
215. *A. Favini and A. Yagi*, Degenerate Differential Equations in Banach Spaces (1999)
216. *A. Illanes and S. Nadler, Jr.*, Hyperspaces: Fundamentals and Recent Advances (1999)
217. *G. Kato and D. Struppa*, Fundamentals of Algebraic Microlocal Analysis (1999)
218. *G. X.-Z. Yuan*, KKM Theory and Applications in Nonlinear Analysis (1999)
219. *D. Motreanu and N. H. Pavel*, Tangency, Flow Invariance for Differential Equations, and Optimization Problems (1999)
220. *K. Hrbacek and T. Jech*, Introduction to Set Theory, Third Edition (1999)
221. *G. E. Kolosov*, Optimal Design of Control Systems (1999)
222. *N. L. Johnson*, Subplane Covered Nets (2000)
223. *B. Fine and G. Rosenberger*, Algebraic Generalizations of Discrete Groups (1999)
224. *M. Väth*, Volterra and Integral Equations of Vector Functions (2000)
225. *S. S. Miller and P. T. Mocanu*, Differential Subordinations (2000)

226. *R. Li et al.*, Generalized Difference Methods for Differential Equations: Numerical Analysis of Finite Volume Methods (2000)
227. *H. Li and F. Van Oystaeyen*, A Primer of Algebraic Geometry (2000)
228. *R. P. Agarwal*, Difference Equations and Inequalities: Theory, Methods, and Applications, Second Edition (2000)
229. *A. B. Kharazishvili*, Strange Functions in Real Analysis (2000)
230. *J. M. Appell et al.*, Partial Integral Operators and Integro-Differential Equations (2000)
231. *A. I. Prilepko et al.*, Methods for Solving Inverse Problems in Mathematical Physics (2000)
232. *F. Van Oystaeyen*, Algebraic Geometry for Associative Algebras (2000)
233. *D. L. Jagerman*, Difference Equations with Applications to Queues (2000)
234. *D. R. Hankerson et al.*, Coding Theory and Cryptography: The Essentials, Second Edition, Revised and Expanded (2000)
235. *S. Dăscălescu et al.*, Hopf Algebras: An Introduction (2001)
236. *R. Hagen et al.*, C*-Algebras and Numerical Analysis (2001)
237. *Y. Talpaert*, Differential Geometry: With Applications to Mechanics and Physics (2001)
238. *R. H. Villarreal*, Monomial Algebras (2001)
239. *A. N. Michel et al.*, Qualitative Theory of Dynamical Systems, Second Edition (2001)
240. *A. A. Samarskii*, The Theory of Difference Schemes (2001)
241. *J. Knopfmacher and W.-B. Zhang*, Number Theory Arising from Finite Fields (2001)
242. *S. Leader*, The Kurzweil-Henstock Integral and Its Differentials (2001)

Additional Volumes in Preparation

THE KURZWEIL-HENSTOCK INTEGRAL AND ITS DIFFERENTIALS

A Unified Theory of Integration on $\mathbb{R}$ and $\mathbb{R}^n$

Solomon Leader
Rutgers University
New Brunswick, New Jersey

CRC Press
Taylor & Francis Group
Boca Raton London New York

CRC Press is an imprint of the
Taylor & Francis Group, an **informa** business

CRC Press
Taylor & Francis Group
6000 Broken Sound Parkway NW, Suite 300
Boca Raton, FL 33487-2742

First issued in paperback 2019

ISBN-13: 978-0-8247-0535-0 (hbk)
ISBN-13: 978-0-367-39715-9 (pbk)

Visit the Taylor & Francis Web site at
http://www.taylorandfrancis.com

and the CRC Press Web site at
http://www.crcpress.com

PREFACE

For many years after the introduction of the generalized Riemann integral by Henstock and Kurzweil the importance of this integral was not widely recognized. Even as the generalized Riemann integral became better known, some analysts dismissed it as just another approach to the integrals of Denjoy and Perron. However, it is not the integral that is of primary importance, but the integration process itself.

Both Kurzweil and Henstock applied their integration process to approximating sums whose summands are much more general than those of the classical form $f\Delta x$. On such general objects of integration (called "summants" here) the integration process induces an equivalence relation: Two summants are equivalent if the absolute value of their difference integrates to zero. Limits in the integration process are unaffected if a summant is replaced by an equivalent one.

The central feature of this book is the novel concept of differential as an equivalence class of summants. Exploiting this concept we present here a thorough exposition of Kurzweil-Henstock integration on the real line. There is also an introductory chapter on the higher dimensional case since it has some subtle aspects that do not appear in dimension one. A final chapter offers some background material.

The text includes 180 theorems with proofs. (This count is slightly inflated since some theorems are reformulations of earlier ones.) Many well known theorems are included, but the versions presented here are often more general than the standard versions. Supplementing the text are over 400 exercises to keep the interested reader engaged.

The book is suitable as a textbook for a graduate course on special topics in analysis, or as a supplementary text for a first-year graduate course in real analysis. As a monograph it can be read independently, offering much of value to readers interested in integration theory and the fundamental concepts of calculus.

Solomon Leader

CONTENTS

Preface   **iii**

Introduction

 §0.1 The Gauge-Directed Integral   1
 §0.2 Differentials   5
 §0.3 Guidance for the Reader   6

Chapter 1. Integration of Summants

 §1.1 Cells, Figures and Partitions   9
 §1.2 Tagged Cells, Divisions, and Gauges   11
 §1.3 The Upper and Lower Integrals of a Summant
   over a Figure   15
 §1.4 Summants with Special Properties   21
 §1.5 Upper and Lower Integrals as Functions on the
   Boolean Algebra of Figures   29
 §1.6 Uniform Integrability and Its Consequences   33
 §1.7 Term-by-Term Integration of Series   40
 §1.8 Applications of Term-by-Term Upper Integration   43
 §1.9 Integration over Arbitrary Intervals   45

Chapter 2.   Differentials and Their Integrals

§2.1 Differential Equivalence and Differentials 53

§2.2 The Riesz Space $\mathbb{D} = \mathbb{D}(K)$ of All Differentials on K 55

§2.3 Differential Norm and Summable Differentials 57

§2.4 Conditionally and Absolutely Integrable Differentials 60

§2.5 The Differential dg of a Function g 66

§2.6 The Total Variation of a Function on a Cell K 68

§2.7 Functions as Differential Coefficients 73

§2.8 The Lebesgue Space $\mathcal{L}_1$ and Convergence Theorems 78

Chapter 3.   Differentials with Special Properties

§3.1 Products Involving Tag-Finite Summants and Differentials 87

§3.2 Continuous Differentials 96

§3.3 Archimedean Properties for Differentials 99

§3.4 Differentials on Open-Ended Intervals 106

§3.5 σ-Nullity of the Union of All σ-Null Cells 119

§3.6 Mappings of Differentials Induced by Lipschitz Functions 120

§3.7 n-Differentials on a Cell K 125

Chapter 4.   Measurable Sets and Functions

§4.1 Measurable Sets 129

§4.2 The Hahn Decomposition for Differentials 133

§4.3 Measurable Functions 137

§4.4 Step Functions and Regulated Functions 145

§4.5 The Radon-Nikodym Theorem for Differentials 160

§4.6 Minimal Measurable Dominators 163

Chapter 5.   The Vitali Covering Theorem Applied to Differentials

§5.1 The Vitali Covering Theorem with some Applications to Upper Integrals 169

§5.2 $\nu(1_E df)$ and Lebesgue Outer Measure of $f(E)$ 175
§5.3 Continuity σ-Everywhere of ρ Given $\rho\sigma = 0$ 181

Chapter 6.   Derivatives and Differentials
§6.1 Differential Coefficients from the Gradient   187
§6.2 Integration by Parts and Taylor's Formula   196
§6.3 A Generalized Fundamental Theorem of Cal-
      culus   209
§6.4 L'Hôpital's Rule and the Limit Comparison Test
      Using Essential Limits   222
§6.5 Differentiation Under the Integral Sign   229

Chapter 7.   Essential Properties of Functions
§7.1 Essentially Bounded Functions   235
§7.2 Essentially Regulated Functions   238
§7.3 Essential Variation   241

Chapter 8.   Absolute Continuity
§8.1 Various Concepts of Absolute Continuity for
      Differentials   249
§8.2 Absolute Continuity for Restricted Classes of
      Differentials   253
§8.3 Absolutely Continuous Functions   257
§8.4 The Vitali Convergence Theorem   262

**Chapter 9.   Conversion of Lebesgue-Stieltjes
      Integrals into Lebesgue Integrals**
§9.1 Banach's Indicatrix Theorem   267
§9.2 A Generalization of the Indicatrix Theorem with
      Applications   270

**Chapter 10.   Some Results on Higher Dimen-
      sions**
§10.1 Integral and Differential on n-Cells   285
§10.2 Direct Products of Summants   293

§10.3 A Fubini Theorem 299
§10.4 Integration on Paths in $\mathbb{R}^n$ 303
§10.5 Green's Theorem 312

Chapter 11. Mathematical Background
§11.1 Filterbases, Lower and Upper Limits 319
§11.2 Metric Spaces 322
§11.3 Norms and Inner Products 327
§11.4 Topological Spaces 331
§11.5 Regular Closed Sets 335
§11.6 Riesz Spaces 339
§11.7 The Inclusion-Exclusion Formula 342

References **347**

Index **351**

INTRODUCTION

§0.1 The Gauge-Directed Integral.

The main defects of the Riemann integral are its restriction to bounded integrands and its feeble convergence properties. The Lebesgue integral was introduced as a remedy for these particular defects. Its development led to measure theory and integration on measure spaces with a multitude of applications in classical analysis, functional analysis, and probability theory.

But both the Riemann and Lebesgue integrals demand absolute integrability; integrability of f requires integrability of $|f|$. So some derivatives fail to be integrable, thereby inhibiting the fundamental theorem of calculus.

Two equivalent integrals, the Denjoy and Perron integrals, were introduced to integrate all derivatives. Although they admit conditional integrability it was simpler to retain the old theory of improper integrals for the elementary cases of conditional integrability encountered in applications.

In another direction the total variation of a function was defined as the supremum of its approximating sums. It is actually a special case of the refinement-directed Stieltjes integral.

Application of Lebesgue theory to the Stieltjes integral on $\mathbb{R}$ has inherent limitations. By demanding absolute integrability the Lebesgue-Stieltjes integral restricts the fundamental algorithm

(1) $$\int_a^b df(t) = f(b) - f(a)$$

to functions f of bounded variation.

But the Riemann-Stieltjes integral gives (1) for *all* f since summation of Δf over any partition of [a, b] collapses identically to $f(b) - f(a)$ by additivity of Δf on abutting closed intervals. Indeed (1) holds for any integral defined as a limit of approximating sums $\sum \Delta f$ no matter what filter is used to define their convergence.

Having abandoned Riemann sums the Lebesgue integral on $\mathbb{R}$ required some tedious preliminaries. One had either to construct Lebesgue measure or extend a linear functional defined by a primitive integral. Such elaborate approaches made access to the Lebesgue integral on $\mathbb{R}$ difficult and time consuming.

But, since there seemed to be no single integration process that could encompass all of the diverse integrals on $\mathbb{R}$, the impressive advances of Lebesgue theory made the Lebesgue integral the standard integral in analysis. The Riemann integral became a useless anachronism relegated to text books in elementary calculus. The prevailing opinion among analysts was forcefully expressed by Jean Dieudonné [5]: "It may be suspected that had it not been for its prestigious name, $\cdots$ [the Riemann integral] $\cdots$ would have been dropped long ago, for (with due reverence for Riemann's genius) it is certainly clear to any working mathematician that nowadays such a 'theory' has at best the importance of a mildly interesting exercise in the general theory of measure and integration. Only the stubborn conservatism of academic tradition could freeze it into a regular part of the curriculum, long after it had outlived its historical importance. $\cdots$ When one needs a more powerful tool $\cdots$ [than the Cauchy integral of regulated functions] $\cdots$ there is no point in stopping halfway, and the general theory of ('Lebesgue') integration is the only sensible answer."

But by 1960 a significant breakthrough had occurred that revived Riemann's approach to integration on $\mathbb{R}$. A crucial modification of Riemann's definition of the integral was introduced independently by Kurzweil [19] and Henstock [11]. It can be described briefly as follows.

To form an approximating sum

$$(2) \qquad \sum_{i=1}^{n} f(t_i)\Delta x(I_i)$$

for the integral of f over $K = [a, b]$ we partition K into nonoverlapping closed intervals $I_1, \cdots, I_n$ and assign to each cell I_i a "tag" t_i belonging to I_i. In Riemann's view I is δ-fine if its length $\Delta x(I)$ is less than δ. In the modified view the positive constant δ is replaced by a positive function $\delta(\cdot)$ on K. In terms of such a "gauge" a tagged cell (I, t) is δ-fine if the length of I is less than $\delta(t)$. The gauge-directed filterbase of approximating sums (2) is much finer than the Riemann filterbase which is directed in effect by the constant gauges. This simple modification of Riemann's definition in which local fineness replaces uniform global fineness yields the generalized Riemann integral whose remarkable properties overcome the defects of both the Riemann and Lebesgue integrals on $\mathbb{R}$ [33]. Moreover, for absolutely integrable functions the generalized Riemann integral corresponds to the Lebesgue integral on K.

But our intention here is to go beyond the generalized Riemann integral by developing a process of integration based on the ideas of Kurzweil and Henstock. This unified approach eliminates the need for specialized integrals. It yields a theory of differentials on $\mathbb{R}$ based directly on the integration process. Differentials as the ultimate objects of integration clarify the exposition of integration on $\mathbb{R}$ and give traditional differential notation a solid foundation.

To transcend the generalized Riemann integral we use three further modifications introduced by Kurzweil and Henstock.

The first extends integration to "summants". The second demands that the tag t for a tagged cell (I, t) be an endpoint of I. The third extends the domain of integration to arbitrary intervals in $[-\infty, \infty]$.

By substituting Δg or $|\Delta g|$ for Δx in (2) we can get the generalized Stieltjes integrals $\int_K f \, dg$ and $\int_k f |dg|$ which include the special case $g = x$ of the generalized Riemann integral. This suggests that for greater generality we can form approximating sums

$$(3) \qquad\qquad \sum_{i=1}^{n} S(I_i, t_i)$$

in place of (2) for *any* function S on the set of all tagged cells (I, t) in K. For such a "summant" S the lower and upper gauge-directed limits of the approximating sums define the lower and upper integrals of S over K,

$$(4) \qquad\qquad -\infty \leq \underline{\int_K} S \leq \overline{\int_K} S \leq \infty.$$

If these two integrals are equal their common value defines the integral $\int_K S$. We call S integrable if $\int_K S$ exists and is finite.

Under gauge-direction, restriction to endpoint-tagged cells significantly refines the filterbase of approximating sums (3). When Stieltjes [40] introduced the integral $\int_K f \, dg$ as a generalization of the Riemann integral $\int_K f \, dx$ his definition was quite vague. He failed to specify the filterbase of the approximating sums $\sum f \Delta g$ whose limit defines the integral. Subsequently two versions of the Stieltjes integral emerged. One version uses constant gauges as Riemann had done. The other gets a finer filterbase using refinement of partition. But an even finer filterbase comes from gauge-direction with endpoint-tagged cells. So we get many more pairs f, g with $f \Delta g$ integrable. We also get the total variation of any function g on K defined directly by the integral $\int_K |dg| \leq \infty$, as well as a generalized Stieltjes integral that includes the Lebesgue-Stieltjes integral as a special case.

Treating approximating sums of the classical form (2) E. J. McShane [34] used gauge direction but allowed the tags to be *any* points in K. So (I, t) is δ-fine for every cell I in K which lies in the δ-neighborhood $(t - \delta(t), t + \delta(t))$ of t whether I contains t or not. McShane's integral is equivalent to the Lebesgue integral on K but has the advantage of avoiding the tedious preliminaries required by the usual approaches to the Lebesgue integral. But, sharing the narrow scope of the Lebesgue integral, McShane's integral precludes conditional integrability. This is a serious disadvantage. End point tagging is essential for a unified approach to integration on $\mathbb{R}$.

Integration over arbitrary intervals is quite simple. We can easily integrate summants over any closed, nondegenerate interval K in $[-\infty, \infty]$ since such a compact interval is topologically equivalent to a closed, bounded interval $[a, b]$ in $\mathbb{R}$. (We shall actually define integration over such K directly by defining the δ-neighborhoods of $\pm\infty$.) To integrate over an arbitrary subinterval L of $[-\infty, \infty]$ we integrate over its closure K in $[-\infty, \infty]$ by setting the summant equal to 0 at any tagged cell in K which contains an endpoint of K that does not belong to L. This effectively subsumes the improper integrals of elementary calculus. From our newly informed point of view the improper integrals required separate treatment in the past only because inadequate definitions of the integral failed to include improper integrals having computable values.

§0.2 Differentials.

The integral presented here has many advantages. Its direct definition as a limit of approximating sums (3) makes its properties readily accessible. Despite its simple formulation it includes equivalents of the integrals of Lebesgue, Stieltjes, and

Denjoy/Perron on $\mathbb{R}$. It makes the Cauchy and Harnack extensions redundant, obviating the need for a patchwork theory of improper integrals. It simplifies the definition and treatment of total variation, absolute continuity, regulated functions, and step functions. It offers improved versions of many basic results in calculus. It also yields new results in real analysis and raises new questions to be answered by researchers.

But the most striking innovation in this exposition is a definition of differential based on the integration of summants. The relation $\int_K |S - S'| = 0$ between summants S and S' on a cell K is an equivalence. A differential on K is just an equivalence class of summants on K. Since the upper and lower integrals are invariant for equivalent summants each differential acquires an upper and lower integral. So we have the concepts of integral and integrability for differentials. Every function f on K induces an integrable differential df, the equivalence class of the summant Δf, which satisfies (1). Moreover, every integrable differential is the differential of a function. In another role functions on K act as multipliers ("differential coefficients") of differentials. So we have the differentials $f\,dg$ on K for all functions f, g on K. Since the differentials on K form a Riesz space we also have the differentials $f|dg|, f(dg)^+$, and $f(dg)^-$. In addition to these we have some novel differentials such as the unit differential ω represented by the constant summant 1.

This concept of differential, introduced and developed by the author ($[21], \cdots, [28]$), is central to our discourse on integration here. Its evident utility in a variety of applications stands in sharp contrast to the uselessness of the old bogus definition $df = f'\Delta x$ foisted on generations of calculus students.

§0.3 Guidance for the reader.

This exposition should be accessible to readers familiar with the fundamentals of real analysis and some elementary concepts of topology. The reader should know about set operations, cardinality, suprema and infima, upper and lower limits, series, convergence, continuity, open and closed sets, connected sets, limit points, closure, and compactness. Prior knowledge of measure theory and Lebesgue integration is not required. Readers with such knowledge may skip over the standard material in §4.1 and §4.3. But they should be prepared for modifications and extensions of familiar concepts.

Chapter 7 may be omitted with no loss of continuity. Chapters 1–9 study the integration process and its associated differentials on 1-dimensional intervals. The n-dimensional case for $n \geq 1$ is treated in Chapter 10.

Chapter 11 is an appendix of selected background material to be used by readers as suits their needs.

Almost every section has an appended set of exercises. Some of these are routine applications of the definitions and theorems in the section. Others are more difficult. Except for occasional hints solutions are left to the reader. Two exercises (7 in §3.3 and 6 in §5.2) pose open questions.

Although this exposition has its roots in the author's earlier works ([21],···,[28]), what appears here from that work has been recast and reworked, notation and terminology have been revised, and much new material has been added. Some theorems make their first appearance here.

Except for a few lemmas all results here are labeled as theorems without regard for their relative importance. The numbering of items in each section is autonomous. So any reference in one section to a numbered item in another section will include the number of the section in which that item appears.

Complete proofs are supplied for most of the theorems presented here. In a few instances a slight gap in a proof is the occasion for an exercise left to the reader. Where a theorem reformulates an earlier version little or no proof is offered, sufficing in all such instances to leave no burden upon the reader.

The absence of solutions to the exercises is deliberate. It removes any temptation for an impatient reader to look at a printed solution prematurely, thereby reducing the value of a challenging exercise.

CHAPTER 1

INTEGRATION OF SUMMANTS

§1.1 Cells, Figures, and Partitions.

A **cell** is a closed, nondegenerate interval $[a, b]$ in $[-\infty, \infty]$. That is, $-\infty \leq a < b \leq \infty$. Although some cells are bounded and others unbounded, the relevant property for the integration process is that $[a, b]$ is an ordered, topological 1-cell with $a < b$. For every cell $[a, b]$ the interior is (a, b) in the sense of invariance of domain. So the boundary of $[a, b]$ consists of the endpoints a, b.

Two cells **overlap** if their intersection has interior points, that is, if their intersection is a cell. Two cells which do not overlap are either disjoint or abutting. Two cells **abut** if they intersect in a single point; the right endpoint of one cell is the left endpoint of the other.

A **figure** is any finite union of cells. Although no cell is empty we do have the empty figure $\emptyset$ as a vacuous union of cells. For later use we summarize the relevant structural properties of figures. (See §11.5 for the theoretical setting.)

A **partition** of a figure A is a finite set $\mathbb{A}$ of nonoverlapping cells whose union is A. Every figure has a partition. Indeed, in our 1-dimensional case every figure is a finite union of disjoint cells, its topological components. Topologically all figures are regular closed sets. That is, every figure A is the closure of its interior, $A = \overline{A^\circ}$. A figure is just a regular closed subset of $[-\infty, \infty]$ with only finitely many components.

9

Given a cell I the set of all figures contained in I is a boolean algebra under the lattice operations $\vee, \wedge$, and $'$ defined as follows. The smallest figure $A \vee B$ containing figures A, B in I is the figure $A \cup B$. The largest figure $A \wedge B$ contained in both A and B is $\overline{(A \cap B)^\circ} = \overline{A^\circ \cap B^\circ}$. The complementary figure A' to A in I is $\overline{I - A}$. A' is characterized by the two properties $A' \wedge A = \emptyset$ and $A' \vee A = I$. Figures A, B, C satisfy the distributive law $A \wedge (B \vee C) = (A \wedge B) \vee (A \wedge C)$.

As a lattice a boolean algebra is a distributive, complemented lattice. The ring-theoretic characterization of boolean algebras is quite useful. A boolean algebra is a commutative ring with unit such that every element A is idempotent, $A^2 = A$. In such a ring $-A = A$, so addition and subtraction are identical. The unit I satisfies the identity $IA = A$. The lattice structure comes from the ring operations through the ordering $A \leq B$ given by $A = AB$. Under this ordering $A \wedge B = AB, A \vee B = A + B + AB = A + (B - AB)$ with $AB \leq B$ and $A(B - AB) = 0, A' = I + A = I - A$, and $0 \leq A \leq I$. (See [9].)

In the other direction the ring operations come from the lattice operations. Specifically $A + B = (A \vee B) \wedge (A \wedge B)'$ and $AB = A \wedge B$. For figures A, B we say A **overlaps** B if $AB \neq \emptyset$. This agrees with the corresponding definition for cells since two figures overlap if and only if their intersection has interior points.

A **boolean field** [9] in a set I is a boolean algebra of subsets of I with zero $\emptyset$, unit I, $A \vee B = A \cup B$, and $A \wedge B = A \cap B$. So A' is just the complement of A in I.

The **indicator** of a subset E of I is the function 1_E on I defined by $1_E(t) = 1$ if $t \in I, 0$ if $t \notin I$. A boolean field can be represented by the indicators of its members. (See Exercise 5.)

Exercises (§1.1).

Prove the following for cells A, B, C in Exercises 1 - 4.

1. The following three conditions are equivalent: $A \cap B \neq \emptyset$; A either abuts or overlaps B; $A \, \breve{\cup} \, B$ is a cell.
2. The following six conditions are equivalent for $A = [a_1, a_2]$, $B = [b_1, b_2]$: A overlaps B; $A \cap B$ is a cell; $A \wedge B \neq \emptyset$; $a_1 < b_2$ and $b_1 < a_2$; $a_1 \vee b_1 < a_2 \wedge b_2$; Either $A = B$ or some interior point of one cell is an endpoint of the other.
3. $A + B$ is a cell if and only if one of the following holds: A abuts B; A and B have a common endpoint and one of the cells is a proper subset of the other.
4. $(A \vee B) \wedge C$ is a cell if and only if one of the following holds: $A \cap B \neq \emptyset$ and C overlaps either A or B; $A \cap B = \emptyset$ and C overlaps exactly one of the cells A, B.
5. Prove the following indicator identities for members D, E of a boolean field of subsets of a set I:

   (i) $1_\emptyset = 0$,   (ii) $1_I = 1$,

   (iii) $1_{D \cap E} = 1_D \wedge 1_E = 1_D 1_E$,

   (iv) $1_{D \cup E} = 1_D \vee 1_E = 1_D + 1_E - 1_D 1_E$,

   (v) $1_{D+E} = |1_D - 1_E|$ where $D + E$ is the symmetric difference,

   (vi) $1_{D'} = 1 - 1_D$.

§1.2 Tagged Cells, Divisions, and Gauges.

A **tagged cell** in a figure K is a pair (I, t) where I is a cell contained in K and t is an endpoint of I. For convenience we shall attribute to (I, t) properties of I, e.g. length.

A **division** of a figure K is a finite set $\mathcal{K}$ of nonoverlapping tagged cells whose union is K. That is, a division is a partition with each member assigned one of its endpoints as its tag.

A **gauge** on a subset A of $[-\infty, \infty]$ is any function δ on A such that $0 < \delta(t) < \infty$ for all t in A. A gauge δ plays a purely topological role by assigning to each point t in its domain A a neighborhood $N_\delta(t)$ of t in $[-\infty, \infty]$. Specifically the

δ-**neighborhood** $N_\delta(t)$ is defined to be the bounded interval $(t-\delta(t), t+\delta(t))$ if $-\infty < t < \infty$, the left halfline $[-\infty, -\frac{1}{\delta(-\infty)})$ if $t = -\infty$, and the right halfline $(\frac{1}{\delta(\infty)}, \infty]$ if $t = \infty$. So smaller gauges give smaller neighborhoods, $N_\alpha(t) \subseteq N_\beta(t)$ for gauges $\alpha(t) \leq \beta(t)$. Moreover, the gauge neighborhoods are basic. That is, if each point t in A is assigned an arbitrary neighborhood $N(t)$ then there exists a gauge δ on A such that $N_\delta(t) \subseteq N(t)$ for all t in A. This yields the following elementary but useful result.

THEOREM 1. *Given a closed subset E of $K = [-\infty, \infty]$ there exists a gauge δ on K such that for all t in K*

$$N_\delta(t) \cap E = \begin{cases} \emptyset \text{ if } t \notin E \\ t \text{ if } t \text{ is an isolated point of } E. \end{cases}$$

PROOF. To each point t in the subset $K - E$ of K assign the neighborhood $N(t) = K - E$ which is open relative to K. To each isolated point t of E assign a neighborhood $N(t)$ that contains no points of E other than t. For t a limit point of E set $N(t) = K$. Take a gauge δ such that $N_\delta(t) \subseteq N(t)$ for all t in K to get the desired conclusion. $\square$

For a gauge δ a tagged cell (I, t) is δ-**fine** if $I \subseteq N_\delta(t)$. Since t is an endpoint of I this means that for t finite the length of I is less than $\delta(t)$. A δ-**division** is a division whose members are δ-fine tagged cells.

Our next result shows that as $\delta \to 0$ the δ-divisions are ultimately finer than any given partition.

THEOREM 2. *Let $\{K_1, \cdots, K_n\}$ be a partition of a figure K. Then there exists a gauge δ on K such that every δ-division $\mathcal{K}$ of K is a union $\mathcal{K} = \mathcal{K}_1 \cup \cdots \cup \mathcal{K}_n$ where $\mathcal{K}_i$ is a δ-division of the cell K_i and each endpoint of K_i is the tag for some member of $\mathcal{K}_i$.*

PROOF. We can dismiss the vacuous case $K = \emptyset$ as trivial. The finite set E of endpoints of $K_1, \cdots, K_n$ is a closed set with no limit points. By Theorem 1 there is a gauge δ such that if (I, t) is a δ-fine tagged cell and I contains a point p belonging to E then $p = t$, an endpoint of I. For $I \subseteq K = K_1 \cup \cdots \cup K_n$ some K_i must overlap I. Since no endpoint of K_i is interior to I, K_i must contain I. Since I contains at most one endpoint of K_i, I is a proper subset of K_i.

Given a δ-division $\mathcal{K}$ of K let $\mathcal{K}_i$ consist of all members of $\mathcal{K}$ which lie in K_i. Since none of the other members of $\mathcal{K}$ can overlap K_i, $\mathcal{K}_i$ partitions K_i. So $\mathcal{K}_i$ is a δ-division of K_i. Since each endpoint of K_i belongs to I for some member (I, t) of $\mathcal{K}_i$ it must be the tag t. $\square$

For integration over a figure K we must have δ-divisions of K for every gauge δ on K. The next result assures us of their existence. Our indirect proof follows the proof of a 2-dimensional version given by Pierre Cousin [4] in 1895.

THEOREM 3. (COUSIN'S LEMMA). *Given a gauge δ on a figure K there exists a δ-division of K. Every δ-division of a subfigure of K can be extended to a δ-division of K.*

PROOF. Since every figure has a partition we need only consider the case in which K is a cell $[a, b]$. Take disjoint δ-fine tagged cells $([a, a_o], a)$ and $([b_o, b], b)$ at the ends of K. To get a δ-division of K we need only adjoin a δ-division of the bounded cell $K_0 = [a_o, b_o]$.

Now suppose there is no δ-division of K_0. We contend this leads to a contradiction. By induction we shall get a nested sequence of bounded cells $K_0 \supset K_1 \cdots \supset K_n \supset \cdots$ converging to a point c with no K_n having a δ-division.

Given a bounded cell K_n having no δ-division, bisect K_n into two cells I_n, J_n abutting at the midpoint of K_n. A least one of the cells I_n, J_n has no δ-division. Otherwise K_n would have one. Take one of the cells I_n, J_n having no δ-division to be K_{n+1}. So $K_{n+1} \subset K_n$ and diam $K_{n+1} = \frac{1}{2}$ diam K_n.

By induction diam $K_n = \frac{1}{2^n}$ diam $K_0 < \infty$ for $n = 0, 1, \cdots$.

So diam $K_n \searrow 0$ as $n \nearrow \infty$. Since $\mathbb{R}$ is complete the K_n's have a unique point of intersection c to which they converge. Take n large enough so that diam $K_n < \delta(c)$. If c is interior to $K_n = [a_n, b_n]$ let $\mathcal{K}_n$ consist of the two tagged cells $([a_n, c], c)$ and $([c, b_n], c)$. If c is an endpoint of K_n let $\mathcal{K}_n$ consist of the single member (K_n, c). In either case $\mathcal{K}_n$ is a δ-division of K_n, a contradiction.

Given a δ-division $\mathcal{A}$ of a subfigure A of K adjoin it to a δ-division $\mathcal{B}$ of the complementary figure $B = \overline{K - A}$ to get a δ-division $\mathcal{K} = \mathcal{A} \cup \mathcal{B}$ of K. $\square$

Exercises §(1.2).

1. Define a gauge δ on $K = [-\infty, \infty]$ such that every δ-division of K is a union of δ-divisions of $[-\infty, 0]$ and $[0, \infty]$.

2. Define a gauge δ on $K = [-\infty, \infty]$ such that every δ-division of K is a union

$$\mathcal{I} \cup \mathcal{K}_{-9} \cup \cdots \cup \mathcal{K}_9 \cup \mathcal{K}_{10} \cup \mathcal{J}$$

where $\mathcal{I}$ is a division of $I = [-\infty, -10], \mathcal{K}_i$ is a division of $K_i = [i-1, i]$ for $i = -9, -8, \cdots, 9, 10$, and $\mathcal{J}$ is a division of $J = [10, \infty]$.

3. Apply Theorem 3 to prove the following:

(i) Every continuous function f on a bounded cell K is uniformly continuous. (Given $\varepsilon > 0$ take a gauge δ on K small enough to ensure that, by continuity of f, $|f(x) - f(t)| < \varepsilon/2$ for all x, t in K such that $|x - t| < 2\delta(t)$. Take a δ-division $\mathcal{K}$ of K and let α be the minimum value of δ on the tags from $\mathcal{K}$. Show that $|f(x) - f(y)| < \varepsilon$ for all x, y in K such that $|x - y| < \alpha$.)

(ii) Every continuous function on $[-\infty, \infty]$ is uniformly continuous on $\mathbb{R}$.

(iii) Every continuous function on a cell K is bounded.

(iv) A function with a positive derivative everywhere on a bounded cell K is increasing on K.

(v) (Heine- Borel) Let $N(t)$ be a given neighborhood of t for each point t in a cell K. Show $K \subseteq N(t_1) \cup \cdots \cup N(t_n)$ for some finite subset $\{t_1, \cdots, t_n\}$ of K. (Take δ small enough so that $N_\delta(t) \subseteq N(t)$ for all t in K).

(For these and other applications of gauged divisions see reference [8].)

4. Let c be a real number. For each $I = [p, q]$ in $K = [-\infty, \infty]$ let $I_c = [p + c, q + c]$. ($t + c = t$ if $t = \pm\infty$.) Given a gauge β on K define the gauge α on K by

$$\alpha(t) = \frac{\beta(t + c)}{1 + |c|\beta(t)}.$$

Given a division $\mathcal{K}$ of K let $\mathcal{K}_c$ be the set of all $(I_c, t + c)$ such that (I, t) belongs to $\mathcal{K}$. Show that if $\mathcal{K}$ is an α-division of K then $\mathcal{K}_c$ is a β-division of K. (Existence of such α for a given β on K holds for any orientation-preserving, topological transformation of K. Translations are a special case of such transformations. It is only for K of dimension 1 that cells, tagged cells, figures, partitions, and divisions in K are purely topological entities. These structures in higher dimensions depend upon the ordering in $\mathbb{R}^n$.)

§1.3 The Upper and Lower Integrals of a Summant over a Figure.

A **summant** S on a figure K is a function on the set of all tagged cells (I, t) in K. To give some examples let $I = [r, s]$ be any cell in K, t be either of the endpoints r, s of I, and f, g be functions on K.

An important class of summants consists of the cell summants $S(I, t) = S(I)$ which do not depend on the tag t. Among these are Δg and $|\Delta g|$ given by $\Delta g(I) = g(s) - g(r)$ and $|\Delta g|(I) = |\Delta g(I)|$. Δg and $|\Delta g|$ are the special case $f = 1$ of the Stieltjes summants $f\Delta g$ and $f|\Delta g|$ given by $(f\Delta g)(I,t) = f(t)\Delta g(I)$ and $(f|\Delta g|)(I,t) = f(t)|\Delta g(I)|$. Another special case of the Stieltjes summants is formed by the classical summants $f\Delta x$ in which g is just the identity function x.

A useful summant is the orientation summant Q given by $Q(I,t) = 1$ if $t = r, -1$ if $t = s$. As a counterpart to the cell summants we have the tag summants $f|Q|$ given by $(f|Q|)(I,t) = f(t)|Q(I,t)| = f(t)$.

Given a nonempty figure K let $\mathcal{D} = \mathcal{D}iv(K)$ be the set of all divisions of K. For each gauge δ on K let $\mathcal{D}_\delta = \delta\text{-}\mathcal{D}iv(K)$ be the set of all δ-divisions of K. The sets $\mathcal{D}_\delta$ corresponding to the gauges δ on K form a filterbase Φ in $\mathcal{D}$. To verify this we note that the members $\mathcal{D}_\delta$ of Φ are directed downward by inclusion since the pointwise minimum $\delta = \alpha \wedge \beta$ of two gauges α, β on K is a gauge for which $\mathcal{D}_\delta = \mathcal{D}_\alpha \cap \mathcal{D}_\beta$. So $\mathcal{D}_\alpha \subseteq \mathcal{D}_\beta$ for $\alpha \leq \beta$. Theorem 3 (§1.2) assures us that no member $\mathcal{D}_\delta$ of Φ is empty. (See §11.1.)

Given a summant S on the nonempty figure K each division $\mathcal{K}$ of K yields an **approximating sum**

$$(1) \qquad \left(\sum S\right)(\mathcal{K}) = \sum_{(I,t)\in\mathcal{K}} S(I,t).$$

The summant S thereby induces a function $\sum S$ on the set $\mathcal{D}$ of all divisions $\mathcal{K}$ of K. This function $\sum S$ carries the filterbase Φ in $\mathcal{D}$ to a filterbase Φ_S in $\mathbb{R}$. Each member $\mathcal{E}_\delta$ of Φ_S is determined by a gauge δ, $\mathcal{E}_\delta$ being the image of $\mathcal{D}_\delta$ under $\sum S$. The upper and lower limits of the filterbase Φ_S in $[-\infty, \infty]$ define the upper and lower integrals of the summant S. To give explicit details we introduce some notation which we shall use repeatedly hereafter.

The **lower δ-sum** of S over K is defined by $S_{(\delta)}(K) =$

inf $\mathcal{E}_\delta$. That is,

(2) $$S_{(\delta)}(K) = \inf_{\mathcal{K} \in \delta\text{-}\mathcal{D}iv K} (\sum S)(\mathcal{K})$$

in terms of (1). The **upper δ-sum** is defined by $S^{(\delta)}(K) = \sup \mathcal{E}_\delta$. Explicitly

(3) $$S^{(\delta)}(K) = \sup_{\mathcal{K} \in \delta\text{-}\mathcal{D}iv K} (\sum S)(\mathcal{K}).$$

(Strictly speaking these are not necessarily sums, but are limits of sums $(\sum S)(\mathcal{K})$.) Since $S_{(\delta)}(K)$ and $S^{(\delta)}(K)$ are lower and upper bounds of a nonempty subset $\mathcal{E}_\delta$ of $\mathbb{R}$

(4) $$-\infty \le S_{(\delta)}(K) \le S^{(\delta)}(K) \le \infty$$

and

(5) $$S_{(\delta)}(K) < \infty, \ S^{(\delta)}(K) > -\infty.$$

For each gauge δ on K the closed interval $[S_{(\delta)}(K), S^{(\delta)}(K)]$ is a terminal set for the filterbase Φ_S in $[-\infty, \infty]$ since it contains the member $\mathcal{E}_\delta$ of Φ_S. Since $\mathcal{E}_\alpha \subseteq \mathcal{E}_\beta$ for gauges $\alpha \le \beta$ their convex closures are similarly nested. That is,

(6) $\quad [S_{(\alpha)}(K), S^{(\alpha)}(K)] \subseteq [S_{(\beta)}(K), S^{(\beta)}(K)]$ for $\alpha \le \beta$.

Explicitly

(7) $\quad S_{(\beta)}(K) \le S_{(\alpha)}(K) \le S^{(\alpha)}(K) \le S^{(\beta)}(K)$ for $\alpha \le \beta$.

The terminal intervals $[S_{(\delta)}(K), S^{(\delta)}(K)]$ indexed by δ meet in a closed, nonempty interval $I = [\sup_\delta S_{(\delta)}(K), \inf_\delta S^{(\delta)}(K)]$ in the compact cell $[-\infty, \infty]$. The endpoints of I are the lower and upper limits of the filterbase Φ_S. (See §11.1).

We define the **lower integral** of S over K to be the lower limit

$$\text{(8)} \qquad \underline{\int_K} S = \sup_\delta S_{(\delta)}(K).$$

We define the **upper integral** to be the upper limit

$$\text{(9)} \qquad \overline{\int_K} S = \inf_\delta S^{(\delta)}(K).$$

From these definitions we get

$$\text{(10)} \qquad -\infty \le \underline{\int_K} S \le \overline{\int_K} S \le \infty$$

by (4) and (6), (7). If the lower and upper integrals are equal, their common value (the limit of the filterbase Φ_S in $[-\infty, \infty]$) defines the **integral** $\int_K S$ of S over K,

$$\text{(11)} \qquad -\infty \le \int_K S = \underline{\int_K} S = \overline{\int_K} S \le \infty.$$

S is **integrable** over K if $\int_K S$ exists and is finite. So integrability of S is just convergence of Φ_S to a finite limit. By the Cauchy criterion such convergence is equivalent to the existence in Φ_S of members $\mathcal{E}_\delta$ having arbitrarily small diameter $S^{(\delta)}(K) - S_{(\delta)}(K)$. By the monotoneity (7) this is just $S^{(\delta)}(K) - S_{(\delta)}(K) = \text{diam } \mathcal{E}_\delta \searrow 0$ as $\delta \searrow 0$. In more primitive terms S is integrable if and only if given $\varepsilon > 0$ in $\mathbb{R}$ there exists a gauge δ on K such that for all δ-divisions $\mathcal{K}, \mathcal{K}'$ of K

$$\text{(12)} \qquad \left| \left(\sum S \right)(\mathcal{K}) - \left(\sum S \right)(\mathcal{K}') \right| < \varepsilon.$$

In such primitive terms convergence of Φ_S to c in $\mathbb{R}$ is explicitly expressed as follows: S is integrable with $\int_K S = c$ if and only if given $\varepsilon > 0$ in $\mathbb{R}$ there exists a gauge δ on K such that

$$(13) \qquad \left| \left(\sum S \right) (\mathcal{K}) - c \right| < \varepsilon$$

for all δ-divisions $\mathcal{K}$ of K.

To apply the integration process it is not necessary for all values of the summant S to be finite as long as all infinite values of S are of the same sign. Then the indeterminate form $\infty - \infty$ cannot occur in the approximating sums (1).

Let S take its values in an interval H that is either $[-\infty, \infty)$ or $(-\infty, \infty]$. In either case addition in H is determinate. So we can form the approximating sums $\left(\sum S \right) (\mathcal{K})$ with values in H. Taking infima and suprema we can define $S_{(\delta)}(K)$ and $S^{(\delta)}(K)$ by (2) and (3) as was done for the case $H = \mathbb{R}$. The inequalities (4) hold just as in the finite case $H = \mathbb{R}$. But one of the inequalities in (5) may fail. As in the finite case the lower and upper integrals are defined by (8) and (9), and the integral by (11) wherever it exists. Of course, for S to be integrable its values must ultimately be finite. That is, there must be a gauge δ on K such that $-\infty < S_{(\delta)}(I) \le S(I, t) \le S^{(\delta)}(I) < \infty$ for all δ-fine (I, t) in $(\mathcal{K})$.

In the absence of an explicit qualification such as $-\infty \le S < \infty$ or $-\infty < S \le \infty$ we implicitly assume that all values of S are finite. We shall give only occasional reminders of this assumption.

By the conventions of logic all empty sums equal zero. So all integrals over the empty figure $\emptyset$ equal 0.

Exercises (§1.3).

1. For S a summant on a figure K and δ a gauge on K prove $(-S)^{(\delta)} = -S_{(\delta)}$ and $\overline{\int_K}(-S) = -\underline{\int_K} S$.

2. For R, S summants on a figure K and δ a gauge on K prove $(R + S)^{(\delta)} \le R^{(\delta)} + S^{(\delta)}$ and $R_{(\delta)} + S_{(\delta)} \le (R + S)_{(\delta)}$.

In the next three exercises we call a summant S on a cell K **refinement integrable** to c if given $\varepsilon > 0$ there exists a partition $\mathbb{K}_\varepsilon$ of K such that (13) holds for every division $\mathcal{K}$ of K that refines $\mathbb{K}_\varepsilon$. We call S **constant-gauge integrable** to c on K if given $\varepsilon > 0$ there exists a constant gauge δ such that (13) holds for every δ-division $\mathcal{K}$ of K.

3. Prove the implications (i) $\Rightarrow$ (ii) $\Rightarrow$ (iii):

   (i) S is constant-gauge integrable to c.

   (ii) S is refinement integrable to c.

   (iii) S is gauge integrable to c. (See the definition associated with (13).)

4. Let S be a summant on K for which given p in K^o and $\varepsilon > 0$ there exists $\delta > 0$ such that $|S(I,t)| < \varepsilon$ for all (I,t) such that I contains p and the length of I is less than δ. Show that for such S (ii) $\Rightarrow$ (i) in Exercise 3.

5. Let S be the summant on $K = [-1,1]$ defined by $S(I,t) = S(I) = 1_I(0)$. Prove:

   (i) S is refinement integrable to 2 on K.

   (ii) S is *not* constant-gauge integrable on K since given any constant $\delta > 0$ there exist δ-divisions $\mathcal{K}_1$ and $\mathcal{K}_2$ of K such that $\left(\sum S\right)(\mathcal{K}_j) = j$ for $j = 1,2$.

6. (Refer to Exercise 4 in §1.2. This exercise is a continuation of that one.)

   Given a summant S on $K = [-\infty, \infty]$ define the summant T on K by $T(I,t) = S(I_c, t + c)$.
   Prove:
   (i) Given a gauge β on K there exists a gauge α on K such that $T^{(\alpha)}(K) \le S^{(\beta)}(K)$.

   (ii) $\overline{\int_K} T \le \overline{\int_K} S$.

(iii) $\overline{\int_K} T = \overline{\int_K} S$. (Apply the preceding with c replaced by $-c$ and S, T interchanged.)

(iv) $\underline{\int_K} T = \underline{\int_K} S$. (Replace S, T above by $-S, -T$.)

7. For the constant summant $S \equiv 1$ on a cell K show that $\int_K S = \infty$.

§1.4 Summants with Special Properties.

A **cell summant** S on a figure K is an extended real-valued summant such that

$$(1) \qquad -\infty \leq S(I, s) = S(I, t) \leq \infty \text{ for all } I = [s, t] \text{ in } K.$$

So a cell summant S is just an extended real-valued function on the set of all cells I in K since the tags are irrelevant for the evaluation of S under (1). Thus we can set $S(I, t) = S(I)$ for cell summants S. Of particular interest are cell summants that are additive, subadditive, or superadditive.

A summant S on K is **additive** if it is a cell summant such that $-\infty < S < \infty$ and

$$(2) \qquad S(I \cup J) = S(I) + S(J)$$

for all abutting cells I, J in K. This is equivalent to the existence of a constant c in $\mathbb{R}$ such that

$$(3) \qquad \left(\sum S \right)(\mathcal{K}) = c$$

for every division $\mathcal{K}$ of K. (See Exercise 2.) So additivity implies integrability with $\int_K S = c$ in (3). Moreover, additivity on K implies additivity on every cell in K. So integration of an additive summant S over any cell in K is redundant. That is,

$$(4) \qquad \int_I S = S(I) \text{ for every cell } I \text{ in } K.$$

Every function g on K induces an additive summant Δg on K defined by

(5) $\Delta g(I) = g(s) - g(r)$ for $I = [r, s]$ in K.

Moreover, every additive summant is induced in this way. Indeed, given S additive on K with components $K_i = [a_i, b_i]$ for $i = 1, \cdots, n$ choose for each i an arbitrary constant c_i and define g on K_i by

(6) $g(t) = \begin{cases} c_i \text{ for } t = a_i \\ c_i + S[a_i, t] \text{ for } a_i < t \le b_i. \end{cases}$

Since both Δg and S are additive on K and are equal at every initial segment $[a_i, t]$ of K_i for $i = 1, \cdots, n$

(7) $\Delta g = S$ on K.

(See Exercise 3.) (7) puts (4) into the form

(8) $\int_I \Delta g = \Delta g(I)$

for every function g on a figure K and every cell I in K.

A cell summant S on a figure K with at least one of the values $\pm\infty$ excluded from its range is **subadditive** if

(9) $S(I \cup J) \le S(I) + S(J)$

for all abutting cells I, J in K, **superadditive** if

(10) $S(I) + S(J) \le S(I \cup J)$

for all abutting cells I, J in K. The sum $S(I) + S(J)$ in (9) and (10) is determinate since the indeterminate form $\infty - \infty$ is excluded by the restriction imposed on the range of S. An additive summant is just a cell summant with all its values finite that is both subadditive and superadditive. Each of these conditions is hereditary. That is, if subadditivity, superadditivity or additivity holds on K it holds on every subfigure of K.

Let S be a summant on K such that $-\infty < S < \infty$. Then for each gauge δ on K the lower sum $S_{(\delta)}$ is subadditive and the upper sum $S^{(\delta)}$ is superadditive. These are the cell summants $-\infty \leq S_{(\delta)} < \infty$ and $-\infty < S^{(\delta)} \leq \infty$ defined in accordance with (2), (3) in §1.3 by

(11)
$$S_{(\delta)}(I) = \inf_{\mathcal{I} \in \delta\text{-Div}(I)} \left(\sum S \right)(\mathcal{I}),$$
$$S^{(\delta)}(I) = \sup_{\mathcal{I} \in \delta\text{-Div}(I)} \left(\sum S \right)(\mathcal{I}).$$

We contend that $\int_K S$ exists if S is either subadditive or superadditive.

THEOREM 1.

(i) *Given S subadditive on K define*

$$S^\star(K) = \sup_{\mathbb{K} \in Ptn(K)} \left(\sum S \right)(\mathbb{K})$$

where $Ptn(K)$ is the set of all partitions $\mathbb{K}$ of K. Then for every gauge δ on K

(12) $$-\infty \leq S^{(\delta)}(K) = S^\star(K) = \int_K S \leq \infty.$$

(ii) *Given S superadditive on K define*

$$S_\star(K) = \inf_{\mathbb{K} \in Ptn(K)} \left(\sum S \right)(\mathbb{K}).$$

Then for every gauge δ on K

(13) $\qquad -\infty \leq S_{(\delta)}(K) = S_*(K) = \int_K S \leq \infty.$

PROOF. We need only prove (i) since application of (i) to $-S$ gives the dual (ii) of (i).

For partitions $\mathbb{K}, \mathbb{K}'$ of K such that $\mathbb{K}'$ refines $\mathbb{K}$ subadditivity of S gives

(14) $\qquad \left(\sum S\right)(\mathbb{K}) \leq \left(\sum S\right)(\mathbb{K}').$

Given $\mathbb{K}$ apply Theorem 2 (§1.2) to get a gauge δ such that every partition $\mathbb{K}'$ induced by a δ-division of K refines $\mathbb{K}$. Then (14) and the definition of lower sum give $\left(\sum S\right)(\mathbb{K}) \leq S_{(\delta)}(K)$. Thus, since $S_{(\delta)}(K) \leq \underline{\int}_K S$ by the definition of lower integral,

(15) $\qquad \left(\sum S\right)(\mathbb{K}) \leq \underline{\int}_K S$ for every partition $\mathbb{K}$ of K.

By the definition of S^* (15) is equivalent to

(16) $\qquad S^*(K) \leq \underline{\int}_K S.$

Now the definitions of upper sum, upper integral, and S^* give

(17) $\qquad \overline{\int}_K S \leq S^{(\delta)}(K) \leq S^*(K)$ for every gauge δ on K.

Thus, since $\underline{\int}_K S \leq \overline{\int}_K S$, equality holds throughout (16) and (17), thereby giving (12). $\square$

THEOREM 2. *For every additive summant R on K the integral of $|R|$ over K exists and*

$$0 \le \int_K |R| = \sup_{\mathbb{K} \in Ptn(K)} \left(\sum |R| \right) (\mathbb{K}) \le \infty.$$

PROOF. Since R is additive $|R|$ is subadditive. Apply (i) in Theorem 1 with $S = |R|$. □

For f a function on a cell K and S a summant on K the summant fS on K is defined by

(18) $(fS)(I, t) = f(t)S(I, t)$

for all tagged cells (I, t) in K. Applied to the unit summant $S \equiv 1$ (18) extends each function f on K to a summant ("tag summant") on K,

(19) $f(I, t) = f(t)$ for every tagged cell (I, t) in K.

So (18) is just a product of two summants. (The abbreviated notation in (19) actually denotes the composition $(f \circ P_2)(I, t)$ where P_2 is the projection $P_2(I, t) = t$.)

If S is an additive summant then $S = \Delta g$ for some function g and (18) gives the Stieltjes summant $f\Delta g$. For g the identity x on a bounded cell K these are just the classical summants $f\Delta x$.

A useful summant Q on $[-\infty, \infty]$ is defined by

(20) $Q(I, t) = \begin{cases} +1 \text{ if } t \text{ is the left endpoint of } I \\ -1 \text{ if } t \text{ is the right endpoint of } I. \end{cases}$

In terms of the endpoints s, t of I, $Q(I, t) = sgn(s - t)$. (See Exercise 12 for an application of Q).

Exercises (§1.4). Prove the following:

1. Let S be a summant on a figure K and δ be a gauge on K such that $S_{(\delta)}(K) = S^{(\delta)}(K)$. Then there is a function g on K such that $S(I,t) = \Delta g(I)$ for all δ-fine (I,t) in K, and $\int_I S = \Delta g(I)$ for every cell I in K.

2. A summant S on a cell K is additive if and only if $(\sum S)(\mathcal{K})$ is constant for all divisions $\mathcal{K}$ of K.

3. If S, T are additive summants on a cell K and p is a point in K such that $S(I) = T(I)$ for every cell I in K having p as an endpoint then $S = T$. (*Hint*: $S - T$ is additive.)

4. S is subadditive if and only if $-S$ is superadditive.

5. If S is subadditive then $S \le S_{(\delta)}$ for every gauge δ. If S is superadditive then $S^{(\delta)} \le S$ for every gauge δ.

6. If S is subadditive on a cell K and $S(I) = -\infty$ for some cell I in K then $S(J) = -\infty$ for every cell J in K which contains I. If S is superadditive on K and $S(I) = \infty$ for some cell I in K then $S(J) = \infty$ for every cell J in K which contains I.

7. Given a summant S on a cell K and gauge δ on K: $S_{(\delta)}$ is the largest subadditive summant T such that $-\infty \le T(I) \le S(I,t)$ for all δ-fine (I,t) in K, $S^{(\delta)}$ is the smallest superadditive summant T such that $S(I,t) \le T(I) \le \infty$ for all δ-fine (I,t) in K.

8. $\int_K |S| = 0$ for a summant S on a cell K if and only if given $\varepsilon > 0$ there exist a gauge δ and superadditive summant T on K such that $T(K) < \varepsilon$ and $|S(I,t)| \le T(I)$ for all δ-fine (I,t) in K.

9. For S a summant on a figure K:

   (i) $S_{(\delta)}(K) \le \int_K S_{(\delta)} \le \underline{\int_K} S \le \overline{\int_K} S \le \int_K S^{(\delta)} \le S^{(\delta)}(K)$ for every gauge δ on K.

(ii) $\int_K S_{(\delta)} \nearrow \underline{\int_K} S$ and $\int_K S^{(\delta)} \searrow \overline{\int_K} S$ as $\delta \searrow 0$.

10. Let $f(t) = 1 - |sgn\, t|$ and $F(t) = 1 + sgn\, t$ for $-\infty \leq t \leq \infty$. Then in terms of (19), $\int_I f = \Delta F(I)$ for every cell I in $[-\infty, \infty]$.

11. Let $p_n \nearrow b$ in $K = [a, b]$ where $a < p_1 < \cdots < p_n < \cdots < b$. Let f be a function on K that vanishes on the complement of $\{a, p_1, p_2, \cdots, b\}$. For Q defined by (20) let $R = -Q$. Then

(i) $\underline{\int_K} fR = \Delta f(K) - \overline{\lim}_{n \to \infty} f^-(p_n)$

(ii) $\overline{\int_K} fR = \Delta f(K) + \overline{\lim}_{n \to \infty} f^+(p_n)$

(iii) fR is integrable over K if and only if $f(p_n) \to 0$, in which case $\int_K fR = \Delta f(K)$.

(iv) If the series $\sum_{n=1}^{\infty} f(p_n) = c$ is convergent then in terms of (19), $\int_K f = f(a) + 2c + f(b)$ and $\int_K |f| = |f(a)| + 2\sum_{n=1}^{\infty} |f(p_n)| + |f(b)| \leq \infty$.

12. For Q defined by (20) on a cell K:

(i) For every function f on K and cell I with endpoints s, t in K

$$(Q\Delta f)(I, t) = f(s) - f(t)$$

and in terms of (19)

$$(f + Q\Delta f)(I, t) = f(s) \text{ with } s = (x + Q\Delta x)(I, t).$$

(ii) For all functions f, g on the cell K the summant identity $\Delta(fg) = (f + Q\Delta f)\Delta g + g\Delta f = f\Delta g + g\Delta f + Q\Delta f\Delta g$ holds on K. (We return to this identity in §6.2.)

13. For functions f, g on a cell K the upper and lower integrals of $f\Delta g$ are unaffected if we relax the restriction of tags to

endpoints and admit tagged cells (I, t) with t any point in I. But this is false for some summants of the form $f|\Delta g|$.

14. If f is continuous on a bounded cell K then $f\Delta x$ is constant-gauge integrable on K. (See Exercise 3 in §1.2 and Exercises 3,4 in §1.3.)

15. If $f\Delta x$ is constant-gauge integrable on a bounded cell K then f is bounded on K.

16. Let $f\Delta x$ be constant-gauge integrable on a bounded cell $K = [0, b]$. Then the function g is well defined on $(0, 1)$ by

(21) $$g(r) = b(1 - r) \sum_{i=1}^{\infty} f(r^i b) r^{i-1} \text{ for } 0 < r < 1.$$

That is, the series converges. Moreover,

(22) $$\int_K f\Delta x = g(1-).$$

(*Hint*: $b(1 - r)r^{i-1} = r^{i-1}b - r^i b$.)

17. For $f(t) = t^n$ on $K = [0, b]$ with n a positive integer (21) reduces to

$$g(r) = b^{n+1} r^n (1 + r + r^2 + \cdots + r^n)^{-1}$$

and (22) gives

$$\int_K f\Delta x = \frac{b^{n+1}}{n + 1}.$$

(As the reader should be aware, Exercises 14-17 pertain to the Riemann integral.)

18. Let Q be defined by (20) on $K = [a, b]$. Given a gauge δ on K there exists by induction a sequence $a = a_0 < a_1 < \cdots < a_i < \cdots$ in $[a, b)$ such that (I_i, a_i) is δ-fine for $I_i =$

$[a_i, a_{i+1}]$, $i = 0, 1, \cdots$. For $J = [a, p]$ with $p = \lim_{i \to \infty} a_i$ the division $\mathcal{J}_n = \{(I_0, a_0), \cdots, (I_n, a_n), (H_n, p)\}$ of J with $H_n = [a_{n+1}, p]$ has $(\sum Q)(\mathcal{J}_n) = n$ and is a δ-division for all sufficiently large n. So $Q^{(\delta)}(J) = \infty$, hence $Q^{(\delta)}(K) = \infty$. Therefore $\overline{\int_K} Q = \infty$. Similarly $\underline{\int_K} Q = -\infty$ which follows by a dual argument.

19. Condition (i) of Theorem 1 holds for the constant summant $S = 1$. So Theorem 1 reaffirms the result $\int_K S = \infty$ of Exercise 7 in §1.3. (This result also follows from Exercise 18 since $|Q| = 1$.)

§1.5 Upper and Lower Integrals as Functions on the Boolean Algebra of Figures.

The next result shows that lower and upper integrals are additive on finite unions of nonoverlapping figures where the indeterminate form $\infty - \infty$ does not occur. But even for the indeterminate case there are appropriate conclusions for finite summants $-\infty < S < \infty$.

THEOREM 1. *Let S be a summant on a union C of nonoverlapping figures A, B. Then the lower and upper integrals are additive in the following sense:*

(i) $\underline{\int_C} S = \underline{\int_A} S + \underline{\int_B} S$ if the right-hand side is determinate.

If the right-hand side is indeterminate then $\underline{\int_C} S = -\infty$.

(ii) $\overline{\int_C} S = \overline{\int_A} S + \overline{\int_B} S$ if the right-hand side is determinate.

If the right-hand side is indeterminate then $\overline{\int_C} S = \infty$.

PROOF. We need only prove (i) since the dual (ii) follows from (i) applied to $-S$.

Take any partitions $\mathbb{A}$ of A and $\mathbb{B}$ of B. Then $\mathbb{A} \cup \mathbb{B}$ is a partition of C since $AB = \emptyset$. By Theorem 2 (§1.2) there is a gauge δ on C such that every δ-division $\mathcal{C}$ of C refines the partition $\mathbb{A} \cup \mathbb{B}$. So $\mathcal{C}$ is the union of δ-divisions $\mathcal{A}$ of A and $\mathcal{B}$ of B. Thus

(1) $$\left(\sum S \right)(\mathcal{C}) = \left(\sum S \right)(\mathcal{A}) + \left(\sum S \right)(\mathcal{B}).$$

Now any δ-divisions $\mathcal{A}$ of A and $\mathcal{B}$ of B unite to form a δ-division $\mathcal{C}$ of C. Since $\mathcal{A}$ and $\mathcal{B}$ can be chosen independently the operation of taking the infimum distributes into (1) to give

(2) $$S_{(\delta)}(C) = S_{(\delta)}(A) + S_{(\delta)}(B) \text{ with } S_{(\delta)} < \infty$$

under (2) and (5) in §1.3. For every figure K, $S_{(\delta)}(K) \nearrow \int_K S$ as $\delta \searrow 0$. So (2) gives (i) in the determinate case. If one of the terms on the right-hand side in (i) equals $-\infty$, say $\int_B S = -\infty$, then $S_{(\delta)}(B) = -\infty$ for every gauge δ. This implies $S_{(\delta)}(C) = -\infty$ for every gauge δ since the sum in (2) is determinate. So $\underline{\int_C} S = -\infty$. $\square$

The next two results follow from Theorem 1.

THEOREM 2. *Let S be a summant on a figure C, and B be a subfigure of C. Then the following hold:*

(i) If $\underline{\int_C} S > -\infty$ then $\underline{\int_B} S > -\infty$.

(ii) If $|\underline{\int_C} S| < \infty$ then $|\underline{\int_B} S| < \infty$.

(iii) If $\overline{\int_C} S < \infty$ then $\overline{\int_B} S < \infty$.

(iv) If $|\overline{\int_C} S| < \infty$ then $|\overline{\int_B} S| < \infty$.

PROOF. Let A be the complementary figure to B in C. That is, $AB = \emptyset$ and $A \cup B = C$. Given $\underline{\int_C} S > -\infty$ (i) in Theorem 1 gives a determinate sum $\underline{\int_A} S + \underline{\int_B} S = \underline{\int_C} S > -\infty$. Hence, neither $\underline{\int_A} S$ nor $\underline{\int_B} S$ can equal $-\infty$. So (i) holds. If, moreover, $\underline{\int_C} S$ is finite then so are $\underline{\int_A} S$ and $\underline{\int_B} S$ by (i) in Theorem 1. This gives (ii). The duals (iii), (iv) of (i), (ii) follow from (i), (ii) applied to $-S$. $\square$

THEOREM 3. *Let S be a summant on a figure C.*
(i) If A, B are nonoverlapping figures whose union is C, and S is integrable over both A and B, then S is integrable over C and

$$\int_C S = \int_A S + \int_B S.$$

(ii) If S is integrable over C then S is integrable over every subfigure B of C.

PROOF. (i) follows directly from Theorem 1. To prove (ii) let S be integrable over C. Let A be the complementary figure to B in C. Then Theorem 1 gives $\int_C S = \int_A S + \int_B S = \overline{\int_A} S + \overline{\int_B} S$ with all these integrals finite by Theorem 2. Therefore, since $\underline{\int_A} S \leq \overline{\int_A} S$ and $\underline{\int_B} S \leq \overline{\int_B} S$ equality must hold in both of these inequalities. In particular, $\underline{\int_B} S = \overline{\int_B} S$ which is finite. That is, S is integrable over B. $\square$

Note that in Theorem 3 $S = S_A + S_B$ ultimately where $S_A = S$ on the tagged cells in $A, 0$ on the tagged cells in B, with a similar definition for S_B. So (i) in Theorem 3 is a special case of the sum of the limits of two functions on a filterbase. Applied to approximating sums $\sum R$ and $\sum T$ of summants R, T on a figure K the general result takes the form: If R and T are integrable summants on a figure K then $R + T$ is integrable and

(3)
$$\int_K (R + T) = \int_K R + \int_K T.$$

In the general case of functions on a filterbase additivity of limits follows from superadditivity of lower limits and subadditivity of upper limits. In the special case of the integration process for summants R, T on a figure K the general case takes the form

(4)
$$\underline{\int_K} R + \underline{\int_K} T \leq \underline{\int_K} (R + T) \leq \overline{\int_K} (R + T) \leq \overline{\int_K} R + \overline{\int_K} T$$

disregarding inequalities involving an indeterminate sum of the form $\infty - \infty$. (See §11.1.)

In the boolean algebra of figures we can easily extend Theorem 3 to include the case of overlapping A, B in (i).

THEOREM 4. *Let S be a summant on a figure C, and A and B figures such that $A \cup B = C$. Then S is integrable over C if and only if S is integrable over both A and B. For S integrable over C*

$$\int_C S = \int_A S + \int_B S - \int_{AB} S.$$

PROOF. In terms of the ring operations on figures $B = (B - AB) + AB$ with $(B - AB)(AB) = \emptyset$. By Theorem 3 $\int_B S = \int_{B-AB} S + \int_{AB} S$. So $\int_B S - \int_{AB} S = \int_{B-AB} S$. Similarly $C = A \cup B = A + (B - AB)$ with $A(B - AB) = \emptyset$. So $\int_C S = \int_A S + \int_{B-AB} S = \int_A S + \int_B S - \int_{AB} S$. □

Exercises (§1.5).

Let S be a summant on $K = [a, b]$.

1. Given $\int_K S > -\infty$ let δ be a gauge on K such that $S_{(\delta)}(K)$ is finite. Define the function G on K by $G(t) = S_{(\delta)}[a, t]$ for all t in K. Prove:

    (i) $\Delta G \leq S_{(\delta)}$.

    (ii) $\Delta G \leq S$ on all δ-fine tagged cells in K.

    (iii) $\Delta G(K) \leq \int_K S$.

2. Given $\overline{\int_K} S < \infty$ let δ be a gauge on K such that $S^{(\delta)}(K)$ is finite. Define the function H on K by $H(t) = S^{(\delta)}[a, t]$ for all t in K. Prove:

    (i) $S^{(\delta)} \leq \Delta H$.

    (ii) $S \leq \Delta H$ on all δ-fine tagged cells in K.

    (iii) $\overline{\int_K} S \leq \Delta H(K)$.

3. Show that S is integrable on K if and only if given $\varepsilon > 0$ there exist a gauge δ and functions G, H on K such that $\Delta G \leq S \leq \Delta H$ on all δ-fine tagged cells in K and moreover $\Delta(H - G)(K) < \varepsilon$.

4. Let f be a function on a bounded cell K such that $\underline{\int_K} f\Delta x$ is finite. Given $\varepsilon > 0$ show that there exists a function G on K with upper derivative $\overline{D}G \leq f$ on K and $\underline{\int_K} f\Delta x - \varepsilon < \Delta G(K) \leq \underline{\int_K} f\Delta x$.

5. Let f be a function on a bounded cell K such that $\overline{\int_K} f\Delta x$ is finite. Given $\varepsilon > 0$ show that there exists a function H on K with lower derivative $\underline{D}H \geq f$ on K and $\overline{\int_K} f\Delta x \leq \Delta H(K) < \overline{\int_K} f\Delta x + \varepsilon$.

6. (The Perron Integral.) For f a function on a bounded cell K show that $f\Delta x$ is integrable if and only if given $\varepsilon > 0$ there exist functions G, H on K such that $\Delta(H - G)(K) < \varepsilon$ and $\overline{D}G \leq f \leq \underline{D}H$. (See [7].)

§1.6 Uniform Integrability and Its Consequences.

We wish now to show that in (ii) of Theorem 3 (§1.5) we have uniform integrability of S over all subfigures of C. This is the result (iii) in Theorem 1.

THEOREM 1. *Let S be a summant on figure C.*

(i) If $\overline{\int_C} S$ is finite then given $\varepsilon > 0$ there exists a gauge δ on C such that $S^{(\delta)}(B) \leq \overline{\int_B} S + \varepsilon$ for every subfigure B of C.

(ii) If $\underline{\int_C} S$ is finite then given $\varepsilon > 0$ there exists a gauge δ on C such that $S_{(\delta)}(B) \geq \underline{\int_B} S - \varepsilon$ for every subfigure B of C.

(iii) If S is integrable over C then given $\varepsilon > 0$ there exists a gauge δ on C such that $|(\sum S)(\mathcal{B}) - \int_B S| \leq \varepsilon$ for every subfigure B of C and δ-division $\mathcal{B}$ of B.

PROOF. By the hypothesis in $(i) - \infty < \overline{\int_C} S = \inf_\delta S^{(\delta)}(C)$. So given $\varepsilon > 0$ there exists a gauge δ on C such that

$$(1) \qquad\qquad S^{(\delta)}(C) \le \overline{\int_C} S + \varepsilon.$$

We contend that (1) holds for every subfigure B of C, thereby proving (i). Let A be the complementary figure in C to a given subfigure B of C. Since $S^{(\delta)}$ is superadditive (1) gives

$$(2) \qquad S^{(\delta)}(A) + S^{(\delta)}(B) \le S^{(\delta)}(C) \le \overline{\int_C} S + \varepsilon$$

with all terms finite since $\overline{\int_C} S$ is finite, and $-\infty < S^{(\delta)} \le \infty$ for $-\infty < S < \infty$. By the additivity of the upper integral given by (ii) in Theorem 1 (§1.5)

$$(3) \qquad \overline{\int_C} S = \overline{\int_A} S + \overline{\int_B} S \le S^{(\delta)}(A) + \overline{\int_B} S.$$

By (2) and (3)

$$(4) \qquad S^{(\delta)}(A) + S^{(\delta)}(B) \le S^{(\delta)}(A) + \overline{\int_B} S + \varepsilon.$$

Since $S^{(\delta)}(A)$ is finite we can subtract it from both sides of (4) to get $S^{(\delta)}(B) \le \overline{\int_B} S + \varepsilon$ which proves (i).

(ii) is just the dual of (i) which is (i) applied to $-S$.

To prove (iii) apply (i) and (ii) with $\int S = \underline{\int} S = \overline{\int} S$ to get for every subfigure B of C and δ-division $\mathcal{B}$ of B, $\int_B S - \varepsilon \le S_{(\delta)}(B) \le (\sum S)(\mathcal{B}) \le S^{(\delta)}(B) \le \int_B S + \varepsilon$. So $-\varepsilon \le (\sum S)(\mathcal{B}) - \int_B S \le \varepsilon$ which proves (iii). $\square$

The next two theorems are consequences of Theorem 1 that will be reformulated in Chapter 2.

THEOREM 2. *Let S be a summant on a figure C.*

(i) If $-\infty < \overline{\int_I} S \leq 0$ for every cell I in C then $\int_C S^+ = 0$.

(ii) If $0 \leq \underline{\int_I} S < \infty$ for every cell I in C then $\int_C S^- = 0$.

(iii) If $\int_I S = 0$ for every cell I in C then $\int_C |S| = 0$.

PROOF. To prove (i) let $\varepsilon > 0$ be given. By (i) in Theorem 1 there exists a gauge δ on C such that $S^{(\delta)}(B) \leq \overline{\int_B} S + \varepsilon$ for every subfigure B of C. By the hypothesis in $(i) -\infty < \overline{\int_B} S \leq 0$ since the figure B is a finite union of nonoverlapping cells in C, and the upper integral is additive. So $S^{(\delta)}(B) \leq \varepsilon$.

Given any δ-division $\mathcal{C}$ of C let $\mathcal{B}$ consist of all members of $\mathcal{C}$ at which $S > 0$. Then $\mathcal{B}$ is a δ-division of a subfigure B of C. So

$$\left(\sum S^+\right)(\mathcal{C}) = \left(\sum S\right)(\mathcal{B}) \leq S^{(\delta)}(B) \leq \varepsilon.$$

That is, $\left(\sum S^+\right)(\mathcal{C}) \leq \varepsilon$ for every δ-division $\mathcal{C}$ of C. This proves implication (i).

Implication (ii) is just the dual of (i).

Since $|S| = S^+ + S^-, (iii)$ follows from (i) and (ii) for $\overline{\int_I} S = \underline{\int_I} S = \int_I S = 0$. $\square$

THEOREM 3. *Given an integrable summant S on a figure K let $\hat{S}$ be the additive summant defined by*

(5) $$\hat{S}(I) = \int_I S \text{ for every cell } I \text{ in } K.$$

Then

(6) $$\int_K |S - \hat{S}| = 0.$$

Conversely, if S is a summant on K and $\hat{S}$ is an additive summant on K such that (6) holds, then S is integrable and (5) holds.

PROOF. The existence and additivity of $\hat{S}$ defined by (5) follow from Theorem 3 (§1.5). Since $\hat{S}$ is additive $\hat{S}(I) = \int_I \hat{S}$

for every cell I in K. So $\int_I (S - \hat{S}) = 0$ for all I. Apply (iii) in Theorem 2 to $S - \hat{S}$ to get (6). The converse is trivial. $\quad\square$

Theorem 3 is the Saks-Henstock Lemma which we shall reformulate as Theorem 1 (§2.5) giving this result its definitive statement.

A set $\mathbb{S}$ of summants on a figure K is **uniformly integrable** on K if each member S of $\mathbb{S}$ is integrable and given $\varepsilon > 0$ there is a gauge δ on K such that

$$(7) \qquad |(\sum S)(\mathcal{K}) - \int_K S| \leq \varepsilon$$

for every δ-division $\mathcal{K}$ of K and every member S of $\mathbb{S}$. In Theorem 1 (iii) gives an equivalent formulation with (7) replaced by

$$(8) \qquad |(\sum S)(\mathcal{B}) - \int_B S| \leq \varepsilon$$

for every figure B in K, every δ-division $\mathcal{B}$ of B, and for every member S of $\mathbb{S}$. Applying (8) with ε replaced by $\varepsilon/2$ we can express uniform integrability in terms of (5) by the conclusion that

$$(9) \qquad |S - \hat{S}|^{(\delta)}(K) \leq \varepsilon$$

for every member S of $\mathbb{S}$.

Let S_i be a summant on K for each i in a nonempty set X where X is equipped with a transitive binary relation $\succ$ such that given j, k in X there exists i in X such that both $i \succ j$ and $i \succ k$. The convergence $S_i \xrightarrow{i} S$ of such a net of summants means that given a tagged cell (I, t) in K and $\varepsilon > 0$ there exists k in X such that $|S_i(I, t) - S(I, t)| < \varepsilon$ for all $i \succ k$.

We can now formulate a convergence theorem for integrable summants. Although it is just a special case of a general result on the interchange of two limits this convergence theorem is widely applicable.

THEOREM 4. *Let* $S_i \xrightarrow[i]{} S$ *be a convergent net of uniformly integrable summants* S_i *on a cell* K. *Then* S *is integrable and*

$$(10) \qquad \int_K S_i \xrightarrow[i]{} \int_K S.$$

PROOF. Let $\varepsilon > 0$ be given. Since the S_i's are uniformly integrable there exists a gauge δ on K such that by (9)

$$(11) \qquad |S_i - \hat{S}_i|^{(\delta)}(K) \le \varepsilon \text{ for all } i \text{ in } X.$$

Consider an arbitrary cell J in K. For all δ-divisions $\mathcal{J}$ of J and all i, j in X we have

$$|\hat{S}_i(J) - \hat{S}_j(J)| \le |(\sum(\hat{S}_i - S_i))(\mathcal{J})| +$$
$$|(\sum(S_i - S_j))(\mathcal{J})| + |(\sum(S_j - \hat{S}_j))(\mathcal{J})|.$$

By (11) this gives

$$(12) \qquad |\hat{S}_i(J) - \hat{S}_j(J)| \le |(\sum(S_i - S_j))(\mathcal{J})| + 2\varepsilon.$$

Now $S_i - S_j \xrightarrow[i,j]{} 0$ since $S_i \xrightarrow[i]{} S$. Thus $(\sum(S_i - S_j))(\mathcal{J}) \xrightarrow[i,j]{} 0$. So (12) gives $\varlimsup_{i,j} |\hat{S}_i(J) - \hat{S}_j(J)| \le 2\varepsilon$. This is the Cauchy criterion for the convergence

$$(13) \qquad \hat{S}_i(J) \xrightarrow[i]{} \hat{S}(J)$$

where the limit $\hat{S}(J)$ for every cell J in K defines an additive summant $\hat{S}$ on K since each $\hat{S}_i$ is additive.

For any δ-division $\mathcal{K}$ of K

$$|\hat{S}(K) - (\sum S)(\mathcal{K})| \le |(\hat{S} - \hat{S}_i)(K)| +$$
$$|(\sum(\hat{S}_i - S_i))(\mathcal{K})| + |(\sum(S_i - S))(\mathcal{K})| \le$$
$$|(\hat{S} - \hat{S}_i)(K)| + 2\varepsilon \text{ for all } i \text{ by (11)}.$$

By (13) with $J = K$ this gives $|\hat{S}(K) - (\sum S)(\mathcal{K})| \leq 2\varepsilon$ for every δ-division $\mathcal{K}$ of K. That is , $\int_K S = \hat{S}(K)$. So S is integrable and (13) gives (10). $\square$

Uniform integrability of S on all figures B in K ((iii) in Theorem 1) does not in general imply uniform integrability of $1_B S$ on K for all figures B in K. (See Exercises 1 and 2.)

Weakening the hypothesis of Theorem 3 from integrability of S to upper integrability we get the following result.

THEOREM 5. *Let S be a summant on a figure C such that $\overline{\int_C} S$ is finite. Let $\overline{S}$ be the additive summant defined by*

$$(14) \qquad\qquad \overline{S}(I) = \overline{\int_I} S$$

for every cell I in C. Then the following hold:

(i) $\int_B \overline{S} = \overline{\int_B} S$ for every figure B on C,

(ii) $\overline{\int_B}(S - \overline{S}) = 0$ for every figure B in C,

(iii) $\int_C (S - \overline{S})^+ = 0$,

(iv) If T is any additive summant on C such that $\int_C (S - T)^+ = 0$ then $T \geq \overline{S}$.

PROOF. Finiteness of $\overline{S}$ follows from (iv) in Theorem 2 (§1.5), additivity from (ii) in Theorem 1 (§1.5). Since B can be partitioned into cells (i) follows from the additivity of (14).

To prove (ii) we have $\overline{\int_B}(S - \overline{S}) = \overline{\int_B} S - \int_B \overline{S} = 0$ by (i).

To prove (iii) use (ii) to apply (i) of Theorem 2 to the summant $S - \overline{S}$.

To prove (iv) let I be any cell in C. By (14) and the conditions imposed on T, $\overline{S}(I) - T(I) = \overline{\int_I} S - \int_I T = \overline{\int_I}(S - T) \leq \overline{\int_C}(S - T)^+ = 0$. So $\overline{S} \leq T$. $\square$

Exercises (§1.6).

1. Show that if f is a continuous function on a cell K then $1_I \Delta f$ is uniformly integrable on K for all cells I in K.

2. Given a bounded cell K prove that $1_A \Delta x$ is not uniformly integrable on K for all figures A. Specifically, show that given any division $\mathcal{K} = \{(I_1, t_1), \cdots, (I_n, t_n)\}$ of K:

   (i) For each $i = 1, \cdots, n$ there is a unique cell J_i in I_i such that $t_i \in J_i$ and $\Delta x(J_i) = \frac{1}{n+1} \Delta x(I_i)$,

   (ii) For $A = J_1 \cup \cdots \cup J_n$, $\int_K 1_A \Delta x = \frac{1}{n+1} \Delta x(K)$, $\Delta x(K) = (\sum 1_A \Delta x)(\mathcal{K})$, $(\sum 1_A \Delta x)(\mathcal{K}) - \int_K 1_A \Delta x \geq \frac{1}{2} \Delta x(K)$.

3. Let $K = [a, b]$ be a bounded cell. Let $F_1, F_2, \cdots$ be a sequence of differentiable functions on K which are equidifferentiable at each point in K. Let $f_n = F_n'$.

   (i) Show that $f_1 \Delta x, f_2 \Delta x, \cdots$ are uniformly integrable on the bounded cell K.

   (ii) Apply Theorem 4 to show that if $F_n(a) = 0$ for all n and $f_n \longrightarrow f$ then $F_n \longrightarrow F$ where $F(a) = 0$ and $F' = f$ on K.

4. Let $S_1, S_2, \cdots$ be uniformly integrable summants on a cell K. Define the additive summant $\hat{S}_n$ by $\hat{S}_n(I) = \int_I S_n$ for all cells I in K. Show that $|S_1|, |S_2|, \cdots$ are uniformly integrable on K if and only if $|\hat{S}_1|, |\hat{S}_2|, \cdots$ are uniformly integrable on K.

5. Let $S_1, S_2, \cdots$ be nonnegative summants on a cell K and $\cup_k E_k = K$ such that given k the summants $1_{E_k} S_1, 1_{E_k} S_2, \cdots$ integrate uniformly to 0. Prove that $S_1, S_2, \cdots$ integrate uniformly to 0 on K. (Given $\varepsilon > 0$ take $\delta_{k+1} \leq \delta_k$ such that $(1_{E_k} S_n)^{(\delta_k)}(K) < \frac{\varepsilon}{2^k}$ for all n, k. Let $\delta(t) = \delta_k(t)$ where k is the smallest integer such that $t \in E_k$. Show that $S_n^{(\delta)}(K) < \varepsilon$ for all n.)

6. Let $T_1, T_2, \cdots$ be nonnegative summants on a cell K which integrate uniformly to 0. Let f be a function on K. Prove that $fT_1, fT_2, \cdots$ integrate uniformly to 0. (Apply Exercise 5 with E_k the set of points where $|f| < k$.) (Theorem 1 in §1.8 follows from Exercise 6. For other results on uniform integrability see Theorem 5 (§2.8) and Exercise 17 (§3.3).)

§1.7 Term-by-Term Integration of Series.

We need two lemmas for Theorem 1, a fundamental result with many useful consequences.

LEMMA A. *Let S be a summant on a figure K such that $S \geq 0$. Let α, β be gauges on K such that every α-fine tagged cell at which $S > 0$ is β-fine. Then $S^{(\alpha)}(K) \leq S^{(\beta)}(K)$.*

PROOF. Given an α-division $\mathcal{K}$ of K let $\mathcal{K}_0$ consist of all members of $\mathcal{K}$ at which $S = 0$, and $\mathcal{K}_1$ consist of all at which $S > 0$. By hypothesis all members of $\mathcal{K}_1$ are β-fine. Let B be the figure partitioned by $\mathcal{K}_0$. Take an arbitrary β-division $\mathcal{B}$ of B and let $\mathcal{K}'$ be the β-division $\mathcal{K}_1 \cup \mathcal{B}$ of K. Then $(\sum S)(\mathcal{K}) = (\sum S)(\mathcal{K}_1) \leq (\sum S)(\mathcal{K}_1) + (\sum S)(\mathcal{B}) = (\sum S)(\mathcal{K}') \leq S^{(\beta)}(K)$. That is, $(\sum S)(\mathcal{K}) \leq S^{(\beta)}(K)$ for every α-division $\mathcal{K}$ of K. Hence, $S^{(\alpha)}(K) \leq S^{(\beta)}(K)$. $\square$

LEMMA B. *Let the summants $T_i \geq 0$ on K for $i = 0, 1, 2, \cdots$ such that given $0 < c < 1$ there exists an integer-valued function $n(t) \geq 1$ on K satisfying*

$$(1) \qquad cT_0(I, t) \leq \sum_{i=1}^{n(t)} T_i(I, t)$$

for every tagged cell (I, t) in K. Then

$$(2) \qquad 0 \leq \overline{\int}_K T_0 \leq \sum_{i=1}^{\infty} \overline{\int}_K T_i \leq \infty.$$

PROOF. Define the summant S_i on K for $i = 1, 2, \cdots$ by letting $S_i(I, t) = T_i(I, t)$ for $i \le n(t), 0$ for $i > n(t)$. Then $cT_0 \le \sum_{i=1}^{\infty} S_i$ by (1). So for every gauge α on K

$$(3) \qquad c\overline{\int}_K T_0 \le cT_0^{(\alpha)}(K) \le \sum_{i=1}^{\infty} S_i^{(\alpha)}(K).$$

Let $\varepsilon > 0$ be given. Since $0 \le S_i \le T_i$ for all $i \ge 1$ we can choose for each such i a gauge β_i on K small enough so that

$$(4) \qquad S_i^{(\beta_i)}(K) \le \overline{\int}_K S_i + \frac{\varepsilon}{2^i} \le \overline{\int}_K T_i + \frac{\varepsilon}{2^i}$$

for $i = 1, 2, \cdots$. Let $\alpha(t)$ be the minimum value of $\beta_i(t)$ for $i = 1, \cdots, n(t)$. Then by Lemma A

$$(5) \qquad S_i^{(\alpha)}(K) \le S_i^{(\beta_i)}(K) \text{ for } i = 1, 2, \cdots$$

since $\alpha(t) \le \beta_i(t)$ if $S_i(I, t) > 0$. By (3), (5) and (4) $c\overline{\int}_K T_0 \le \sum_{i=1}^{\infty} \overline{\int}_K T_i + \varepsilon$. Letting $\varepsilon \to 0+$ and then $c \to 1-$ we get (2). □

THEOREM 1. *Let S be a summant on a figure K such that $S \ge 0$. For $i = 0, 1, \cdots$ let v_i be a function on K such that $0 \le v_i(t) < \infty$ and $v_o(t) \le \sum_{i=1}^{\infty} v_i(t) \le \infty$ for all t in K. Then*

$$(6) \qquad 0 \le \overline{\int}_K v_o S \le \sum_{i=1}^{\infty} \overline{\int}_K v_i S \le \infty.$$

If moreover $\int_K v_i S$ exists for $i = 1, 2, \cdots$ and $v_o = \sum_{i=1}^{\infty} v_i$ then

$$(7) \qquad 0 \le \int_K v_o S = \sum_{i=1}^{\infty} \int_K v_i S \le \infty.$$

PROOF. We shall apply Lemma B with

$$(8) \qquad T_i = v_i S \text{ for } i = 0, 1, \cdots .$$

Given $0 < c < 1$ consider any point t in K. If $v_o(t) > 0$ then $cv_o(t) < v_o(t) \leq \sum_{i=1}^{\infty} v_i(t)$. So there is a smallest positive integer $n(t)$ such that

$$(9) \qquad cv_o(t) \leq \sum_{i=1}^{n(t)} v_i(t).$$

If on the other hand $v_o(t) = 0$ then (9) holds for $n(t) = 1$. For any cell I with endpoint t, multiplying (9) through by $S(I, t)$ gives (1) in Lemma B under (8). So Lemma B gives (6) by (2) and (8).

The final statement in Theorem 1 follows from (6) and the superadditivity of lower integrals in (2) of §1.5. □

Theorem 1 plays a leading role in our exposition. The next section treats some of it immediate consequences.

Exercises (§1.7).

1. Show that in Lemma A we can conclude that $S^{(\alpha)}(H) \leq S^{(\beta)}(H)$ for every figure H in K.

2. Show that Lemma B remains valid if instead of demanding that (1) hold on all tagged cells in K we require only the existence of a gauge δ on K such that (1) holds on all δ-fine tagged cells in K.

3. Let $S \geq 0$ be a summant on a cell K and $E_n \nearrow K$ for some monotone sequence of subsets of K such that $\int_K 1_{E_n} S$ exists for $n = 1, 2, \cdots$. Show that $\int_K S$ exists and $\int_K 1_{E_n} S \nearrow \int_K S$. (If $1_{E_n} S$ is integrable for all n apply Theorem 1 with $v_o = 1$ and $v_i = 1_{E_i} - 1_{E_{i-1}}$ for $i \geq 1$ and $E_o = \emptyset$.)

§1.8 Applications of Term-by-Term Upper Integration.

The next two theorems are corollaries of Theorem 1 (§1.7) that will be useful in Chapter 2 with elegant reformulations.

THEOREM 1. *Let T be a summant on a figure K such that $\int_K |T| = 0$. Then $\int_K |fT| = 0$ for every function f on K.*

PROOF. Apply Theorem 1 (§1.7) with $v_o = |f|, v_i = 1$ for $i = 1, 2, \cdots$, and $S = |T|$ to get $\int_K |fT| = \sum_{i=1}^{\infty} \int_K |T| = 0$ according to (7)(§1.7). $\square$

Recall that a function on a set K is the indicator 1_E of a subset E of K if $1_E(t) = 1$ for t in $E, 0$ for t in $K - E$.

THEOREM 2. *Let f be a function on a figure K. Let E be the set of points t in K such that $f(t) \neq 0$. Then for every summant T on K, $\int_K |fT| = 0$ if and only if $\int_K |1_E T| = 0$.*

PROOF. Let E_i consist of all t such that $|f(t)| > \frac{1}{i}$. Then $E_1 \cup E_2 \cup \cdots = E$, so $1_E \leq \sum_{i=1}^{\infty} 1_{E_i}$. Since $1_{E_i} \leq i|f|, |1_{E_i}T| \leq i|fT|$. So $\int_K |fT| = 0$ implies $\int_K |1_{E_i}T| = 0$. Hence, Theorem 1(§1.7) applied with $S = |T|, v_o = 1_E$, and $v_i = 1_{E_i}$ for $i = 1, 2, \cdots$ gives $\int_K |1_E T| = \sum_{i=1}^{\infty} \int_K |1_{E_i}T| = 0$ by (7) (§1.7).

Conversely $\int_K |1_E T| = 0$ implies $\int_K |f1_E T| = 0$ by Theorem 1. That is, $\int_K |fT| = 0$ since $f1_E = f$. $\square$

Hereafter we may revert to the traditional notation $\int_a^b S$ for $\int_{[a,b]} S$ with the convention that $\int_b^a S = -\int_a^b S$ and $\int_a^a S = 0$. This applies also to the upper and lower integrals. In the next section we shall introduce integration over open or half open intervals. We shall denote the integral over (a, b) by $\int_{(a,b)} S$ restricting the notation $\int_a^b S$ to $\int_{[a,b]} S$ for $-\infty \leq a < b \leq \infty$.

The next consequence of Theorem 1 is a primitive monotone convergence theorem.

THEOREM 3. *Let $S \geq 0$ be a summant on $K = [a, b]$. Then*

(i) $\overline{\int_a^r} S \nearrow \overline{\int_a^b} 1_{[a,b)} S$ *as* $r \nearrow b$ *in* (a, b)
and
(ii) $\overline{\int_r^b} S \nearrow \overline{\int_a^b} 1_{(a,b]} S$ *as* $r \searrow a$ *in* (a, b).

PROOF. We shall prove (i). The proof of (ii) is similar and is left as an exercise. Additivity of the upper integral ((ii) in Theorem 1, §1.5) over abutting cells will be used repeatedly.

For $q < r$ in K, $\overline{\int_a^r} S = \overline{\int_a^q} S + \overline{\int_q^r} S \geq \overline{\int_a^q} S$ since $S \geq 0$. For $r < b$, $\overline{\int_a^r} S = \overline{\int_a^r} 1_{[a,b)} S \leq \overline{\int_a^b} 1_{[a,b)} S$. So for some $c \leq \infty$,

$$(1) \qquad \overline{\int_a^r} S \nearrow c \leq \overline{\int_a^b} 1_{[a,b)} S.$$

Since (i) is trivial if $c = \infty$ in (1) we need only consider the case $c < \infty$.

Take a sequence $r_0 < r_1 < \cdots$ in (a, b) such that $r_n \nearrow b$. By the monotone convergence in (1)

$$(2) \qquad \overline{\int_a^{r_n}} S \nearrow c < \infty.$$

Now since the upper integral is additive over abutting cells,

$$(3) \qquad \overline{\int_a^{r_n}} S = \overline{\int_a^{r_0}} S + \sum_{i=1}^{n} \overline{\int_{r_{i-1}}^{r_i}} S$$

By (2) and (3) we get the convergent series

$$(4) \qquad \overline{\int_a^{r_0}} S + \sum_{i=1}^{\infty} \overline{\int_{r_{i-1}}^{r_i}} S = c < \infty.$$

Now for all $n \geq 1$,

$$(5) \qquad 1_{[a,b)} \leq 1_{[a,r_n)} + \sum_{i=n}^{\infty} 1_{(r_{i-1}, r_{i+1})}.$$

Applying Theorem 1 to (5) and invoking (2) we get

$$\overline{\int_a^b} 1_{[a,b)} S \leq \overline{\int_a^{r_n}} S + \sum_{i=n}^{\infty} \overline{\int_{r_{i-1}}^{r_{i+1}}} S \leq$$

$$c + \sum_{i=n}^{\infty} \left(\overline{\int_{r_{i-1}}^{r_i}} S + \overline{\int_{r_i}^{r_{i+1}}} S \right) \leq c + 2 \sum_{i=n}^{\infty} \overline{\int_{r_{i-1}}^{r_i}} S.$$

As $n \to \infty$ this gives

(6)
$$\overline{\int_a^b} 1_{[a,b)} S \leq c$$

by the convergence of the series in (4). Finally, (i) follows from (1) and (6).   □

Exercises (§1.8).

1. Let g be a continuous function on a cell K such that $\Delta g \geq 0$. Show that $\int_K f(\Delta g)^2 = 0$ for every function f on K.

2. Let $S \geq 0$ be a summant on $K = [0,1]$ such that $\int_K 1_0 S = 0$ and S is integrable on $[\varepsilon, 1]$ for all ε in $(0,1)$. Show that $\int_0^1 S$ exists, $0 \leq \int_0^1 S \leq \infty$, and $\int_\varepsilon^1 S \nearrow \int_0^1 S$ as $\varepsilon \searrow 0$.

3. Apply Theorem 1 to show that if S and S' are summants on a cell K such that $\int_K |S - S'| = 0$ then $\int_K |fS - fS'| = 0$ for every function f on K. (The importance of this result will become evident in §2.7.)

§1.9 Integration over Arbitrary Intervals.

Integration over figures in a cell K is easily subsumed by integration over K. Given a summant S_A on a figure A in K extend it to a summant S on K by letting $S(I,t) = S_A(I,t)$ if $I \subseteq A$, and $S(I,t) = 0$ otherwise. Let B be the complementary figure to A in K. Then by the additivity given by Theorem 1 ($\S1.5$), $\overline{\int_K} S = \overline{\int_A} S + \overline{\int_B} S = \overline{\int_A} S_A + 0 = \overline{\int_A} S_A$ and similarly $\underline{\int_K} S = \underline{\int_A} S_A$. Thus, if it were convenient we could treat all integration in one dimension as integration over $[-\infty, \infty]$ with no loss of generality.

A similar technique can be used to define integration over an arbitrary interval L in $[-\infty, \infty]$ which may be open at one or both ends. The details are as follows.

Let L be a subinterval of a cell $K = [a, b]$ such that $\overline{L} = K$. That is, L is one of the four intervals $[a, b), (a, b], (a, b)$, or $[a, b]$. A summant S_L on L is a function on the set of all tagged cells in L. We extend such a summant S_L to a summant S on K by defining

$$(1) \qquad S(I, t) = \begin{cases} S_L(I, t) & \text{if } I \subseteq L \\ 0 & \text{otherwise .} \end{cases}$$

(Note that (1) actually defines S as a summant on $[-\infty, \infty]$.) We then define integration of S_L over L by

$$(2) \qquad \overline{\int_L} S_L = \overline{\int_K} S \text{ and } \underline{\int_L} S_L = \underline{\int_K} S$$

with concomitant definitions of $\int_L S_L = \int_K S$ whenever the latter integral exists and integrability of S_L over L whenever S is integrable over K.

Given a summant S on K its restriction to the tagged cells in L is a summant S_L on L. In this case the extension defined by (1) is just the summant $\mathbf{1}_L S$ on K where $\mathbf{1}_L$ is the cell summant on K that indicates the cells in L,

$$(3) \qquad \mathbf{1}_L(I) = 1 \text{ if } I \subseteq L, 0 \text{ otherwise .}$$

Consider any gauge small enough so that any δ-fine tagged cell in K which contains an endpoint a, b of K must have that endpoint as its tag. For such a gauge δ, $1_L(I) = 1_L(t)$ for every δ-fine tagged cell (I, t) in K. So for the restriction S_L of S to the tagged cells in L we can express (2) in the form

$$\overline{\int_L} S_L = \overline{\int_K} 1_L S = \overline{\int_K} 1_L S$$

(4)   and

$$\underline{\int_L} S_L = \underline{\int_K} 1_L S = \underline{\int_K} 1_L S$$

which is valid for any interval L such that $\overline{L} = K$.

For S a summant on $K = [a, b]$, and $L = [a, b), S = 1_L S + 1_b S$ since $1_L + 1_b = 1$ on K. This suggests that we must study the behavior of $1_p S$ at the endpoints p of K in order to understand the behavior of $1_L S$ in (4). The next result gives explicit expressions for the upper and lower integrals of $1_p S$ for any point p in K.

THEOREM 1. *For S a summant on $K = [a, b]$*
(i) $\overline{\int_K} 1_b S = \overline{\lim_{r \to b-}} S([r, b], b)$ *and* $\underline{\int_K} 1_b S = \underline{\lim_{r \to b-}} S([r, b], b)$

and

(ii) $\overline{\int_K} 1_a S = \overline{\lim_{r \to a+}} S([a, r], a)$ *and* $\underline{\int_K} 1_a S = \underline{\lim_{r \to a+}} S([a, r], a)$.
Moreover, if $a < p < b$ then
(iii) $\overline{\int_K} 1_p S = \overline{\lim_{r \to p-}} S([r, p], p) + \overline{\lim_{r \to p+}} S([p, r], p)$ *with value* ∞ *if the right-hand side is indeterminate,*
and
(iv) $\underline{\int_K} 1_p S = \underline{\lim_{r \to p-}} S([r, p], p) + \underline{\lim_{r \to p+}} S([p, r], p)$ *with value* $-\infty$ *if the right-hand side is indeterminate. Finally, for all p in K and all cells I in K with p as an endpoint,*
(v) $\overline{\lim_{(I,p) \to p}} |S(I, p)| \leq \overline{\int_K} 1_p |S| \leq 2 \overline{\lim_{(I,p) \to p}} |S(I, p)|$.

PROOF. For a sufficiently small gauge δ every δ-division $\mathcal{K}$ of K has a final member with tag b, and every δ-fine tagged cell in K with tag b is the final member of some δ-division $\mathcal{K}$. For such $\mathcal{K}$ $(\sum 1_b S)(\mathcal{K}) = S([r, b], b)$ where $a \leq r < b$ and $r \in N_\delta(b)$. Hence,

$$(1_b S)^{(\delta)}(K) = \sup_{r \in [a,b) \cap N_\delta(b)} S([r, b], b)$$

and

$$(1_b S)_{(\delta)}(K) = \inf_{r \in [a,b) \cap N_\delta(b)} S([r, b], b).$$

As $\delta \to 0$ this gives (i).

A similar proof gives (ii). To get (iii) and (iv) apply (i) and (ii) to $[a, p]$ and $[p, b]$ respectively. Then use Theorem 1 (§1.5) to get additivity over $[a, b] = [a, p] \cup [p, b]$. (v) follows from $(i), (ii), (iii)$ applied to $|S|$. $\square$

The next result will enable us to characterize integrability over $[a, b)$.

THEOREM 2. *A summant S on $K = [a, b]$ is integrable over K if and only if*
(i) S is integrable over $[a, r]$ for all r in (a, b)
and
(ii) $\lim_{r \to b-} \left[\int_a^r S + S([r, b], b) \right]$ exists and is finite. Moreover, for S integrable over K the limit in (ii) equals $\int_K S$.

PROOF. Let S be integrable over K. Then (i) follows from (ii) in Theorem 3 (§1.5). To prove (ii) we have $\int_r^b S = \int_a^r S + \int_r^b S$. So $\left[\int_a^r S + S[r, b], b) \right] - \int_a^b S = S([r, b], b) - \int_r^b S$. Therefore, to prove that $\int_a^b S$ equals the limit in (ii) we need only prove

$$(5) \qquad S([r, b], b) - \int_r^b S \to 0 \text{ as } r \to b-\,.$$

Let $\varepsilon > 0$ be given. By the uniform integrability condition
(iii) in Theorem 1 (§1.6) there exists a gauge δ on K such
that $|(\sum S)(\mathcal{B}) - \int_B S| \leq \varepsilon$ for every subfigure B of K and
δ-division $\mathcal{B}$ of B. Thus, for r in (a, b) such that $([r, b], b)$ is
δ-fine, this single tagged cell forms a δ-division $\mathcal{B}$ of $B = [r, b]$.
So $|S([r, b], b) - \int_r^b S| \leq \varepsilon$ for all r in $(a, b) \cap N_\delta(b)$. This proves
(5). So (ii) holds and the limit in (ii) equals $\int_K S$.

Conversely let (i) and (ii) hold. Let c be the finite limit in
(ii). Define the function f on K by

(6)
$$f(t) = \begin{cases} \int_a^t S \text{ for } a \leq t < b \\ c \text{ for } t = b. \end{cases}$$

Since Δf is an additive summant on K we need only prove

(7)
$$\int_K |S - \Delta f| = 0$$

to conclude by Theorem 3 (§1.6) that S is integrable. Indeed,
by that theorem and (i), $\int_a^r |S - \Delta f| = 0$ for all r in (a, b). By
Theorem 3 (§1.8) $\int_a^r |S - \Delta f| \nearrow \overline{\int_K} 1_{[a,b)} |S - \Delta f|$ as $r \nearrow b$. So
$\int_K 1_{[a,b)} |S - \Delta f| = 0$. Since $1_{[a,b)} + 1_b = 1_K$ this implies

(8)
$$\overline{\int_K} |S - \Delta f| = \overline{\int_K} 1_b |S - \Delta f|.$$

Now by (i) in Theorem 1

(9)
$$\overline{\int_K} 1_b |S - \Delta f| = \overline{\lim_{r \to b-}} |S - \Delta f|([r, b], b).$$

In terms of (6), $|S - \Delta f|([r, b], b) = |S([r, b], b) - \Delta f[r, b]| =$
$|S([r, b], b) + f(r) - f(b)| = |S([r, b], b) + \int_a^r S - c| \to 0$ as
$r \to b-$ by the definition of c as the limit in (ii). That is,

(10) $|S - \Delta f|([r, b], b) \to 0$ as $r \to b-$.

Finally, (7) follows from (8), (9), and (10). $\square$

Theorem 2 readily gives a characterization of integrability
of S_L over $[a, b)$ which is just integrability of S in (1) over $[a, b]$
according to (2).

THEOREM 3. *A summant S on $L = [a, b)$ is integrable over L if and only if*

(i) S is integrable over $[a, r]$ for all r in (a, b)
and
(ii) $\lim_{r \to b-} \int_a^r S$ exists and is finite.
If S is integrable over L then $\int_{[a,b)} S = \lim_{r \to b-} \int_a^r S$.

PROOF. Apply Theorem 2 to $\overline{S}$ defined on $[a, b]$ by (1). □

Results similar to those in Theorem 2 and Theorem 3 hold at the left endpoint of $[a, b]$. The proofs of these results and the following generalization of Theorem 3 are left as exercises.

THEOREM 4. *A summant S on a nondegenerate interval L in $[-\infty, \infty]$ is integrable over L if and only if*

(i) S is integrable over every cell J contained in L, and for the set of those cells directed upward by containment,

(ii) $\lim_{J \to L} \int_J S$ exists and is finite.
Moreover, for S integrable over L the limit in (ii) equals $\int_L S$.
Finally,
(iii) If $\int_I S = 0$ for every cell in L then $\int_L |S| = 0$.

((iii) comes from (iii) in Theorem 2 (§1.6).)

Exercises (§1.9).

1. Given a function f on $L = [a, b)$ define for each r in L the function f_r on $[a, b]$ by $f_r(t) = f(t \wedge r)$ for all t in $[a, b]$. Prove:

   (i) $\Delta f_r(I) = 0$ for every cell I in $[r, b]$.

   (ii) $\int_L \Delta f_r = \int_a^b \Delta f_r = \int_a^r \Delta f = f(r) - f(a)$.

   (iii) The following three conditions are equivalent:

   (A) Δf is integrable on L.

   (B) $f(b-)$ exists and is finite.

(C) $1_L \Delta f_r$ is uniformly integrable on the cell $[a, b]$ for all r in L.

(Use Theorem 4 (§1.6) and Theorem 3 (§1.9).)

2. Show that $\int_{\mathbb{R}} (\Delta x)^2 = 0$ for the identity function x on $\mathbb{R}$.

3. Let $f(t) = sgn\, t$ for all t in $\mathbb{R}$. Show that

(i) $\int_K \Delta f = 1_a(0) + 1_b(0) + 2\, 1_{(a,b)}(0)$ and $\int_K \Delta |f| = 1_a(0) - 1_b(0)$ for every cell $K = [a, b]$ in $\mathbb{R}$.

(ii) $\int_L \Delta f = 2\, 1_L(0)$ and $\int_L \Delta |f| = 0$ for every open interval L in $\mathbb{R}$.

4. Given a function F on $\mathbb{R}$ and a point p in $\mathbb{R}$ prove:
(i) The Dini derivates (the lower and upper left derivatives D_- and D^-, the lower and upper right derivatives D_+ and D^+) are given by

$$\underline{\int_{\mathbb{R}}} 1_p Q^- \frac{\Delta F}{\Delta x} = D_- F(p), \overline{\int_{\mathbb{R}}} 1_p Q^- \frac{\Delta F}{\Delta x} = D^- F(p),$$

$$\underline{\int_{\mathbb{R}}} 1_p Q^+ \frac{\Delta F}{\Delta x} = D_+ F(p), \overline{\int_{\mathbb{R}}} 1_p Q^+ \frac{\Delta F}{\Delta x} = D^+ F(p).$$

(ii) F has finite left and right derivatives at p if and only if $1_p \frac{\Delta F}{\Delta x}$ is integrable on $\mathbb{R}$.

(iii) F is differentiable at p if and only if $1_p \frac{\Delta F}{\Delta x}$ is integrable on $\mathbb{R}$ and its integrals over $(-\infty, p]$ and $[p, \infty)$ are equal.

5. Given a function f on $\mathbb{R}$ show that $\int_{\mathbb{R}} |\Delta f| = 0$ if and only if f is constant.

6. Let $\mathcal{R}$ be the set of all functions f on $\mathbb{R}$ such that Δf is integrable over every nondegenerate interval in $\mathbb{R}$. Prove:
(a) If f, g belong to $\mathcal{R}$ then so do $f + g$, fg, and $|f|$.

(b) Every uniform limit of functions in $\mathcal{R}$ belongs to $\mathcal{R}$.

(c) Every bounded, monotone function on $\mathbb{R}$ belongs to $\mathcal{R}$.

(This exercise gives a preview of §4.4.)

7. Let g be a continuous, decreasing function on $\mathbb{R}$ whose limit at ∞ is 0. Let S be the summant whose value at (I, t) is $g(t) \sin t \Delta x(I)$. Show that S is integrable over $[0, \infty)$. (This proves in effect that the improper Riemann integral $\int_0^\infty g(x) \sin x \, dx$ is finite. In the next chapter we shall justify this traditional differential notation for integrals.)

8. Let S be a summant on $\mathbb{R}$ such that $0 \leq \underline{\int_I S} < \infty$ for every cell I in $\mathbb{R}$.
   Show that $\int_{\mathbb{R}} S^- = 0$. (Use Theorem 2 in §1.6.)

CHAPTER 2

DIFFERENTIALS AND THEIR INTEGRALS

§2.1 Differential Equivalence and Differentials.

We are now ready to make the transition from summants to differentials, the ultimate objects of integration. We accomplish this by annihilating the ideal of summants S on a figure K such that $\int_K |S| = 0$.

For summants S, T on a figure K we define **differential equivalence** $S \sim T$ to be $\int_K |S - T| = 0$. This relation is obviously reflexive and symmetric. That is, $S \sim S$, and $S \sim T$ implies $T \sim S$. Transitivity follows from the triangle inequality, $\int_K |R - T| \leq \int_K |R - S| + \int_K |S - T|$, which implies $R \sim T$ if both $R \sim S$ and $S \sim T$.

A **differential** σ on K is an equivalence class of summants on K under differential equivalence. That is, $\sigma = [S]$ where $[S]$ is the set of all summants S' on K such that $S' \sim S$. We define the **lower integral** and **upper integral** of a differential $\sigma = [S]$ respectively by

(1) $$\underline{\int_K} \sigma = \underline{\int_K} S \quad \text{and} \quad \overline{\int_K} \sigma = \overline{\int_K} S \ .$$

To show that these definitions are effective we must prove that they are independent of the particular representative S of the differential σ.

THEOREM 1. *If $S' \sim S$ on K then $\underline{\int}_K S' = \underline{\int}_K S$ and $\overline{\int}_K S' = \overline{\int}_K S$.*

PROOF. Since $-|S' - S| \leq S' - S \leq |S' - S|$ the equivalence $\int |S' - S| = 0$ implies $\int (S' - S) = 0$. So $\underline{\int} S' = \underline{\int}(S' - S) + \underline{\int} S = 0 + \underline{\int} S = \underline{\int} S$ and $\overline{\int} S' = \int (S' - S) + \overline{\int} S = 0 + \overline{\int} S = \overline{\int} S$. (All integrals here are over K.)   □

We can attribute to differentials any summant property that is invariant under differential equivalence. Conversely every property of a differential reverts to each of its representative summants. In the former direction we define the **integral** of a differential σ on K by

$$(2) \qquad \int_K \sigma = \underline{\int}_K \sigma = \overline{\int}_K \sigma$$

whenever the lower and upper integrals in (1) are equal. That is, $\int_K \sigma$ exists and equals $\int_K S$ whenever $\int_K S$ exists for some (hence, for all) S belonging to σ. σ is **integrable** on K if $\int_K \sigma$ exists and is finite.

For each differential σ and function f on K we define the differential $f\sigma$ on K by

$$(3) \qquad f\sigma = [fS] \quad \text{for} \quad \sigma = [S]$$

where fS is the summant defined by (18) in §1.4. Effectiveness of (3) follows from Theorem 1 (§1.8) which asserts that $T \sim 0$ implies $fT \sim 0$. Applied to $T = S - S'$ for any summants S, S' representing σ this gives $fS \sim fS'$. So (3) is valid.

Note: We shall study (3) further in §2.7.

Under the extension of summants introduced in §1.9 we have differentials on any nondegenerate interval L in $[-\infty, \infty]$ defined as differentials on the cell $K = \bar{L}$.

Exercises (§2.1).

1. Show that $S \sim T$ on a cell K if and only if $\bar{\int}_I (S - T) = \bar{\int}_I (T - S) = 0$ on every cell I in K. (Apply (iii) in Theorem 2 (§1.6).)

2. Show that $S \sim T$ on K implies $S^+ \sim T^+$, $S^- \sim T^-$, and $|S| \sim |T|$ on K.

3. (i) Show that $\rho + \sigma = [R + S]$ is an effective definition for the sum of differentials $\rho = [R]$ and $\sigma = [S]$ on a cell K. (ii) Show that if $\rho = [0]$ then $\rho + \sigma = \sigma$.

4. Verify that under definition (3) and the definition of $\rho + \sigma$ in Exercise 3 the following distributive laws hold:

   (i) $f(\rho + \sigma) = f\rho + f\sigma$,

   (ii) $(f + g)\sigma = f\sigma + g\sigma$.

5. Using (3) show that if $\tau = 0$ (i.e., $\tau = [0]$) on a cell K then $f\tau = 0$ for every function f on K. (This is a reformulation of Theorem 1 (§1.8).)

§2.2 The Riesz Space $\mathbb{D} = \mathbb{D}(K)$ of All Differentials on K.

A summant on a figure K is just a function on the set of all tagged cells in K. So the summants on K form a Riesz space $\mathbb{Y}$ under the pointwise operations and ordering for real-valued functions on a set. In the Riesz space $\mathbb{Y}$ the summants $S \sim 0$ form a Riesz ideal $\mathbb{Z}$. That is, $\mathbb{Z}$ is a linear subspace of $\mathbb{Y}$ such that for all S, T in $\mathbb{Y}$

(1)   If $T \in \mathbb{Z}$ and $|S| \leq |T|$ then $S \in \mathbb{Z}$. ($\mathbb{Z}$ is **solid**).

To verify that $\mathbb{Z}$ is a linear space we have for S, T belonging to $\mathbb{Z}$, $\overline{\int}_K |S+T| \leq \overline{\int}_K |S| + \overline{\int}_K |T| = 0$ and $\overline{\int}_K |cS| = |c| \overline{\int}_K |S| = 0$

for any constant c. To verify (1) we infer from $|S| \leq |T|$ that $\overline{\int}_K |S| \leq \overline{\int}_K |T|$. So $\int_K |S| = 0$ if $\int_K |T| = 0$.

Since $\mathbb{Z}$ is a Riesz ideal in the Riesz space $\mathbb{Y}$ the Riesz space $\mathbb{Y}/\mathbb{Z}$ is the homomorph of $\mathbb{Y}$ with the kernel $\mathbb{Z} = [0]$ just the zero differential represented by the summant 0. The elements of $\mathbb{Y}/\mathbb{Z}$ are just the differentials on K. That is, $\mathbb{Y}/\mathbb{Z} = \mathbb{D}$ where $\mathbb{D} = \mathbb{D}(K)$ is the set of all differentials on K. The Riesz space structure of $\mathbb{Y}$ is transfered homomorphically to $\mathbb{D}$.

So for differentials $\rho = [R]$ and $\sigma = [S]$ on K we have $\rho + \sigma = [R + S]$, $c\rho = [cR]$ for every real constant c, $\rho \vee \sigma = [R \vee S]$, $\rho \wedge \sigma = [R \wedge S]$, $|\rho| = [|R|]$, $\rho^+ = [R^+]$, and $\rho^- = [R^-]$.

In Riesz space $\mathbb{D}$ we have the general properties $\sigma^+ = \sigma \vee 0$, $\sigma^- = -(\sigma \wedge 0) = (-\sigma)^+$ with $\sigma = \sigma^+ - \sigma^-$, $\sigma^+ \wedge \sigma^- = 0$, and $|\sigma| = \sigma \vee (-\sigma) = \sigma^+ + \sigma^- = \sigma^+ \vee \sigma^-$. We also have $\rho + \sigma = \rho \vee \sigma + \rho \wedge \sigma$ and $(\rho - \sigma)^+ = \rho - \rho \wedge \sigma = \rho \vee \sigma - \sigma = (\sigma - \rho)^-$. (See §11.6.)

Exercises (§2.2).

1. Let $\mathbb{W}$ be a Riesz ideal in $\mathbb{Y}$ such that $\underline{\int}_K S = 0$ for every member S of $\mathbb{W}$. Show that $\mathbb{W} \subseteq \mathbb{Z}$.

2. The Riesz space $\mathbb{Y}$ of all summants on a cell K is a ring since the product ST belongs to $\mathbb{Y}$ for all S, T in $\mathbb{Y}$.

   (i) Show that $\mathbb{Z}$ is not a ring ideal by finding S in $\mathbb{Y}$ and T in $\mathbb{Z}$ such that ST does not belong to $\mathbb{Z}$.

   (ii) Prove that if S belongs to $\mathbb{Y}$, T belongs to $\mathbb{Z}$, and $\int_K |1_p S| < \infty$ for all p in K then ST belongs to $\mathbb{Z}$.

   Hint: Use Theorem 1 in §1.8 and Theorem 1 in §1.9.

3. Let $\mathbb{W}$ be the smallest Riesz ideal in $\mathbb{Y}$ such that the constant summant 1 on K belongs to $\mathbb{W}$, and $\mathbb{W} \supseteq \mathbb{Z}$. Show that $\mathbb{W}$ consists of all summants S on K such that $\overline{\int}_K |1_p S|$ is

uniformly bounded for all p in K. (See Theorem 1 in §1.9.) So $\mathbb{W}/\mathbb{Z}$ consists of all σ in $\mathbb{D}$ such that $\sup\limits_{p \in K} \int_K |1_p \sigma| < \infty$.

4. Let $\sigma_1, \cdots, \sigma_n$ be disjoint differentials on a cell K. That is, $|\sigma_i| \wedge |\sigma_j| = 0$ for $i \neq j$. Show that there exist summants $S_1, \cdots, S_n$ representing $\sigma_1, \cdots, \sigma_n$ respectively such that $S_i S_j = 0$ for $i \neq j$. (By disjointness $|\sigma_i| = |\sigma_i| - |\sigma_i| \wedge |\sigma_j| = (|\sigma_i| - |\sigma_j|)^+$ if $i \neq j$. So $|\sigma_i| = \wedge_{\substack{j=1 \\ (j \neq i)}}^n (|\sigma_i| - |\sigma_j|)^+$. Given $[S_i] = \sigma_i$ let $\bar{S}_i = (\operatorname{sgn} S_i) \wedge_{\substack{j=1 \\ (j \neq i)}}^n (|S_i| - |S_j|)^+$. Show that $\bar{S}_i \sim S_i$ for all i and $\bar{S}_i \bar{S}_j = 0$ for $i \neq j$.)

§2.3 Differential Norm and Summable Differentials.

We define an extended Riesz norm ν on $\mathbb{D}$ by

$$(1) \quad \nu(\sigma) = \overline{\int_K} |\sigma| \leq \infty \quad \text{for every differential } \sigma \text{ on } K.$$

The following properties of ν are obvious:

(i) $0 \leq \nu(\sigma) \leq \infty$,

(ii) $\nu(\sigma) = 0$ if and only if $\sigma = 0$,

(iii) $\nu(c\sigma) = |c|\nu(\sigma)$ for every constant c,

(iv) $\nu(\rho + \sigma) \leq \nu(\rho) + \nu(\sigma)$,

(v) $\nu(\rho) \leq \nu(\sigma)$ if $|\rho| \leq |\sigma|$.

We call ν the **differential norm** on $\mathbb{D}$. Although ν is not a proper norm because it may attain the value ∞ it nevertheless defines **differential convergence** in $\mathbb{D}$ by

$$(2) \qquad \sigma_n \to \sigma \quad \text{in} \quad \mathbb{D} \quad \text{if} \quad \nu(\sigma_n - \sigma) \to 0.$$

A differential σ is **summable** if $\nu(\sigma) < \infty$. The reader can verify that the summable differentials form a Riesz subspace $\mathbb{S}$ of $\mathbb{D}$ with ν a finite Riesz norm on $\mathbb{S}$. We contend that $\mathbb{S}$ is a Banach lattice under ν. We need only prove completeness which we formulate in terms of series convergence. $\sum_{i=1}^{\infty} \rho_i = \sigma$ means that for $\sigma_n = \sum_{i=1}^{n} \rho_i$, $\sigma_n \to \sigma$ as defined by (2).

THEOREM 1. *Let* $\rho_1, \rho_2, \cdots, \rho_i, \cdots$ *be a sequence of summable differentials on a figure* K *such that* $\sum_{i=1}^{\infty} \nu(\rho_i) < \infty$. *Then there exists a summable differential* σ *on* K *such that* $\sigma = \sum_{i=1}^{\infty} \rho_i$. *Moreover,* $\nu(\sigma) \leq \sum_{i=1}^{\infty} \nu(\rho_i)$.

PROOF. We may assume each $\rho_i \neq 0$. Choose a summant R_i representing ρ_i for $i = 1, 2, \cdots$. Then choose a decreasing sequence of gauges $\delta_i > \delta_{i+1}$ on K such that

$$(3) \qquad |R_i|^{(\delta_i)}(K) < 2\nu(\rho_i).$$

This is possible because $0 < \nu(\rho_i) < \infty$. Let the summant P_i indicate the δ_i-fine tagged cells in K. That is, $P_i(I, t) = 1$ if (I, t) is δ_i-fine, 0 otherwise. According to (3),

$$(4) \qquad \left(\sum |P_i R_i| \right)(\mathcal{K}) < 2\nu(\rho_i)$$

for every division $\mathcal{K}$ of K. So

$$(5) \qquad |P_i R_i|(I, t) < 2\nu(\rho_i)$$

for every tagged cell (I, t) in K. We can therefore define $S(I, t) = \sum_{i=1}^{\infty} (P_i R_i)(I, t)$ since this series is absolutely convergent by (5) because $\sum_{i=1}^{\infty} \nu(\rho_i) < \infty$. Let $\sigma = [S]$. Every δ_n-division $\mathcal{K}$ of K is a δ_i-division for $i = 1, \cdots, n$ since

$\delta_n \leq \delta_i$ for such i. For any δ_n-division $\mathcal{K}$ of K, $(\sum |S -$
$\sum_{i=1}^{n} R_i|)(\mathcal{K}) = (\sum |S - \sum_{i=1}^{n} P_i R_i|)(\mathcal{K}) = (\sum | \sum_{i=n+1}^{\infty} P_i R_i|)(\mathcal{K}) \leq$
$(\sum (\sum_{i=n+1}^{\infty} |P_i R_i|))(\mathcal{K}) = \sum_{i=n+1}^{\infty} (\sum |P_i R_i|)(\mathcal{K}) < 2 \sum_{i=n+1}^{\infty} \nu(\rho_i)$
by (4). So $\nu(\sigma - \sum_{i=1}^{n} \rho_i) \leq 2 \sum_{i=n+1}^{\infty} \nu(\rho_i) \to 0$ as $n \to \infty$.

Hence, $\sigma = \sum_{i=1}^{\infty} \rho_i$.

By triangle inequality (iv), $\nu(\sigma) \leq \nu(\sigma - \sum_{1}^{n} \rho_i) + \sum_{1}^{n} \nu(\rho_i)$.
As $n \to \infty$ this gives $\nu(\sigma) \leq \sum_{i=1}^{\infty} \nu(\rho_i) < \infty$. This proves that
σ is summable. $\square$

By definition completeness means that of every Cauchy sequence converges:
$\sigma_m - \sigma_n \to 0$ as $m, n \to \infty$ implies $\sigma_n \to \sigma$ for some σ. The equivalence of this condition and its series formulation in Theorem 1 for normed linear spaces is a familiar exercise in functional analysis. (See Exercises 2 and 3 in §11.3.) Theorem 1 has the following important corollary.

THEOREM 2. *The Riesz space* $\mathbb{D}$ *of all differentials on a figure* K *is complete under* ν-*convergence defined by (1), (2).*

PROOF. Let $\sigma_1, \sigma_2, \cdots$ be a Cauchy sequence in $\mathbb{D}$. That is, $\nu(\sigma_m - \sigma_n) \to 0$ as $m, n \to \infty$. This implies that $\nu(\sigma_m - \sigma_n) < \infty$ ultimately. So we can choose N large enough so that $\tau_k = \sigma_{N+k} - \sigma_N$ is summable for $k = 0, 1, 2, \cdots$. Now $\tau_j - \tau_k = \sigma_{N+j} - \sigma_{N+k} \to 0$ as $j, k \to \infty$. So $\tau_1, \tau_2, \cdots$ is a Cauchy sequence of summable differentials on K. By Theorem 1, $\tau_k \to \tau$ for some differential τ on K. So $\sigma_{N+k} = \tau_k + \sigma_N \to \tau + \sigma_N$ as $k \to \infty$. Hence, $\sigma_n \to \sigma$ where $\sigma = \tau + \sigma_N$. $\square$

All of the Riesz space operations are continuous with respect to the convergence defined by (2). That is, if $\sigma_n \to \sigma$ in $\mathbb{D}$ then

$|\sigma_n| \to |\sigma|$, $\sigma_n^+ \to \sigma^+$, $\sigma_n^- \to \sigma^-$, $c\sigma_n \to c\sigma$ for every constant c, and $\nu(\sigma_n) \to \nu(\sigma)$ in $[0, \infty]$. Moreover, let $\sigma_n \to \sigma$ and $\tau_n \to \tau$ in $\mathbb{D}$. Then $\sigma_n + \tau_n \to \sigma + \tau$, $\sigma_n \vee \tau_n \to \sigma \vee \tau$, and $\sigma_n \wedge \tau_n \to \sigma \wedge \tau$. Hence, $\sigma \leq \tau$ if $\sigma_n \leq \tau_n$ for all n. These properties are valid under $\nu \leq \infty$ because $\nu(\sigma_n - \sigma) \to 0$ implies $\nu(\sigma_n - \sigma) < \infty$ ultimately, although we may have $\nu(\sigma_n) = \nu(\sigma) = \infty$ for all n.

Theorem 2 is useful since to prove a subspace of $\mathbb{D}$ is complete we need only prove it is closed in the complete space $\mathbb{D}$ of all differentials on K.

Exercises (§2.3).

1. Verify the properties (i)—(v) of ν under definition (1).

2. Show that σ is summable if $\rho \leq \sigma \leq \tau$ with both ρ and τ summable.

3. Show that the linear space of all integrable differentials on a figure K is complete (i.e., closed in $\mathbb{D}$).

4. Given σ, τ in $\mathbb{D}$ such that $\tau \geq 0$ and τ is integrable show that $\sigma \leq \tau$ if and only if $\nu(\sigma \vee \tau) = \nu(\tau)$.

5. Prove that a differential σ on a cell K is summable if and only if $|\sigma| \leq \tau$ for some integrable differential τ on K. (See Theorem 5 in §1.6.)

§2.4 Conditionally and Absolutely Integrable Differentials.

A differential σ on a figure K is **absolutely integrable** over K if both σ and $|\sigma|$ are integrable over K. σ is **conditionally integrable** if σ is integrable but $|\sigma|$ is not integrable.

THEOREM 1. *(i) σ is absolutely integrable if and only if both σ^+ and σ^- are integrable.*

(ii) If σ is integrable over K then $\int_K |\sigma|$ exists with $0 \leq \int_K |\sigma| \leq \infty$.

(iii) σ is absolutely integrable if and only if it is both integrable and summable.

(iv) If σ is conditionally integrable over K then $\int_K \sigma^+ = \int_K \sigma^- = \int_K |\sigma| = \infty$.

PROOF. (i) follows from the Riesz space identities $\sigma = \sigma^+ - \sigma^-$, $|\sigma| = \sigma^+ + \sigma^-$, $\sigma^+ = \frac{1}{2}(|\sigma| + \sigma)$, and $\sigma^- = \frac{1}{2}(|\sigma| - \sigma)$ along with the linearity ((3) in §1.5) of the integral on integrable differentials.

By Theorem 3 (§1.6) an integrable σ is of the form $\sigma = [\hat{S}]$ for some additive summant $\hat{S}$. Theorem 2 (§1.4) with $R = \hat{S}$ gives (ii).

If σ is integrable and summable, then σ is absolutely integrable by (ii). Since the converse is trivial we have (iii).

For σ conditionally integrable (ii) implies $\int_K |\sigma| = \infty$. So since the integral of σ is finite we can conclude that $\int_K \sigma^+ = \frac{1}{2}(\int_K |\sigma| + \int_K \sigma) = \frac{1}{2}(\infty + \int_K \sigma) = \infty$ and similarly $\int_K \sigma^- = \frac{1}{2}(\int_K |\sigma| - \int_K \sigma) = \frac{1}{2}(\infty - \int_K \sigma) = \infty$. This proves (iv). $\square$

In addition to continuity of the Riesz space operations on the space $\mathbb{D}$ of differentials on K we must investigate continuity of the integral as a functional on those members of $\mathbb{D}$ for which the integral exists. We examine first the limiting behavior of the lower and upper integrals acting on $\mathbb{D}$.

THEOREM 2. *The following hold:*

(i) $-\nu(\rho) \leq \underline{\int_K} \rho \leq \overline{\int_K} \rho \leq \nu(\rho)$ for all ρ in $\mathbb{D}$.

(ii) If $\rho_n \to 0$ in $\mathbb{D}$ then $\underline{\int_K} \rho_n \to 0$ and $\overline{\int_K} \rho_n \to 0$.

(iii) If $\sigma_n \to \sigma$ in $\mathbb{D}$ then $\underline{\int_K} \sigma_n \to \underline{\int_K} \sigma$ and $\overline{\int_K} \sigma_n \to \overline{\int_K} \sigma$. Moreover, if $\underline{\int_K} \sigma_n$ is finite for all n then $\underline{\int_K} \sigma$ is finite. If $\overline{\int_K} \sigma_n$ is finite for all n then $\overline{\int_K} \sigma$ is finite.

PROOF. $-|\rho| \leq \rho \leq |\rho|$ for all ρ in $\mathbb{D}$. So $-\overline{\int_K}|\rho| = \underline{\int_K}(-|\rho|) \leq \underline{\int_K}\rho \leq \overline{\int_K}\rho \leq \overline{\int_K}|\rho|$ which proves (i). Application of (i) to ρ_n with $\nu(\rho_n) \to 0$ yields (ii).

To prove the generalization (iii) of (ii) let $\rho_n = \sigma - \sigma_n$. Then $\sigma = \sigma_n + \rho_n$ and $\sigma_n = \sigma - \rho_n$. So

$$(1) \quad \underline{\int_K}\sigma_n + \underline{\int_K}\rho_n \leq \underline{\int_K}\sigma \quad \text{and} \quad \underline{\int_K}\sigma + \underline{\int_K}(-\rho_n) \leq \underline{\int_K}\sigma_n.$$

Since $\rho_n \to 0$ (1) gives the convergence of the lower integrals in (iii) by (ii). (1) also gives the finiteness of $\underline{\int_K}\sigma$ if $\underline{\int_K}\sigma_n$ is finite for all n. Apply this to $-\sigma_n \to -\sigma$ to get the dual results for the upper integral. $\square$

We can now easily prove continuity properties of the integral.

THEOREM 3. *Let* $\sigma_n \to \sigma$ *in* $\mathbb{D}$. *Then*
(i) If $\int_K \sigma_n$ *exists for all* n *then* $\int_K \sigma$ *exists and* $\int_K \sigma_n \to \int_K \sigma$ *in* $[-\infty, \infty]$.
(ii) If σ_n *is integrable for all* n *then* σ *is integrable and* $\int_K \sigma_n \to \int_K \sigma$ *in* $\mathbb{R}$.
(iii) If σ_n *is absolutely integrable for all* n *then* σ *is absolutely integrable and both* $\int_K \sigma_n \to \int_K \sigma$ *and* $\int_K |\sigma_n| \to \int_K |\sigma|$ *in* $\mathbb{R}$.

PROOF. (i) follows directly from (iii) in Theorem 2. So does (ii). Since $\sigma_n \to \sigma$ implies $|\sigma_n| \to |\sigma|$, (iii) follows from (ii) applied to both σ and $|\sigma|$. $\square$

By Theorem 3 the integrable differentials on K, and the absolutely integrable differentials on K, form closed linear subspaces of $\mathbb{D}$. So these subspaces are complete since $\mathbb{D}$ is complete by Theorem 2 (§2.3). In particular, the absolutely integrable differentials form a Banach lattice, a closed subspace of the Banach lattice $\mathbb{S}$ of summable differentials on K.

For integrable differentials we can prove a monotone convergence theorem.

THEOREM 4. *Let $\sigma_0 \leq \sigma_1 \leq \sigma_2 \leq \cdots$ be a monotone sequence of integrable differentials on a figure K such that for some constant $c < \infty$, $\int_K \sigma_n \leq c$ for $n = 1, 2, \cdots$. Then there exists an integrable differential σ on K such that $\sigma \geq \sigma_n$ for all n, $\sigma_n \to \sigma$, and $\int_K \sigma_n \to \int_K \sigma$. If, moreover, σ_0 is absolutely integrable then so are $\sigma_1, \sigma_2, \cdots$ and σ.*

PROOF. Let $\rho_i = \sigma_i - \sigma_{i-1}$ for $i = 1, 2, \cdots$. Then $\rho_i \geq 0$ and $\sum_{i=1}^{n} \rho_i = \sigma_n - \sigma_0$ for $n = 1, 2, \cdots$. Therefore, since all these differentials are integrable, $\sum_{i=1}^{n} \int_K \rho_i = \int_K \sigma_n - \int_K \sigma_0 \leq c - \int_K \sigma_0$. Since $\int_K \rho_i = \nu(\rho_i)$, $\sum_{i=1}^{\infty} \nu(\rho_i) \leq c - \int_K \sigma_0 < \infty$. By completeness of the summable differentials in Theorem 1 (§2.3) we get a summable differential $\rho = \sum_{i=1}^{\infty} \rho_i$. That is, $\sigma_n - \sigma_0 = \sum_{i=1}^{n} \rho_i \to \rho$ as $n \to \infty$. So $\sigma_n \to \sigma$ where $\sigma = \rho + \sigma_0$. Integrability of σ and the convergence $\int_K \sigma_n \to \int_K \sigma$ follow from (ii) in Theorem 3.

By continuity of the lattice operations $\sigma_n \leq \sigma$ for all n. So $\sigma_0 \leq \sigma_n \leq \sigma$. Hence $0 \leq \sigma^- \leq \sigma_n^- \leq \sigma_0^-$ which implies $\nu(\sigma^-) \leq \nu(\sigma_n^-) \leq \nu(\sigma_0^-)$. So summability (equivalent to absolute integrability for integrable differentials) of σ_0 implies that property for σ_n and σ by the contrapositive of implication (iv) in Theorem 1.   □

All four theorems in §1.5 as well as Theorem 3 (§1.9) can be reformulated in terms of differentials by replacing "S" by "σ" and "summant" by "differential." But Theorem 2 (§1.6) can be carried beyond its literal translation to give the following reformulation.

THEOREM 5. *Let σ be a differential on a figure K.*

(i) If $-\infty < \overline{\int_I}\sigma \leq 0$ for every cell I in K then $\sigma \leq 0$ on the figure K.

(ii) If $0 \leq \underline{\int_I}\sigma < \infty$ for every cell I in K then $\sigma \geq 0$ on K.

(iii) If $\int_I \sigma = 0$ for every cell I in K then $\sigma = 0$ on K.

PROOF. Apply Theorem 2 (§1.6) to S for $[S] = \sigma$. Since $\sigma^+ = [S^+]$, $\nu(\sigma^+) = \overline{\int_K} S^+$. So $\overline{\int_K} S^+ = 0$ implies $\sigma^+ = 0$, that is, $\sigma = -\sigma^- \leq 0$. Thus, (i) in Theorem 2 (§1.6) yields (i) in Theorem 5. (ii) is just (i) applied to $-\sigma$. (iii) follows from (i) and (ii). □

For σ integrable its integrals over figures determine the integrals of σ^+, σ^-, and $|\sigma|$ as follows.

THEOREM 6. *Let σ be an integrable differential on a cell K. Let $\mathcal{F}$ be the set of all figures in K. Then*

(i) $0 \leq \sup\limits_{A \in \mathcal{F}} \int_A \sigma = \int_K \sigma^+ \leq \infty$,

(ii) $0 \leq -\inf\limits_{B \in \mathcal{F}} \int_B \sigma = \int_K \sigma^- \leq \infty$,

and

(iii) $0 \leq \sup\limits_{A,B \in \mathcal{F}} [\int_A \sigma - \int_B \sigma] = \int_K |\sigma| \leq \infty$.

PROOF. By Theorem 3 (§1.6) $\sigma = [S]$ for the additive summant S given by $S(I) = \int_I \sigma$. For every partition $\mathbb{K}$ of K $(\Sigma S^+)(\mathbb{K}) = \int_A \sigma$ for A the union of those members I of $\mathbb{K}$ such that $S(I) > 0$. Thus, since $\sigma^+ = [S^+]$, $\int_K \sigma^+ \leq \sup\limits_{A \in \mathcal{F}} \int_A \sigma$. The reverse inequality follows from $\sigma \leq \sigma^+$ and $0 \leq \sigma^+$ which imply $\int_A \sigma \leq \int_A \sigma^+ \leq \int_K \sigma^+$ for every figure A in K. So (i) holds. (ii) is just (i) applied to $-\sigma = \sigma^- - \sigma^+$ with $(-\sigma)^+ = \sigma^-$. (iii) is the sum of (i) and (ii). □

Using Theorem 4 we can get a rudimentary form of Lebesgue decomposition in terms of Riesz space concepts.

THEOREM 7. *Let σ, τ be absolutely integrable differentials on a cell K. Then there exist unique absolutely integrable differentials ρ, θ on K such that: (i) $\tau = \rho + \theta$, (ii) $|\theta| \wedge |\sigma| = 0$, and (iii) $\phi \wedge |\rho| = 0$ for every differential ϕ on K such that $\phi \wedge |\sigma| = 0$.*

PROOF. We treat only the case $\sigma, \tau \geq 0$ since this can be applied in the general case to $|\sigma|$, τ^+ and to $|\sigma|$, τ^-. (See Exercise 3.)

Given integrable $\sigma, \tau \geq 0$ let $\rho_n = \tau \wedge (n\sigma)$ for $n = 1, 2, \cdots$. Then $0 \leq \rho_n \leq \rho_{n+1} \leq \tau$. So Theorem 4 yields an integrable ρ such that $\rho_n \nearrow \rho$ with $0 \leq \rho \leq \tau$. Let $\theta = \tau - \rho \geq 0$ to get (i).

Since $\rho_n \leq \rho$, $\theta \leq \tau - \rho_n = \tau - \tau \wedge (n\sigma) = (\tau - n\sigma)^+$. Now $0 \leq (\sigma - \frac{1}{n}\tau)^+ \leq n(\sigma - \frac{1}{n}\tau)^+ = (n\sigma - \tau)^+ = (\tau - n\sigma)^-$. Hence, since $(\tau - n\sigma)^+ \wedge (\tau - n\sigma)^- = 0, \theta \wedge (\sigma - \frac{1}{n}\tau)^+ = 0$. As $n \nearrow \infty$, $\frac{1}{n}\tau \searrow 0$. So $(\sigma - \frac{1}{n}\tau)^+ \nearrow \sigma^+ = \sigma$. Therefore, $0 = \theta \wedge (\sigma - \frac{1}{n}\tau)^+ \nearrow \theta \wedge \sigma$ which implies $\theta \wedge \sigma = 0$. This proves the disjointness condition (ii).

To prove (iii) let $\phi \wedge \sigma = 0$. Then since $0 \leq \rho_n \leq n\sigma$, $0 \leq \phi \wedge \rho_n \leq \phi \wedge (n\sigma) \leq n(\phi \wedge \sigma) = 0$. So $\phi \wedge \rho_n = 0$ for all n. Therefore, since $\rho_n \nearrow \rho$, $0 = \phi \wedge \rho_n \nearrow \phi \wedge \rho$ which implies $\phi \wedge \rho = 0$. This proves (iii). Note that (ii) and (iii) imply $\theta \wedge \rho = 0$.

To prove uniqueness of the decomposition (i) satisfying (ii), (iii) let ρ', θ' also be positive, integrable differentials satisfying (i), (ii), (iii). Then $\theta \wedge \rho' = 0$ by (ii), (iii). So $\theta + \rho' = \theta \vee \rho' + \theta \wedge \rho' = \theta \vee \rho' \leq \tau$. Hence, $\rho' \leq \tau - \theta = \rho$. Since ρ, θ may be interchanged respectively with ρ', θ' we can conclude that $\rho' = \rho$. So $\theta' = \tau - \rho' = \tau - \rho = \theta$. $\square$

Exercises (§2.4).

1. Show that in (iii) of Theorem 6 we may restrict A, B to pairs of: (a) nonoverlapping figures, (b) complementary figures. (Use Theorems 3,4 of §1.5.)

2. For σ a differential on $K = [a, b]$ prove:

   (a) If $\int_a^{\bar t} \sigma = 0$ for all t in K then $\sigma \leq 0$.

   (b) If σ is integrable on K and $\int_a^t \sigma = 0$ for all t in K then $\sigma = 0$.

3. Show that Theorem 7 holds if it holds for *positive*, integrable σ, τ and ρ, θ.

4. Show that for integrable $\rho_0 \geq \rho_1 \geq \rho_2 \geq \cdots \geq 0$ on a cell K there exists an integrable $\rho \geq 0$ such that $\rho_n \searrow \rho$ and $\int_K \rho_n \searrow \int_K \rho$.

Hint: Apply Theorem 4 to $\sigma_n = \rho_0 - \rho_n$.

§2.5 The Differential dg of a Function g.

Every function g on a cell K yields a differential dg on K defined by

(1) $$dg = [\Delta g]$$

where Δg is the additive summant $\Delta g[r, s] = g(s) - g(r)$ according to (5) in §1.4. Thus, the basic algorithm takes the form

(2) $$\int_I dg = \Delta g(I) \quad \text{for every cell } I \text{ in } K.$$

That every integrable differential on a cell K is the differential of some function on K is the conclusion of the following elegant version of the Saks-Henstock Lemma.

THEOREM 1. *Given an integrable differential σ on a cell $K = [a, b]$ define*

(3) $$g(t) = \int_a^t \sigma \quad \text{for} \quad a \leq t \leq b.$$

Then $dg = \sigma$.

PROOF. (3) implies $\Delta g(I) = \int_I \sigma$ for every cell I in K. That is, $\Delta g = \hat{S}$ where $\hat{S}$ is the additive summant in Theorem 3 (§1.6) defined by $\hat{S}(I) = \int_I \sigma$ for every cell I in K. So $dg = [\hat{S}] = \sigma$ by Theorem 3 (§1.6). □

Theorem 1 supports the standard technique for evaluating integrals. Given an integrable differential σ on a cell K find a function g on K such that $dg = \sigma$. Then compute $\Delta g(K)$ to evaluate $\int_K \sigma$.

Just as Theorem 1 is a reformulation of Theorem 3 (§1.6) in terms of differentials, the next theorem is a reformulation of Theorem 5 (§1.6).

THEOREM 2. *Given a differential σ on $K = [a,b]$ such that $\overline{\int}_K \sigma$ is finite define g on K by*

(4)
$$g(t) = \overline{\int_a^t} \sigma \quad \text{for} \quad a \le t \le b.$$

Then the following hold:

(i) $\Delta g(I) = \int_I dg = \overline{\int}_I \sigma$ for every cell I in K,

(ii) $\overline{\int}_I (\sigma - dg) = 0$ for every cell I in K,

(iii) $dg \ge \sigma$ on K,

(iv) If h is any function on K such that $dh \ge \sigma$ then it must also satisfy $dh \ge dg$.

Exercises (§2.5).

1. Given a function g on a cell K show the equivalence of the following:

   (i) $dg \ge 0$ on K.

   (ii) $\Delta g \ge 0$ on K.

(iii) $g(s) \leq g(t)$ for all $s \leq t$ in K.

2. Given an integrable differential σ on a cell K, a point p in K, and a real number c show that there is a unique function g on K such that $dg = \sigma$ and $g(p) = c$.

3. Using Theorem 2 state and prove an analogous theorem for the lower integral.

Hint: $\underline{\int_K} \sigma = -\overline{\int_K}(-\sigma)$.

4. Show that Theorem 2 yields Theorem 1 by applying Theorem 5 (§2.4).

5. Verify that Theorem 2 follows from Theorem 5 (§1.6).

§2.6 The Total Variation of a Function on a Cell K.

We define the **total variation** of a function g on a cell $K = [a, b]$ to be $\nu(dg) = \int_K |dg| \leq \infty$. Since dg is integrable over K the existence of $\int_K |dg|$ with value in $[0, \infty]$ is assured by (ii) of Theorem 1 (§2.4). g is of **bounded variation** if $\int_K |dg| < \infty$, **unbounded variation** if $\int_K |dg| = \infty$. So according to Theorem 1 (§2.5) a differential σ on a cell K is absolutely integrable if and only if $\sigma = dg$ for some function g of bounded variation on K. σ is conditionally integrable if and only if $\sigma = dg$ for some function g of unbounded variation on K.

The differentials of functions of bounded variation on a cell K form the Banach lattice of all absolutely integrable differentials on K. (In §2.4 see Theorem 1 and the remarks after the proof of Theorem 3.) This Banach lattice is the appropriate setting for the Jordan Decomposition Theorem which we now present.

THEOREM 1. *Given a function g of bounded variation on a cell K there exist functions u, v on K such that $g = u - v$, $du = (dg)^+$, and $dv = (dg)^-$.*

PROOF. Since $|dg|$ is integrable over K Theorem 1 (§2.5) yields a function w on K such that $dw = |dg|$. Define $u =$

$\frac{1}{2}(w + g)$ and $v = \frac{1}{2}(w - g)$. Then $u - v = g$. Also $du = \frac{1}{2}d(w + g) = \frac{1}{2}(dw + dg) = \frac{1}{2}(|dg| + dg) = (dg)^+$. Similarly $dv = \frac{1}{2}(|dg| - dg) = (dg)^-$. □

Note that $w = u + v$ and $dw \geq 0$, $du \geq 0$, $dv \geq 0$. So u, v, w are monotone. Hence, a function g of bounded variation on K has finite unilateral limits $g(t-)$ at t in $(a, b]$ and $g(t+)$ at t in $[a, b)$. Also g has at most countably many points of discontinuity.

For any function g on a figure C and $\mathcal{F}$ the set of all figures in C Theorem 6 (§2.4) applied to the components of C gives

(1) $\int_C (dg)^+ = \sup_{A \in \mathcal{F}} \int_A dg \leq \infty,$

(2) $\int_C (dg)^- = - \inf_{B \in \mathcal{F}} \int_B dg \leq \infty,$

(3) $\int_C |dg| = \sup_{A, B \in \mathcal{F}} (\int_A dg - \int_B dg) \leq \infty.$

According to Exercise 1 in §2.4 (3) also holds with the restriction that A, B be complementary figures in C. In this form (3) yields the following characterization of bounded variation. (In essence it is about boundedness of an additive function on a boolean algebra.)

THEOREM 2. *A function g on a cell K is of bounded variation on K if and only if $\int_{A_n} dg \to 0$ for every sequence $A_1, A_2, \cdots$ of nonoverlapping figures in K.*

PROOF. Let $\int_K |dg| < \infty$. Then for any sequence $A_1, A_2, \cdots$ of nonoverlapping figures in K, $\sum_{i=1}^{n} |\int_{A_i} dg| \leq \sum_{i=1}^{n} \int_{A_i} |dg| = \int_{A_1 + \cdots + A_n} |dg| \leq \int_K |dg| < \infty$. So $\sum_{i=1}^{\infty} |\int_{A_i} dg| < \infty$ which implies $\int_{A_i} dg \to 0$ as $i \to \infty$.

Conversely let $\int_K |dg| = \infty$. Let $B_0 = K$. Having chosen for some positive integer n a figure B_{n-1} with $\int_{B_{n-1}} |dg| = \infty$

apply (3) to get nonoverlapping A_n, B_n such that $A_n + B_n = B_{n-1}$ and

$$(4) \qquad |\int_{A_n} dg - \int_{B_n} dg| > 2 + |\int_{B_{n-1}} dg|.$$

Now $\int_{A_n} |dg| + \int_{B_n} |dg| = \int_{B_{n-1}} |dg| = \infty$. So at least one of the lefthand integrals must be infinite. Since A_n, B_n may be interchanged in (4) we may assume that $\int_{B_n} |dg| = \infty$. Since $\int_{B_{n-1}} dg = \int_{A_n} dg + \int_{B_n} dg$, (4) implies

$$|\int_{A_n} dg - \int_{B_n} dg| > 2 + |\int_{A_n} dg + \int_{B_n} dg|.$$

The reader can verify that this yields the conclusion (Exercise 6)

$$(5) \qquad |\int_{A_n} dg| > 1 \quad \text{for } n = 1, 2, \cdots.$$

Now the inductively chosen sequence $A_1, A_2, \cdots$ of figures is nonoverlapping. To verify this recall that $A_n + B_n = B_{n-1}$ with $A_n B_n = \emptyset$ for $n = 1, 2, \cdots$. Thus $B_0 \supseteq B_1 \supseteq B_2 \supseteq \cdots$. So for $1 \le m < n, B_m \supseteq B_{n-1}$ which implies $A_m A_n \subseteq A_m B_{n-1} \subseteq A_m B_m = \emptyset$. Hence $A_m A_n = \emptyset$ for $m \ne n$.

Clearly (5) implies that the convergence condition $\int_{A_n} dg \to 0$ fails. (S. Bochner showed the author this proof in a more general context in 1952.) $\quad \square$

Exercises (§2.6). Prove the following:

1. For f any bounded function on a cell K let $\|f\| = \sup_{t \in K} |f(t)|$.

The following implications hold:

(i) If f, g are bounded functions on K then the differential inequality $|d(fg)| \leq \|f\| \, |dg| + \|g\| \, |df|$ holds.

(ii) If f and g are of bounded variation on K then so is their product fg.

2. The following conditions are equivalent for any function g on a cell K:

(i) g is of bounded variation ($\int_K |dg| < \infty$).

(ii) $\sum\limits_{n=1}^{\infty} |\Delta g(I_n)| < \infty$ for every sequence of nonoverlapping cells $I_1, I_2, \cdots$ in K.

(iii) (Same as (ii) with "nonoverlapping" replaced by "disjoint".)

(iv) $\sum\limits_{n=1}^{\infty} |g(a_{n+1}) - g(a_n)| < \infty$ for every monotone sequence $a_1, a_2, \cdots$ of points in K.

(v) There is a constant $c < \infty$ such that $|\int_A dg| < c$ for all figures A in K.

3. For g a function on a cell K let $h = |g|$. Then

(i) $|dh| \leq |dg|$ so the total variation of $|g|$ is at most the total variation of g.

(ii) There exists a (discontinuous) function g on K such that $|dh| < |dg|$ on every cell in K.

(Consider the case $h = 1$ for which $dh = 0$.)

(iii) If g is continuous then $|dh| = |dg|$, so $|g|$ and g have the same total variation.

(There exists a gauge δ on K such that $|\Delta h| = |\Delta g|$ at every δ-fine (I, t) in K. Consider each of the cases: $g(t) > 0$, $g(t) < 0$, $g(t) = 0$.)

4. For $\mathbb{K}$ any partition of a bounded cell K let $\|\mathbb{K}\|$ be the maximum length of the cells belonging to $\mathbb{K}$. For any continuous function g on K and any sequence $\langle \mathbb{K}_n \rangle$ of partitions of K such that $\|\mathbb{K}_n\| \to 0$,

$$\left(\sum |\Delta g| \right)(\mathbb{K}_n) \to \int_K |dg| \leq \infty.$$

5. For any function g on a bounded cell K the following two conditions are equivalent:

(i) g is continuous and of bounded variation,

(ii) There is a sequence $\langle g_n \rangle$ of polygonal (i.e., continuous, piecewise linear) functions on K converging uniformly to g with $\langle |\Delta g_n| \rangle$ uniformly integrable.

(To prove (i) implies (ii) let $\mathbb{K}_n$ partition K into 2^n cells of equal length. Take g_n on K linear on each member I of $\mathbb{K}_n$ and equal to g at the endpoints of I. Prove:

(a) $g_n \to g$ uniformly. (Use uniform continuity of g.)

(b) $\int_K |dg_n| \nearrow \int_K |dg|$ as $n \nearrow \infty$. (Apply Exercise 4.)

(c) For every refinement $\mathbb{K}$ of $\mathbb{K}_n$ and all $i = 1, 2, \cdots$

(6) $0 \leq \int_K |dg_i| - \left(\sum |\Delta g_i| \right)(\mathbb{K}) \leq \int_K |dg| - \int_K |dg_n|.$

(For $i \leq n$ the first inequality in (6) is equality.)

(d) Let δ_n be a gauge for which every δ_n-division $\mathcal{K}$ of K refines $\mathbb{K}_n$. For such $\mathcal{K}$ (6) holds with $\mathbb{K}$ replaced by $\mathcal{K}$, thereby giving uniform integrability of $\langle |\Delta g_i| \rangle$.)

6. Given x, y in $\mathbb{R}$ such that $|x - y| - |x + y| > 2$ prove $|x| > 1$. (*Hint:* Square both sides of the given inequality.)

§2.7 Functions as Differential Coefficients.

For σ a differential and f a function on a cell K we have defined

$$(1) \qquad f\sigma = [fS] \quad \text{for } \sigma = [S]$$

in (3) of §2.1. For fS to be a viable summand the function f must be defined and finite everywhere on K. But we want to relax this restriction on the differential coefficient f in the differential $f\sigma$. Specifically we demand only that f be defined and finite "σ-everywhere" on K. We introduce this concept as follows.

For σ a differential on K a subset E of K is σ-**null** if $1_E\sigma = 0$ under (1) with $f = 1_E$. If E is σ-null then so is every subset D of E since $|1_D\sigma| \leq |1_E\sigma| = 0$. If $E_1, E_2, \ldots$ is a sequence of σ-null sets then their union E is σ-null. This follows from Theorem 1 (§1.7) which gives $\nu(1_E\sigma) \leq \sum_{i=1}^{\infty} \nu(1_{E_i}\sigma) = 0$ for $1_E \leq \sum_{i=1}^{\infty} 1_{E_i} \leq \infty$. In summary, every differential σ on K has a hereditary sigma-ring consisting of all σ-null subsets of K.

For the unit differential ω represented by the summant of constant value 1 the only ω-null set is the empty set $\emptyset$. At the other extreme, every subset of K is σ-null if and only if $\sigma = 0$ on K.

For a differential σ on K a property holds σ-**everywhere** on K if it holds at all points in K except for those in some σ-null subset of K. That is, the property holds at all points of a subset D of K such that $1_D\sigma = \sigma$. We shall also use the term "at σ-all t in K" for "σ-everywhere on K."

We can now give a concise reformulation of Theorem 2 (§1.8) as follows.

THEOREM 1. *Let g be a function on a cell K, and σ be a differential on K. Then $g\sigma = 0$ on K if and only if $g = 0$ σ-everywhere on K.*

To extend definition (1) let F be a function that is defined and finite σ-everywhere on K. (On a σ-null subset of K , F may be undefined, or defined with infinite values.) We define the differential $F\sigma$ on K in terms of (1) by

$$(2) \qquad\qquad F\sigma = f\sigma$$

where f is any function defined and finite everywhere on K such that $f = F$ σ-everywhere. To verify that this definition is effective consider any functions f and f' on K each equal to F σ-everywhere. Then $f = f'$ σ-everywhere. So Theorem 1 applied to $g = f - f'$ gives $g\sigma = 0$. That is, $f\sigma - f'\sigma = (f - f')\sigma = 0$. So $f\sigma = f'\sigma$. Thus, the definition of $F\sigma$ in (2) does not depend on the particular choice of the function f that equals F σ-everywhere. So $F\sigma$ is a viable differential.

Note that under definition (2) Theorem 1 is valid for any function g that is defined and finite σ-everywhere on K.

If f and g are defined and finite σ-everywhere on K then so are $f+g$, fg, $f\vee g$, and $f\wedge g$. Moreover, $(f+g)\sigma = f\sigma+g\sigma$ and $(fg)\sigma = f(g\sigma) = g(f\sigma)$. Also, for $\sigma \geq 0$,, $(f\vee g)\sigma = (f\sigma)\vee(g\sigma)$ and $(f \wedge g)\sigma = (f\sigma) \wedge (g\sigma)$. If f is defined both σ-everywhere and τ-everywhere then $f(\sigma + \tau) = f\sigma + f\tau$.

We can now reformulate Theorem 1 (§1.7) in terms of differentials.

THEOREM 2. *Let σ be a differential on a cell K. For $i = 0, 1, 2, \cdots$ let v_i be a function such that both of the inequalities $0 \leq v_i < \infty$ and $v_0 \leq \sum_{i=1}^{\infty} v_i \leq \infty$ hold σ-everywhere on K. Then*

$$(3) \qquad\qquad 0 \leq \nu(v_0\sigma) \leq \sum_{i=1}^{\infty} \nu(v_i\sigma) \leq \infty.$$

If, moreover, $\int_K v_i|\sigma|$ exists for $i = 1, 2, \cdots$ and $v_0 = \sum_{i=1}^{\infty} v_i$
σ-everywhere, then

(4) $$0 \leq \int_K v_0|\sigma| = \sum_{i=1}^{\infty} \int_K v_i|\sigma| \leq \infty.$$

From Theorem 2 we get the following version of Lebesgue's Monotone Convergence Theorem.

THEOREM 3. *Let $\sigma \geq 0$ be a differential on a cell K. Let $f_0, f_1, \cdots$ be a sequence of functions defined σ-everywhere on K such that $f_0 \leq f_1 \leq \cdots$ σ-everywhere, $f_n\sigma$ is integrable for all n, and $\int_K f_n\sigma \nearrow c < \infty$. Then there is a function f on K such that $f_n \nearrow f$ σ-everywhere and $\int_K f\sigma = c$. So $f_n\sigma \to f\sigma$.*

PROOF. By setting $f_n(t) = 0$ on a σ-null set of points t in K we may assume that $f_{n-1}(t) \leq f_n(t) < \infty$ for all t in K, and for all $n = 1, 2, \cdots$. Moreover, replacing f_n by $f_n - f_0$ we may assume that $f_0 = 0$. So we have $0 \leq f(t) \leq \infty$ defined for all t in K by $f_n(t) \nearrow f(t)$.

To show that $f < \infty$ σ-everywhere let A be the set of all t where $f(t) = \infty$. For $i = 1, 2, \cdots$ define $v_i = f_i - f_{i-1}$. So $f_n = v_1 + \cdots + v_n$ and $v_i\sigma$ is integrable with $\int_K v_i\sigma = \int_K f_i\sigma - \int_K f_{i-1}\sigma$. So $\int_K f_n\sigma = \int_K v_1\sigma + \cdots + \int_K v_n\sigma$. Since $f = \infty$ on A and $f \geq 0$, $m1_A \leq f = \sum_{i=1}^{\infty} v_i$ for every positive integer m. So (3) in Theorem 2 gives $m\nu(1_A\sigma) \leq \sum_{i=1}^{\infty} \int_K v_i\sigma = c$. Therefore, since $c < \infty$ and m is arbitrary, $1_A\sigma = 0$. That is, $f < \infty$ σ-everywhere.

By (4) in Theorem 2, $\int_K f\sigma = \sum_{i=1}^{\infty} \int_K v_i\sigma = \lim_{n\to\infty} \int_K f_n\sigma = c$. So $\nu(f\sigma - f_n\sigma) = \int_K (f - f_n)\sigma = c - \int_K f_n\sigma \to 0$ as $n \to \infty$. That is, $f_n\sigma \to f\sigma$. □

For the case $S = \Delta g$, Theorem 1 (§1.9) has the following reformulation.

THEOREM 4. *Let g be a function on a cell $K = [a, b]$. Let p be a point in K. Then under the convention $g(a-) = g(a)$ and $g(b+) = g(b)$ at the endpoints of K,*

$$(5) \qquad \overline{\int_a^b} 1_p dg = \overline{\lim_{r \to p+}}\, g(r) - \varliminf_{r \to p-}\, g(r)$$

with value ∞ if the right-hand side is indeterminate, and

$$(6) \qquad \underline{\int_a^b} 1_p dg = \varliminf_{r \to p+}\, g(r) - \varlimsup_{r \to p-}\, g(r)$$

with value $-\infty$ if the right-hand side is indeterminate. So $1_p dg$ is integrable over K if and only if $g(p+)$ and $g(p-)$ exist and are finite, in which case

$$(7) \qquad \int_a^b 1_p dg = g(p+) - g(p-).$$

Moreover, if $g(p+)$ and $g(p-)$ exist with values in $[-\infty, \infty]$ and the right-hand side of (7) is determinate then (7) holds.

For $S = |\Delta g|$ Theorem 1 (§1.9) yields the following result.

THEOREM 5. *Let g be a function on $K = [a, b]$, and p be a point in K. Then under the convention $g(a-) = g(a)$ and $g(b+) = g(b)$ at the endpoints of K,*

$$(8) \qquad \overline{\int_a^b} 1_p |dg| = \varlimsup_{r \to p-}\, |g(p) - g(r)| + \varlimsup_{r \to p+}\, |g(r) - g(p)|.$$

Consequently as $r \to p$ in K,

$$(9) \qquad \varlimsup_{r \to p}\, |g(r) - g(p)| \le \int_K 1_p |dg| \le 2\varlimsup_{r \to p} |g(r) - g(p)|.$$

Differentials of the form $f\sigma$ with σ integrable are just the Stieltjes differentials $f\,dg$. For g the identity function $x(t) = t$ these are the classical differentials $f\,dx$. (Stieltjes [40] invoked the restriction that g be monotone but seems to have waived this restriction when he treated integration-by-parts.)

Exercises (§2.7).

Let ρ, σ be differentials on a cell K and f, g be functions on K.

1. Prove $f\rho \wedge g\rho = f \wedge g\,\rho$ and $f\rho \vee g\rho = f \vee g\,\rho$ if $\rho \geq 0$.

2. Prove $f\rho \wedge f\sigma = f\,\rho \wedge \sigma$ and $f\rho \vee f\sigma = f\,\rho \vee \sigma$ if $f \geq 0$.

3. Prove $f\rho \wedge g\sigma \leq f \vee g\,\rho \wedge \sigma$ and $f\rho \wedge g\sigma \leq f \wedge g\,\rho \vee \sigma$ if $f, g \geq 0$ and $\rho, \sigma \geq 0$.

4. Prove $f\rho \wedge g\sigma = 0$ if either $\rho \wedge \sigma = 0$ and $f, g \geq 0$, or $f \wedge g = 0$ and $\rho, \sigma \geq 0$.

5. Given a figure A in a cell K let $g_A = -1_{A^\circ}$ where A° is the absolute interior of A. Show that for every function f on K, $\int_K f\,dg_A = \int_A df$. (For a cell I in K, $\int_K f\,dg_I = \Delta f(I)$.)

6. Let D_n, E_n be complementary subsets of a cell K for $n = 1, 2, \cdots$ such that $\bigcap\limits_{n=1}^{\infty} D_n = \emptyset$ ($\bigcup\limits_{n=1}^{\infty} E_n = K$). Let ρ be a differential on K.

Prove:

(i) If $1_{E_n}\rho \geq 0$ for all n then $\rho \geq 0$.
Hint: Each E_n is ρ^--null.

(ii) If $\rho \geq 0$ then $\bigwedge\limits_{n=1}^{\infty}(1_{D_n}\rho) = 0$.
Hint: $0 \leq \sigma \leq 1_{D_n}\rho$ implies E_n is σ-null.

(iii) If $\rho \geq 0$ then $\rho = \bigvee\limits_{n=1}^{\infty}(1_{E_n}\rho)$.

7. Given differentials $\rho, \sigma \geq 0$ on a cell K prove the equivalence of the following conditions:

   (i) $\rho \leq \varepsilon \sigma$ for every $\varepsilon > 0$ in $\mathbb{R}$.

   (ii) $n\rho \leq \sigma$ for every positive integer n.

   (iii) $w\rho \leq \sigma$ for every function w on K.

   (iv) $\rho \leq u\sigma$ for every function u such that $u > 0$ everywhere on K.

   (To prove (ii) implies (iii) apply (iii) in Exercise 6 to $w^+\rho$ with $E_n = (w < n)$.)

8. Show that for f a function on $K = [a, b]$ and p a point in K:

   (i) $\int_a^p f1_p\omega = f(p)1_{(a,b]}(p)$,

   (ii) $\int_p^b f1_p\omega = f(p)1_{[a,b)}(p)$,

   (iii) $\int_a^b f1_p\omega = f(p)[2 - 1_a(p) - 1_b(p)]$.

9. For E a subset of a cell K prove that $1_E\omega$ is integrable if and only if E is finite.

10. By (2) show that Theorem 1 holds for g defined and finite σ-everywhere on K.

§2.8 The Lebesgue Space $\mathcal{L}_1$ and Convergence Theorems.

Given an integrable differential $\sigma \geq 0$ on a cell K the set of all absolutely integrable differentials of the form $f\sigma$ is the **Lebesgue space** $\mathcal{L}_1 = \mathcal{L}_1(K, \sigma)$. It is easy to see that $\mathcal{L}_1$ is a Riesz subspace of the Banach lattice of all absolutely integrable differentials on K. (See Theorem 3 (§2.4) and the ensuing discussion there.) To conclude that $\mathcal{L}_1$ is a Banach lattice under the norm $\nu(f\sigma) = \int_K |f|\sigma$ we must prove that $\mathcal{L}_1$ is complete. The series formulation of completeness is again convenient.

THEOREM 1. *Let $\sigma \geq 0$ be a differential on a cell K. Let $g_1, \cdots, g_n, \cdots$ be a sequence of functions on K such that $g_n \sigma$ is absolutely integrable for all n and we have convergence of the series $\sum\limits_{n=1}^{\infty} \int_K |g_n| \sigma < \infty$. Then there is a function f on K such that $f(t) = \sum\limits_{n=1}^{\infty} g_n(t)$ at σ-all t in K, $f\sigma$ is absolutely integrable on K, and $\int_K f\sigma = \sum\limits_{n=1}^{\infty} \int_K g_n \sigma$.*

PROOF. Since the hypothesis holds with g_n replaced there by either g_n^+ or g_n^- we need only treat the case where $g_n \geq 0$ for all n. For this case apply Theorem 3 (§2.7) on monotone convergence with $f_n = g_1 + \cdots + g_n$ to get $f_n \nearrow f$, σ-everywhere with $\sum\limits_{i=1}^{n} \int_K g_i \sigma = \int_K f_n \sigma \nearrow \int_K f\sigma$. □

For any differential ρ on K the linear space of all integrable differentials of the form $f\rho$ is complete. This follows from the next theorem.

THEOREM 2. *Let ρ be a differential on a cell K and $f_1, \cdots$ be a sequence of functions on K such that*

(1) *$(f_m - f_n)\rho \to 0$ as $m, n \to \infty$ with all the differentials $(f_m - f_n)\rho$ integrable. Then there exists a function f on K and an increasing sequence of positive integers $N_i \nearrow \infty$ such that*

(2) *$f_{N_i} \to f$ ρ-everywhere on K*

and

(3) *$f_n \rho \to f\rho$ as $n \to \infty$.*

PROOF. Using the Cauchy criterion (1) we can choose N_i for $i = 1, 2, \cdots$ such that $N_{i+1} > N_i$ and

(4) $$\int_K |(f_m - f_n)\rho| < \frac{1}{2^i} \quad \text{for all } m, n \geq N_i.$$

Let

(5) $$g_i = f_{N_{i+1}} - f_{N_i} \text{ for } i = 1, 2, \cdots .$$

Then

(6) $$f_{N_1} + \sum_{i=1}^{k} g_i = f_{N_{k+1}}.$$

By (4) and (5) $\int_K |g_i \rho| < \frac{1}{2^i}$. So Theorem 1 yields a function g on K such that $g = \sum_{i=1}^{\infty} g_i$ ρ-everywhere. In terms of (6) this gives (2) for $f = f_{N_1} + g$. Since $f - f_{N_k} = \sum_{i=k}^{\infty} g_i$ ρ-everywhere, application of Theorem 1 to these tail series with $\sigma = |\rho|$ gives $\int_K |(f - f_{N_k}) \rho| \le \sum_{i=k}^{\infty} \int_K |g_i \rho| < \sum_{i=k}^{\infty} \frac{1}{2^i} = \frac{1}{2^{k-1}}$. So $f_{N_k} \rho \to f \rho$ which implies (3) under the Cauchy condition (1). □

The next theorem shows that $f\sigma$ belongs to $\mathcal{L}_1(K, \sigma)$ for every continuous function f on K and integrable $\sigma \ge 0$. It is an elementary result that will be extended later. (See Theorem 6 (§4.3).)

THEOREM 3. *Let f be a continuous function on a cell K, and σ an absolutely integrable differential on K. Then $f\sigma$ is absolutely integrable on K.*

PROOF. By considering f^+, f^- and σ^+, σ^- separately we need only treat the case of continuous $f \ge 0$ and integrable $\sigma \ge 0$.

Given $\varepsilon > 0$ continuity of f yields a gauge δ on K such that $|f(s) - f(t)| < \varepsilon$ for all s in the δ-neighborhood $N_\delta(t)$ of t.

Consider any δ-division $\mathcal{K}$ of K. For each member (I, t) of $\mathcal{K}$ we have $f(s) < f(t) + \varepsilon$ for all s in I. So

(7) $$\overline{\int_I} f\sigma \le (f(t) + \varepsilon) \int_I \sigma \text{ for all } (I, t) \text{ in } \mathcal{K}.$$

Now $\sigma = dv = [\Delta v] \geq 0$ for some function v on K. So summation of (7) over $\mathcal{K}$ gives $\overline{\int_K} f\sigma \leq (\sum f\Delta v)(\mathcal{K}) + \varepsilon \int_K \sigma < \infty$. Since this holds for every δ-division $\mathcal{K}$ of K, $\overline{\int_K} f\sigma \leq \underline{\int_K} f\sigma + \varepsilon \int_K \sigma$. Therefore, since $\varepsilon > 0$ is arbitrary, $0 \leq \overline{\int_K} f\sigma \leq \underline{\int_K} f\sigma < \infty$. Hence, $\int_K f\sigma$ exists and is finite. $\square$

Given a monotone sequence $f_1 \leq f_2 \leq \cdots$ of functions on a cell K, $f_n \nearrow f$ where $-\infty < f(t) \leq \infty$ for all t in K. If $\sigma \geq 0$ is a differential on K such that $\int_K f_n\sigma \nearrow \infty$ it makes sense to set $\int_K f\sigma = \infty$ even though f may fail to be a viable differential coefficient because $f^{-1}(\infty)$ is not σ-null. This convention gives a convenient extension of the monotone convergence theorem: If $\sigma \geq 0$ and $f_n \nearrow f$ with $f_n\sigma$ integrable then $\int_K f_n\sigma \nearrow \int_K f\sigma \leq \infty$. (See the remarks at the end of §1.3 concerning integration of summants S such that $0 \leq S \leq \infty$.)

One of the most useful tools in integration theory is the Dominated Convergence Theorem. The following version based on Theorem 3 (§2.7) (Monotone Convergence) does not require σ to be integrable and applies to some cases of conditionally integrable $f_n\sigma$.

THEOREM 4. *Given $\sigma \geq 0$ on K, let $h, f, f_1, f_2, \ldots$ be functions on K such that $f_n \to f$ σ-everywhere, $|f_n - f_1| \leq h$ σ-everywhere for all n, and $h\sigma$ and all $f_n\sigma$ are integrable. Then $f_n\sigma \to f\sigma$. So $f\sigma$ is integrable and $\int_K f_n\sigma \to \int_K f\sigma$.*

PROOF. Since we can replace f_n by $f_n - f_1$ and f by $f - f_1$ we may in effect assume $f_1 = 0$. Since $f_n\sigma$ is integrable and $|f_n\sigma| \leq h\sigma$ this makes $f_n\sigma$ absolutely integrable. Since it suffices to prove $f_n^+\sigma \to f^+\sigma$ and $f_n^-\sigma \to f^-\sigma$ we may also assume $f_n \geq 0$ for all n. Finally we may assume "σ-everywhere" in the hypothesis is "everywhere" on K.

In the Banach lattice of absolutely integrable differentials on K we have integrability of $(f_m\sigma) \vee (f_n\sigma) = (f_m \vee f_n)\sigma$

and $(f_m\sigma) \wedge (f_n\sigma) = (f_m \wedge f_n)\sigma$. So for each n we can apply Theorem 3 (§2.7) to the monotone convergence $f_n \vee f_{n+1} \vee \cdots \vee f_{n+k} \nearrow \bar{f}_n$ as $k \nearrow \infty$, where $\bar{f}_n = f_n \vee f_{n+1} \vee \cdots \leq h$. By Theorem 3(§2.7) $\bar{f}_n\sigma$ is integrable with $c_n = \int_K \bar{f}_n\sigma \leq \int_K h\sigma < \infty$. Apply Theorem 3 (§2.7) to $h - \bar{f}_n \nearrow h - f$ with $\int_K (h - \bar{f}_n)\sigma \leq \int_K h\sigma$ to conclude that $(h - \bar{f}_n)\sigma \to (h - f)\sigma$. This implies

$$(8) \qquad\qquad \bar{f}_n\sigma \to f\sigma.$$

Similarly Theorem 3 (§2.7) applied to $h - f_n \wedge \cdots \wedge f_{n+k} \nearrow h - \underline{f}_n$ as $k \nearrow \infty$, where $\underline{f}_n = f_n \wedge f_{n+1} \wedge \cdots$, gives the integrability of $(h - \underline{f}_n)\sigma$, hence of $\underline{f}_n\sigma$. So Theorem 3 (§2.7) applied to $\underline{f}_n \nearrow f$ gives

$$(9) \qquad\qquad \underline{f}_n\sigma \to f\sigma.$$

Since $\underline{f}_n \leq f_n \leq \bar{f}_n$ and $\underline{f}_n \leq f \leq \bar{f}_n$, $|f_n - f| \leq \bar{f}_n - \underline{f}_n$. So $|f_n - f|\sigma \leq (\bar{f}_n - \underline{f}_n)\sigma \to 0$ by (8) and (9). Hence we have the convergence $f_n\sigma \to f\sigma$. $\square$

Uniform integrability of a sequence of summants is an inherent property of certain convergence conditions for integrals. Consider $f_n \to f$ on a cell K with $f_n dg$ integrable for all n. By Theorem 4 (§1.6) uniform integrability of $f_n \Delta g$ for all n implies that $f\,dg$ is integrable and $\int_K f_n dg \to \int_K f\,dg$. For the stronger convergence condition $f_n\,dg \to f\,dg$ with g of bounded variation we have the following converse.

THEOREM 5. *Let g be a function of bounded variation on a cell K. Let $f_n \to f$ be a pointwise convergent sequence of functions on K such that $f_n dg$ is integrable for all n and $f_n dg \to f\,dg$. Then the summants $f_1\Delta g, f_2\Delta g, \cdots$ are uniformly integrable on K.*

PROOF. By Theorem 3 (§2.4) and Theorem 1 (§2.5) the hypothesis gives F_n and F on K such that $dF_n = f_n dg$ for all n, $dF = f\,dg$, and $dF_n \to dF$. Since

$$|\Delta F_n - f_n \Delta g| \le |\Delta(F_n - F) - (f_n - f)\Delta g| + |\Delta F - f\Delta g|$$

we need only prove uniform integrability of $(f_n - f)\Delta g$. Replacing f_n by $f_n - f$ and F_n by $F_n - F$ we in effect may assume that $f = F = 0$.

Given $\varepsilon > 0$ it suffices to prove that there exists a gauge δ on K such that

$$(10) \qquad (\sum |\Delta F_n - f_n \Delta g|)(\mathcal{K}) < \varepsilon(2 + \int_K |dg|)$$

for every δ-division $\mathcal{K}$ of K and all n. To prove this choose for each n a gauge δ_n on K such that

$$(11) \qquad |\Delta F_n - f_n \Delta g|^{(\delta_n)}(K) < \varepsilon \text{ for all } n.$$

This is possible because $dF_n = f_n dg$. Since $dF_n \to 0$ we can choose an integer N such that

$$(12) \qquad \int_K |dF_n| < \varepsilon \quad \text{for all } n > N.$$

Take a gauge δ on K such that

$$(13) \qquad \delta \le \delta_n \quad \text{for } n = 1, \cdots, N$$

and for each t in K

$$(14) \qquad \delta(t) \le \delta_n(t) \quad \text{for all } n \text{ such that } |f_n(t)| > \varepsilon.$$

Condition (14) can be met because $f_n(t) \to 0$. Let $\mathcal{K}$ be any δ-division of K. Clearly (10) holds for $n \le N$ by (11) and (13).

We contend it also holds for $n > N$. To prove this define the complementary subsets A_n, B_n of K by

(15) $$A_n = \{t : |f_n(t)| \leq \varepsilon\}$$

and

(16) $$B_n = \{t : |f_n(t)| > \varepsilon\}.$$

Then

$$|\Delta F_n - f_n \Delta g| = (1_{A_n} + 1_{B_n})|\Delta F_n - f_n \Delta g|$$
$$\leq 1_{A_n}|\Delta F_n| + 1_{A_n}|f_n \Delta g| + 1_{B_n}|\Delta F_n - f_n \Delta g|$$
$$\leq |\Delta F_n| + \varepsilon|\Delta g| + 1_{B_n}|\Delta F_n - f_n \Delta g|$$

by (15). Summation over $\mathcal{K}$ gives

(17)
$$\left(\sum |\Delta F_n - f_n \Delta g|\right)(\mathcal{K})$$
$$\leq \int_K |dF_n| + \varepsilon \int_K |dg| + \left(\sum 1_{B_n}|\Delta F_n - f_n \Delta g|\right)(\mathcal{K}).$$

The nonzero terms of the last sum in (17) are on (I, t) in $\mathcal{K}$ with t in B_n. Such (I, t) are δ-fine since $\mathcal{K}$ is a δ-division, hence δ_n-fine by (14) and (16). So these (I, t) all belong to some δ_n-division $\mathcal{K}_n$ of K. Hence,

(18)
$$\left(\sum 1_{B_n}|\Delta F_n - f_n \Delta g|\right)(\mathcal{K}) \leq \left(\sum |\Delta F_n - f_n \Delta g|\right)(\mathcal{K}_n)$$
$$\leq |\Delta F_n - f_n \Delta g|^{(\delta_n)}(K).$$

Finally (17) gives (10) for $n > N$ by (12), (18), and (11). $\square$

Theorem 5 is valid under a weaker hypothesis than bounded variation of g. (See Exercise 6 (§3.3).)   Theorem 5 shows that uniform integrability plays a ubiquitous role in Kurzweil-Henstock integration theory.

Exercises (§2.8).

1. Let $F_1, F_2, \cdots$ be a sequence of functions of bounded variation on a cell K such that $dF_n \to dF$ for some function F on K. Prove that $|\Delta F_1|, |\Delta F_2|, \cdots$ are uniformly integrable.

2. Let g indicate the set of all rationals in $\mathbb{R}$, and h indicate the complementary set of all irrationals. Let $K = [a, b]$ in $\mathbb{R}$ and $J = [a, c]$ where $a \leq c < b$. Let f indicate J. Prove:
   (i) $\overline{\int_K} f dg = h(a)$, (ii) $\underline{\int_K} f \, dg = -g(a)$, (iii) $\overline{\int_K} f \, dh = g(a)$,
   (iv) $\underline{\int_K} f \, dh = -h(a)$.

3. Let J, K be intervals in $\mathbb{R}$ and f be a function on $J \times K$ such that:
   (i) $f(s, t)dt$ is integrable on K for each s in J,

   (ii) There exists a function h on K such that $h(t)dt$ is integrable on K and for dx-all t in K

   $$|f(r, t) - f(s, t)| \leq h(t) \quad \text{for all } r, s \text{ in } J.$$

   Define F on J by $F(s) = \int_K f(s, t)dt$ for all s in J. Apply Theorem 4 to prove $F(\cdot)$ is continuous wherever $f(\cdot, t)$ is continuous for dx-all t in K.

4. Verify that in Theorem 4 we may conclude that $\int_K |f_n| \sigma \to \int_K |f| \sigma$ in $[0, \infty]$.

5. Verify that Theorem 4 applies on $K = [0, 1]$ with $\sigma = \omega$ to the sequence $f_1, f_2, \cdots$ given by

$$f_n(t) = \begin{cases} \left(\dfrac{1}{2}\right)^{n+i} + \dfrac{(-1)^i}{i} & \text{if } \dfrac{1}{t} \text{ is a positive integer } i, \\ 0 & \text{otherwise.} \end{cases}$$

6. Let $g_1, g_2, \cdots$ be a sequence of functions of bounded variation on a cell K such that $dg_n \to dg$ for some function g on K. Let $f_1, f_2, \cdot$ be a uniformly bounded sequence of continuous functions on K such that $f_n \to f$ for some function f on K.

   Prove $f_n dg_n \to f dg$ and $\int_K f)n dg_n \to \int_K f dg$. (Under the additional assumption that f is continuous and $g_n \to g$ one can also conclude that $\int_K g_n df_n \to \int_K g df$ by applying intergration by parts which is treated in §6.2)

CHAPTER 3

DIFFERENTIALS WITH
SPECIAL PROPERTIES

§3.1 Products Involving Tag-Finite Summants and Differentials.

As a generalization of the product $f\tau$ of a function f and a differential τ we can define the product $S\tau$ as a viable differential for a restricted class of summants S. With restrictions on both σ and τ we can define the differential $\sigma\tau$. The restriction we need is "tag-finiteness."

A differential σ on a cell K is **tag-finite** if $1_p\sigma$ is summable for every point p in K. As a differential property tag-finiteness reverts to the summants S representing σ. We want to express tag-finiteness explicitly in terms of S. As a first step, application of Theorem 1 (§1.9) to $|S|$ has the following formulation.

THEOREM 1. *Let* $\sigma = [S]$ *be a differential on* $K = [a, b]$. *Then for all points* p *in* K

$$\nu(1_p\sigma) = \varlimsup_{r \to p-} |S([r, p], p)| + \varlimsup_{r \to p+} |S([p, r], p)|$$

with the convention that all limits at $a-$ *and* $b+$ *equal* 0.

The next result is essentially a corollary.

THEOREM 2. *Let σ be a tag-finite differential on a cell K. Then there exists a function v on K such that*

(1) $\nu(1_p\sigma) < v(p) < \infty$ *for all p in K.*

(i) Given v satisfying (1) and a summant S representing σ there exists a gauge δ on K such that

(2) $|S(I,t)| \leq v(t)$

for every δ-fine tagged cell (I,t) in K.

(ii) Given v satisfying (1) there exists a summant S representing σ such that (2) holds for every tagged cell (I,t) in K.

PROOF. The existence of v satisfying (1) is assured because $\nu(1_p\sigma) < \infty$ for all p in K by tag-finiteness of σ. By Theorem 1 we have

$$\varlimsup_{(I,p)\to p} |S(I,p)| \leq \nu(1_p\sigma)$$

which together with (1) yields (i). Annihilating S at all tagged cells (I,t) where (2) fails to hold gives an equivalent summant satisfying (ii). $\square$

For S a tag-finite summant and τ any differential on a cell K we define the differential $S\tau$ on K by

(3) $S\tau = [ST]$ for $\tau = [T]$.

For $S\tau$ to be a viable differential we must show that the differential $[ST]$ in (3) does not depend on the particular representative T of τ. Given $T' \sim T$ we must verify that $ST' \sim ST$ if S is tag-finite. By (i) in Theorem 2 $|ST - ST'| = |S|\,|T - T'| \leq v|T - T'|$ at all δ-fine tagged cells. Now $v|T - T'| \sim 0$ by Theorem 1 (§1.8) applied to $T - T' \sim 0$. So $ST - ST' \sim 0$. That is, $ST \sim ST'$. So (3) is effective for S tag-finite.

If S and S' represent a tag-finite differential σ there may exist a differential τ such that $S\tau \neq S'\tau$. For example, take the identity function x on a bounded cell K and let $S = (\Delta x)^2$ and $T = \frac{1}{\Delta x}$. Then $S \sim 0$ but $ST = \Delta x$ and $0T = 0$ which are not equivalent since $dx \neq 0$. In this example the zero differential represented by both 0 and S is obviously tag-finite. But $\tau = [T]$ is not tag-finite. Indeed, $\nu(1_p\tau) = \infty$ at every point p in K. Both σ and τ must be tag-finite to yield a viable differential product $\sigma\tau$ which we shall now define.

Given tag-finite differentials $\sigma = [S]$ and $\tau = [T]$ on a cell K define the differential $\sigma\tau$ on K by

$$(4) \qquad\qquad \sigma\tau = [ST].$$

To prove effectiveness of this definition we must show $[ST] = [S'T']$ if $S' \sim S$ and $T' \sim T$. The differential $[ST] - [S'T'] = [ST - S'T'] = S[T - T'] + T'[S - S'] = [0]$ since $S[T - T'] = S[0] = [0]$ and $T'[S - S'] = T'[0] = [0]$ under (3) for S, T' tag-finite. So $[ST] = [S'T']$. Thus, $\sigma\tau$ is well defined by (4) for all tag-finite σ, τ on K. Clearly $\sigma\tau = \tau\sigma$.

THEOREM 3. *If f is a function on K and σ is a tag-finite differential on K then $f\sigma$ is tag-finite.*

PROOF. Since $1_p(t)f(t) = f(p)1_p(t)$ for all p, t in K, $1_p f\sigma = f(p)1_p\sigma$. So $\nu(1_p f\sigma) = |f(p)|\nu(1_p\sigma) < \infty$ for σ tag-finite.   □

THEOREM 4. *Let ρ, σ be differentials on K such that $|\rho| \leq |\sigma|$ and σ is tag-finite. Then ρ is tag-finite.*

PROOF. $1_p|\rho| \leq 1_p|\sigma|$ since $|\rho| \leq |\sigma|$. So $\nu(1_p\rho) \leq \nu(1_p\sigma) < \infty$ for σ tag-finite.   □

THEOREM 5. *If σ and τ are tag-finite differentials on K then so is their product $\sigma\tau$.*

PROOF. Using (ii) in Theorem 2 choose S representing σ such that $|S| \leq v$ for some function v on K. Choose any T

representing τ. By (3) and (4), $\sigma\tau = [ST] = S\tau$. So $|\sigma\tau| = |S\tau| = |S||\tau| \leq v|\tau|$. That is, $|\sigma\tau| \leq |v\tau|$. $v\tau$ is tag-finite by Theorem 3. So $\sigma\tau$ is tag-finite by Theorem 4.

It is an easy exercise to verify that if σ, τ are tag-finite differentials on K then so are $\sigma + \tau$, $|\sigma|$, and $c\sigma$ for c in $\mathbb{R}$. So the set $\mathbb{D}_1$ of all tag-finite differentials on K is a Riesz subspace of the space $\mathbb{D}$ of all differentials on K. Theorem 4 says that $\mathbb{D}_1$ is solid in $\mathbb{D}$. By Theorem 5 $\mathbb{D}_1$ is a commutative algebra. By Theorem 3 functions on K act as linear operators on $\mathbb{D}_1$. We contend now that $\mathbb{D}_1$ is complete.

THEOREM 6. *The Riesz space $\mathbb{D}_1$ of all tag-finite differentials on a cell K is complete under differential convergence.*

PROOF. $\mathbb{D}_1$ is a subspace of $\mathbb{D}$. By Theorem 2 (§2.3) $\mathbb{D}$ is complete. So we need only prove that $\mathbb{D}_1$ is closed in $\mathbb{D}$ to conclude that $\mathbb{D}_1$ is complete.

Given $\sigma_n \to \sigma$ in $\mathbb{D}$ with σ_n in $\mathbb{D}_1$ for all n we contend that σ belongs to $\mathbb{D}_1$. For every point p in K, $1_p\sigma_n \to 1_p\sigma$. Now $1_p\sigma_n$ is summable. That is, $1_p\sigma_n$ belongs to the space $\mathbb{S}$ of all summable differentials on K. By Theorem 1 (§2.3) $\mathbb{S}$ is complete. So $1_p\sigma$ belongs to $\mathbb{S}$. Since this holds for all p in K, σ belongs to $\mathbb{D}_1$.   □

For σ integrable on a cell K we have $\sigma = dg$ for some function g on K by Theorem 1 (§2.4). We contend now that tag-finiteness of σ is just boundedness of g.

THEOREM 7. *Let g be a function on a cell K. Then dg is tag-finite on K if and only if g is bounded on K.*

PROOF. Given dg tag-finite apply Theorem 2 to $\sigma = dg$ with $S = \Delta g$ to get a function v and gauge δ on K for which (2) gives $|g(s) - g(t)| \leq v(t)$ for all s and t in K such that s lies in the neighborhood $N_\delta(t)$ of t. So $|g(s)| \leq v(t) + |g(t)|$ for all s in $K \cap N_\delta(t)$. Hence g is bounded on $K \cap N_\delta(t)$ for each t in

K. Since K is compact, finitely many δ-neighborhoods cover K. So the local boundedness of g implies boundedness on K.

Conversely let $|g(r)| \le c < \infty$ for all r in K. Then $|g(r) - g(p)| \le |g(r)| + |g(p)| \le 2c$ for all r, p in K. Hence, by (9) in Theorem 5 (§2.7), $\nu(1_p dg) \le 2\overline{\lim}_{r \to p} |g(r) - g(p)| \le 4c < \infty$ for all p in K. So dg is tag-finite. $\square$

The constant summant 1 defines the **unit differential** $\omega = [1]$ on K. ω is tag-finite on $K = [a, b]$ since $\nu(1_p\omega) = \int_K 1_p\omega = 1$ if $p = a$ or $p = b$, 2 if $a < p < b$. ω is the unit in the commutative algebra $\mathbb{D}_1$ of tag-finite differentials on K. That is,

(5) $\qquad\qquad \omega\sigma = \sigma \quad$ for every tag-finite σ on K.

This follows from the definition of $\omega\sigma$ given by (4), $\omega\sigma = [1][S] = [1S] = [S] = \sigma$ for S representing σ.

We can characterize the tag-finite differentials in terms of the unit differential ω.

THEOREM 8. *A differential σ on a cell K is tag-finite if and only if $|\sigma| \le v\omega$ for some function $v \ge 0$ on K.*

PROOF. Given σ tag-finite apply Theorem 2 to get a summant S representing σ and a function v such that $|S| \le v1$ by (2). So $|\sigma| \le v\omega$ since $[|S|] = |\sigma|$ and $[v1] = v[1] = v\omega$.

Since ω is tag-finite the converse follows from Theorem 3 and Theorem 4. $\square$

The set $\mathbb{D}_1$ of all tag-finite differentials on a cell K is a commutative ring with unit ω. Now the idempotents $\sigma = \sigma^2$ in any ring with unit form a boolean algebra in which the greatest lower bound of idempotents σ, τ is taken to be $\sigma\tau$. Does this agree with the ordering in the Riesz space $\mathbb{D}_1$? Does $\sigma\tau = \sigma \wedge \tau$ in $\mathbb{D}_1$ if σ, τ are idempotent? Is $\sigma \wedge \tau$ idempotent if σ, τ are idempotent in $\mathbb{D}_1$? We shall answer these questions affirmatively.

A **damper** on K is a function u on K such that $u(t) > 0$ for all t in K.

THEOREM 9. *If σ, τ are tag-finite differentials on a cell K and $|\sigma| \wedge |\tau| = 0$ then $\sigma\tau = 0$.*

PROOF. We may assume $\sigma, \tau \geq 0$. By Theorem 8 there exists a damper u on K such tht $u\sigma \leq \omega$ and $u\tau \leq \omega$. (Set $u = 1/(v+1)$ in Theorem 8.) The first inequality gives $u\sigma\tau \leq \omega\tau = \tau$. The second gives $u\sigma\tau \leq \sigma$. So $0 \leq u\sigma\tau \leq \sigma \wedge \tau = 0$. That is, $u\sigma\tau = 0$. Multiply by $1/u$ to get $\sigma\tau = 0$. $\square$

(The converse is false according to (ii) in Exercise 8.)

We call a differential σ on a cell K **idempotent** if σ is tag-finite and $\sigma = \sigma^2$. The next result is a lemma for Theorem 11. Indeed it follows from Theorem 11.

THEOREM 10. *If σ is an idempotent differential then $0 \leq \sigma \leq \omega$.*

PROOF. $0 \leq \sigma^2 = \sigma$. So $0 \leq \sigma$. Similarly $0 \leq (\omega - \sigma)^2 = \omega^2 - 2\omega\sigma + \sigma^2 = \omega - 2\sigma + \sigma = \omega - \sigma$. So $\sigma \leq \omega$. $\square$

THEOREM 11. *For any differential σ on a cell K the following conditions are equivalent:*

(i)   *σ is idempotent,*

(ii)   *$\sigma = [S]$ for some indicator summant $S = S^2$,*

(iii)   *$\sigma \wedge (\omega - \sigma) = 0$.*

PROOF. (i) $\Rightarrow$ (ii). By Theorem 10 we may assume that given (i)

(6) $$\sigma = [S_0] \quad \text{with} \quad 0 \leq S_0 \leq 1.$$

Let S indicate $(S_0 > \frac{1}{2})$, that is, $(1 - S_0 < \frac{1}{2})$. Let $T = 1 - S$ which indicates $(2S_0 \leq 1)$. So $2TS_0 \leq T$ which yields the differential inequality $2T\sigma \leq T\omega$. Hence, since $0 \leq \sigma = \sigma^2$, $0 \leq 2T\sigma = 2T\sigma^2 \leq T\omega\sigma = T\sigma$ which implies $T\sigma = 0$.

That is, since $S + T = 1$,

$$(7) \qquad\qquad \sigma = S\sigma.$$

By Theorem 10 this implies

$$(8) \qquad\qquad \sigma \leq S\omega.$$

Now $S(1 - S_0) \leq \frac{1}{2}S \leq S_0$ which implies

$$(9) \qquad\qquad S(\omega - \sigma) \leq \sigma.$$

By Theorem 10 $\omega - \sigma \geq 0$. With $0 \leq S \leq 1$ this gives

$$(10) \qquad\qquad 0 \leq S(\omega - \sigma) \leq \omega - \sigma.$$

By (7), $\cdots$, (10) $0 \leq S\omega - \sigma = S\omega - S\sigma = S(\omega - \sigma) = S^2(\omega - \sigma)^2 \leq \sigma(\omega - \sigma) = \sigma - \sigma^2 = 0$. So $S\omega - \sigma = 0$. That is, $\sigma = S\omega$ which gives (ii).

(ii) $\Rightarrow$ (iii). By (ii) $\sigma \wedge (\omega - \sigma) = (S\omega) \wedge (\omega - S\omega) = (S\omega) \wedge ((1 - S)\omega) = (S \wedge (1 - S))\omega = 0\omega = 0$. That is, (iii) holds.

(iii) $\Rightarrow$ (i). Set $\tau = \omega - \sigma$ in Theorem 9 to get $\sigma(\omega - \sigma) = 0$ from (iii). That is, $\sigma - \sigma^2 = 0$ which gives (i). (σ is tag-finite since $0 \leq \sigma \leq \omega$ by (iii).) $\square$

THEOREM 12. *If σ, τ are idempotent differentials on a cell K then $\sigma \wedge \tau = \sigma\tau$.*

PROOF. $(\sigma \wedge \tau) \wedge (\omega - \sigma \wedge \tau) = \sigma \wedge \tau \wedge ((\omega - \sigma) \vee (\omega - \tau)) \leq (\sigma \wedge (\omega - \sigma)) \vee (\tau \wedge (\omega - \tau)) = 0$ by Theorem 11. So Theorem 11 gives $\sigma \wedge \tau = (\sigma \wedge \tau)^2$. Since $0 \leq \sigma \wedge \tau \leq \sigma$ and $0 \leq \sigma \wedge \tau \leq \tau$, $(\sigma \wedge \tau)^2 \leq \sigma\tau$. That is, $\sigma \wedge \tau \leq \sigma\tau$. To reverse this inequality we have $0 \leq \sigma$ and $\tau \leq w$ from Theorem 10. Thus $\sigma\tau \leq \sigma w = \sigma$. That is, $\sigma\tau \leq \sigma$. Similarly $\sigma\tau \leq \tau$. So $\sigma\tau \leq \sigma \wedge \tau$. □

Exercises (§3.1).

1. For ρ, σ, τ tag-finite differentials on a cell K and f, g functions on K prove:

   (i) $\rho(\sigma + \tau) = \rho\sigma + \rho\tau$,

   (ii) $\rho(\sigma\tau) = (\rho\sigma)\tau$,

   (iii) $\rho^2 \geq 0$,

   (iv) $(f\sigma)\tau = \sigma(f\tau) = f(\sigma\tau)$,

   (v) $(f\sigma)(g\tau) = (fg)(\sigma\tau)$,

   (vi) $(\rho|\tau|)\wedge(\sigma|\tau|) = (\rho\wedge\sigma)|\tau|$ and $(\rho|\tau|)\vee(\sigma|\tau|) = (\rho\vee\sigma)|\tau|$.

2. Verify that $\sigma = Sw$ for tag-finite $\sigma = [S]$.

3. Show that if $0 \leq \sigma \leq cw$ for some $c < 1$ and $\sigma^2 = \sigma$ then $\sigma = 0$.

4. Given an idempotent σ on $K = [a, b]$ prove:

   (i) $|w - 2\sigma| = w$,

   (ii) The set C of all σ-null points p in K is σ-null,

   (iii) Each of the following integrals has value 0 or 1 for all points p in K:

$$\overline{\int_a^p} 1_p\sigma, \quad \underline{\int_a^p} 1_p\sigma, \quad \overline{\int_p^b} 1_p\sigma, \quad \underline{\int_p^b} 1_p\sigma.$$

5. For f a function on a cell K show that $fw = 0$ on K if and only if $f = 0$. (So the only w-null set is $\emptyset$.)

6. Let $t_1, t_2, \cdots$ be a monotone sequence of distinct points in (a, b). Let f be a function on $K = [a, b]$ that vanishes on the complement of $\{t_1, t_2, \cdots\}$. Prove: (i) fw is integrable on K if and only if the series $f(t_1) + f(t_2) + \cdots$ is convergent,

(ii) If $f\omega$ is integrable then its integral $\int_K f\omega = 2(f(t_1) + f(t_2) + \cdots)$.

7. Show that the number of points in a subset E of $K = [a, b]$ is given by $\int_K 1_E g\omega$ where $g = 1 - \frac{1}{2}1_{(a,b)}$.

8. (i) For differentials σ, τ on a cell K prove that $|\sigma| \wedge |\tau| = 0$ if and only if there exist complementary indicator summants V, W such that $V\tau = W\sigma = 0$.

(ii) Find tag-finite σ, τ on a cell K such that $\sigma\tau = 0$ but $|\sigma| \wedge |\tau| > 0$.

(Consider the case $\sigma = \tau > 0$.)

(iii) Prove Theorem 9 using Exercise 4 (§2.2).

9. Prove that in the Riesz space $\mathbb{D}$ of all differentials on a cell K the set $\mathbb{D}_1$ of all tag-finite members of $\mathbb{D}$ is the smallest Riesz ideal $\mathbb{E}$ in $\mathbb{D}$ such that $\omega \in \mathbb{E}$, and if $1_p\sigma \in \mathbb{E}$ for some σ in $\mathbb{D}$ and all p in K then $\sigma \in \mathbb{E}$.

10. Let h be a function of bounded variation on a cell K such that h is continuous at the endpoints of K and $h(t) = (h(t-)+h(t+))/2$ at every interior point t of K. Let Q be the summant defined by (20) in §1.4. Show that $\int_K Qf(dh)^2 = 0$ for every function f on K.

11. Given a point p in $K = [a, b]$, and a subset E of K prove:

(i) $\int_K 1_p Q^+\omega = 1_{[a,b)}(p)$, $\quad \int_K 1_p Q^-\omega = 1_{(a,b]}(p)$,

(ii) $\int_K 1_E Q^+\omega$ equals the number of points in $E \cap [a, b)$, and $\int_K 1_E Q^-\omega$ the number in $E \cap (a, b]$.

12. (See Theorem 11.) Given an idempotent differential σ on a cell K prove:

(i) σ is summable if and only if $\sigma = 1_D\sigma$ for some finite subset D of K,

(ii) σ is integrable if and only if $\sigma = (1_A Q^+ + 1_B Q^-)\omega$ for some finite subsets A, B of K.

(For more about integrable idempotents see Exercise 22 in §4.4 of the next chapter.)

13. Let f be a function on $K = [a, b]$ such that $\sum_{s \in K} |f(s)| < \infty$. Let $D_n = \{t_1, \cdots, t_n\}$ where $t_1, t_2, \cdots$ is a sequential arrangement of the points at which $f \neq 0$. Prove:

  (i) $1_{K - D_n} f\omega \to 0$ as $n \to \infty$,

  (ii) $1_{D_n} Q^+ f\omega \to Q^+ f\omega$, $1_{D_n} Q^- f\omega \to Q^- f\omega$,

  (iii) $\int_a^t 1_{D_n} Q^+ f\omega = \sum_{i=1}^n 1_{[a,t)}(t_i) f(t_i)$,
$\int_a^t 1_{D_n} Q^- f\omega = \sum_{i=1}^n 1_{(a,t]}(t_i) f(t_i)$,

  (iv) $\int_a^t Q^+ f\omega = F(t)$ and $\int_a^t Q^- f\omega = G(t)$ for $F(t) = \sum_{s \in [a,t)} f(s)$, $G(t) = \sum_{s \in (a,t]} f(s)$,

  (v) $f\omega = d(F + G)$.

14. For σ a differential on a cell K prove the equivalence of the following:

  (i) $\sigma = [S]$ for some bounded S,

  (ii) $|\sigma| \leq c\omega$ for some $c > 0$ in $\mathbb{R}$,

  (iii) σ is tag-finite and $\sigma\tau$ is summable for every summable differential τ on K.

   (That (ii) implies (iii) is easy to prove. The converse is much harder. For an indirect proof show that if (ii) is false and σ is tag-finite then $\sigma\tau$ fails to be summable for $\tau = f\omega$ for some $f \geq 0$ with $\sum_{t \in K} f(t) < \infty$.)

15. As an extension of Theorem 3 verify that if σ is tag-finite and F is a function defined and finite σ-everywhere on K then $F\sigma$ as defined by (2) in §2.7 is tag-finite.

16. Show that the only differential ρ on a cell K satisfying the condition $\rho \wedge \omega = 0$ is $\rho = 0$.

§3.2 Continuous Differentials.

   A differential σ on a cell K is **continuous** at a point p in K if $1_p \sigma = 0$, that is, if p is σ-null. σ is **continuous** on a set E contained in K if σ is continuous at every point in E. If σ is continuous on K then σ is clearly tag-finite. The following result shows that the term "continuous" is appropriate. It is a direct consequence of (9) in Theorem 5 (§2.7).

THEOREM 1. *For g a function on a cell K the differential dg is continuous at a point p in K if and only if the function g is continuous at p.*

A more general result is the following.

THEOREM 2. *Let $\sigma = [S]$ be a differential on a cell K and p be a point in K. Then $1_p\sigma = 0$ if and only if $S(I,p) \to 0$ as $(I,p) \to p$ in K.*

PROOF. Let $h(p) = \overline{\lim}|S(I,p)|$ as $(I,p) \to p$ in K. By (v) in Theorem 1 (§1.9), $h(p) \le \nu(1_p\sigma) \le 2h(p)$. So $\nu(1_p\sigma) = 0$ if and only if $h(p) = 0$.   □

We can reformulate Theorem 2 in terms of the unit differential $\omega = [1]$ as follows.

THEOREM 3. *A differential σ on a cell K is continuous on K if and only if*

$$(1) \qquad\qquad |\sigma| \le \varepsilon\omega \quad \text{for all } \varepsilon > 0 \text{ in } \mathbb{R}.$$

PROOF. Let (1) hold. Then $1_p|\sigma| \le \varepsilon 1_p\omega$ for all p in K and all $\varepsilon > 0$. So $\nu(1_p\sigma) \le \varepsilon\nu(1_p\omega) \le 2\varepsilon$ for all $\varepsilon > 0$. Hence, $\nu(1_p\sigma) = 0$. So $1_p\sigma = 0$ for all p in K.

Conversely let $1_p\sigma = 0$ for all p. Given S representing σ and $\varepsilon > 0$, Theorem 2 yields a gauge δ on K such that $|S(I,t)| < \varepsilon$ for all δ-fine (I,t) in K. So $|\sigma| = [|S|] \le \varepsilon[1] = \varepsilon\omega$.   □

It is (1) that makes continuous differentials appear to be "ghosts of departed quantities." From the modern point of view the existence of nonzero differentials σ satisfying (1) just means that ω is non-archimedean. This leads to the topic of the next section.

Exercises (§3.2).

1. Show that the set $\mathbb{C}$ of all continuous differentials on a cell K is a complete, solid Riesz subspace of the Riesz space $\mathbb{D}$ of all differentials on K.

2. Given σ belonging to $\mathbb{C}$ prove:

   (i) $f\sigma$ belongs to $\mathbb{C}$ for every function f on K,

   (ii) $\rho\sigma$ belongs to $\mathbb{C}$ for every tag-finite differential ρ on K.

3. For σ a differential on a cell K prove the equivalence of the following:

   (i) $(|\sigma| - \varepsilon\omega)^+$ is continuous for every $\varepsilon > 0$ in $\mathbb{R}$,

   (ii) σ is continuous on K,

   (iii) Every countable subset of K is σ-null,

   (iv) σ is tag-finite and $\rho\sigma = 0$ for every summable differential ρ on K,

   (v) $1_I\sigma = 1_{I^\circ}\sigma$ for every cell I in K,

   (vi) $1_L\sigma = 1_{\bar{L}}\sigma$ for every interval L in K,

   (vii) $1_A\sigma = 1_{A^\circ}\sigma$ for every figure A in K,

   (viii) σ is continuous σ-everywhere on K.

4. For σ an integrable differential on a cell K prove the equivalence of the following:

   (i) σ is continuous on K,

   (ii) $1_I\sigma$ is integrable and $\int_K 1_I\sigma = \int_I \sigma$ for every cell I in the cell K,

   (iii) $1_A\sigma$ is integrable and $\int_K 1_A\sigma = \int_A \sigma$ for every figure A in the cell K.

5. Prove that if a differential σ is summable on a cell K and continuous on a subset E of K then $1_E\sigma^2 = 0$.

6. Show that 0 is the only continuous, idempotent differential.

§3.3 Archimedean Properties for Differentials.

Many non-summable differentials are useful in analysis. To classify these we introduce some new concepts. Recall (§3.1) that a damper on a cell K is a function u on K such that $0 < u(t) < \infty$ for all t in K. (A gauge has the same definition. But we shall use dampers as differential coefficients while gauges are used only to produce neighborhoods.)

A differential σ on K is **dampable** if $u\sigma$ is absolutely integrable for some damper u on K. σ is **damper-summable** if $u\sigma$ is summable for some damper u on K. That is, $u|\sigma| \leq \tau$ for some damper u and some integrable τ on K. (See Theorem 2, §2.5.) A summant S on K is **tame** if $u|S| \leq T$ for some damper u and some additive summant T on K. σ is damper-summable if and only if $\sigma = [S]$ for some tame S. (This is the equivalence of (i) and (iii) in Exercise 15.)

Clearly every dampable differential is damper-summable, as is every summable differential. But not every summable differential is dampable, as the following example shows.

On $K = [0, 1]$ let $\sigma = 1_0 dg$ where $g(t) = 0$ if $t = 0$, $\sin \frac{1}{t}$ if $t > 0$. By (8) in Theorem 5 (§2.7), $\nu(\sigma) = \int_K 1_0 |dg| = 1$. So σ is summable. For any damper u on K, $u\sigma = u(0)\sigma$. So $\int_K u\sigma = u(0)$ and $\int_{\underline{K}} u\sigma = -u(0)$ since $\int_K \sigma = 1$ and $\int_{\underline{K}} \sigma = -1$ by (5), (6) in Theorem 4 (§2.7). Therefore, since $u(0) > 0$, $u\sigma$ is not integrable. (See Exercise 6 (§4.3) for an example with σ continuous.)

A differential σ on K is **archimedean** if the only differential ρ on K such that $|\rho| \leq \varepsilon|\sigma|$ for all $\varepsilon > 0$ in $\mathbb{R}$ is $\rho = 0$. This is a Riesz space concept. (For alternative definitions of archimedean differential see Exercise 7 (§2.7).) σ is **weakly archimedean** if $P\sigma = 0$ for every continuous summant P on K. (P is continuous whenever the differential $[P]$ is continuous.)

THEOREM 1. *Every damper-summable differential σ on a cell K is archimedean.*

PROOF. Take a damper u on K such that $\nu(u\sigma) < \infty$. Let ρ be a differential on K such that $|\rho| \le \varepsilon|\sigma|$ for all $\varepsilon > 0$ in $\mathbb{R}$. Then $|u\rho| \le \varepsilon|u\sigma|$, so $\nu(u\rho) \le \varepsilon\nu(u\sigma) < \infty$ for all $\varepsilon > 0$. Hence $\nu(u\rho) = 0$. That is, $u\rho = 0$. So $\rho = \frac{1}{u}(u\rho) = 0$. $\square$

THEOREM 2. *Every archimedean differential σ on a cell K is weakly archimedean.*

PROOF. Given a continuous summant P on K we contend that $P\sigma = 0$. Given $\varepsilon > 0$ apply Theorem 2 (§3.2) to get a gauge δ on K such that $|P| < \varepsilon$ at every δ-fine tagged cell in K. At such tagged cells $|PS| \le \varepsilon|S|$ for $[S] = \sigma$. Therefore, $|P\sigma| \le \varepsilon|\sigma|$. Since this holds for all $\varepsilon > 0$ and σ is archimedean, $P\sigma = 0$. $\square$

For the next theorem we need an indirect proof.

THEOREM 3. *Every weakly archimedean differential σ on a cell K is tag-finite.*

PROOF. We may assume $\sigma \ge 0$, so $\sigma = [S]$ for some $S \ge 0$. Suppose σ is not tag-finite. That is, $\nu(1_p\sigma) = \infty$ for some point p in K. So there exists a sequence of tagged cells in K with tag p such that $(I_n, p) \to p$ and

$$(1) \qquad\qquad S(I_n, p) \to \infty \text{ as } n \to \infty.$$

Define $\rho = [R]$ on K by

$$(2) \quad R(I, t) = \begin{cases} \sqrt{S(I,t)} & \text{if } (I,t) = (I_n, p) \text{ for some } n \\ 0 & \text{otherwise.} \end{cases}$$

Then define the summant P by

$$(3) \qquad\qquad P = \begin{cases} \frac{1}{R} & \text{if } R > 0 \\ 0 & \text{if } R = 0 \end{cases}.$$

By (1),(2),(3) P is a continuous summant on K. By (2), (3) $PS = R$. So $P\sigma = \rho$. By (2), $\rho = 1_p\rho$ and $\nu(\rho) = \nu(1_p\rho) = \infty$ by (2), (1). So $\rho \neq 0$. Hence, σ cannot be weakly archimedean. $\square$

In summary we have the following chain of implications for a differential σ on K: σ absolutely integrable $\Rightarrow$ σ summable $\Rightarrow$ σ damper-summable $\Rightarrow$ σ archimedean $\Rightarrow$ σ weakly archimedean $\Rightarrow$ σ tag-finite. The first three conditions are easily seen to be distinct. The last two are also distinct since the unit differential ω is tag-fnite but not weakly archimedean. Indeed, $\rho\omega = \rho$ for every tag-finite ρ. So $\rho\omega \neq 0$ if ρ is continuous and nonzero. (That σ damper-summable, σ archimedean, and σ weakly archimedean are distinct conditions is only conjectural. Perhaps all three are equivalent. See Exercise 7.)

The following characterization of weakly archimedean differentials is a corollary of Theorem 3. Its proof is left as an exercise. (Compare with (iii) in Exercise 3 in §3.2.)

THEOREM 4. *A differential σ on a cell K is weakly archimedean if and only if σ is tag-finite and $\rho\sigma = 0$ for every continuous differential ρ on K.*

For a more primitive characterization define $R = o(S)$ for summants R, S on K to mean that given $\varepsilon > 0$ in $\mathbb{R}$ there exists a gauge δ on K such that $|R| \leq \varepsilon|S|$ at all δ-fine tagged cells in K.

THEOREM 5. *A differential σ on a cell K is weakly archimedean if and only if $R = o(S)$ for some S representing σ implies $R \sim 0$.*

PROOF. Let $\sigma = [S]$ be weakly archimedean and $R = o(S)$. Define $P = R/S$ if $S \neq 0$, $P = 0$ if $S = 0$. P is continuous since $R = o(S)$. So $P\sigma = 0$ since σ is weakly archimedean. That is, $R \sim PS \sim 0$.

Conversely given P continuous let $R = PS$. Then $R = o(S)$ so $R \sim 0$. That is $P\sigma = [PS] = [R] = 0$. $\square$

We can extend the decomposition given by Theorem 7 (§2.4) for absolutely integrable differentials to dampable differentials. We call the reader's attention again to the fact that the conditions (i), (ii), and (iii) are purely Riesz space concepts.

THEOREM 6. *Let σ, τ be dampable differentials on a cell K. Then there exist unique dampable differentials ρ, θ on K such that: (i) $\tau = \rho + \theta$, (ii) $|\theta| \wedge |\sigma| = 0$, (iii) $\phi \wedge |\rho| = 0$ for every differential ϕ on K such that $\phi \wedge |\sigma| = 0$.*

PROOF. There exist dampers v, w and absolutely integrable σ_0, τ_0 such that $\sigma = v\sigma_0$ and $\tau = w\tau_0$. Apply Theorem 7 (§2.4) to get absolutely integrable ρ_0, θ_0 such that (i), (ii), (iii) hold for $\sigma_0\tau_0, \rho_0, \theta_0$ in place of $\sigma, \tau, \rho, \theta$ respectively. Let $\rho = w\rho_0$ and $\theta = w\theta_0$. So ρ, θ are dampable. Also (i) holds since $\tau = w\tau_0$ and $\tau_0 = \rho_0 + \theta_0$. (ii) follows from $|\theta| \wedge |\sigma| = |w\theta_0| \wedge |v\sigma_0| \leq (w \vee v)(|\theta_0| \wedge |\sigma_0|) = 0$ since (ii) holds for θ_0, σ_0.

To prove (iii) for σ, ρ from (iii) for σ_0, ρ_0 let $\phi \wedge |\sigma| = 0$. Then $\phi \wedge |\sigma_0| = \phi \wedge |\frac{1}{v}\sigma| \leq (1 + \frac{1}{v})(\phi \wedge |\sigma|) = 0$. So $\phi \wedge |\rho_0| = 0$ by (iii) for σ_0, ρ_0. Therefore, $0 \leq \phi \wedge |\rho| = \phi \wedge |w\rho_0| \leq (1+w)(\phi \wedge |\rho_0|) = 0$ which proves (iii).

To prove uniqueness let ρ, θ and ρ_1, θ_1 satisfy (i), (ii), (iii). By (ii) both θ and θ_1 are disjoint from σ, $|\theta| \wedge |\sigma| = |\theta_1| \wedge |\sigma| = 0$. So by (iii) ρ and ρ_1 are disjoint from both θ and θ_1, $|\rho| \wedge |\theta| = |\rho| \wedge |\theta_1| = |\rho_1| \wedge |\theta| = |\rho_1| \wedge |\theta_1| = 0$. Disjointness of the summands in (i), $\tau = \rho + \theta = \rho_1 + \theta_1$, gives $\tau^+ = \rho^+ \vee \theta^+ = \rho_1^+ \vee \theta_1^+$. Hence, $\rho^+ = \rho^+ \wedge (\rho_1^+ \vee \theta_1^+) = (\rho^+ \wedge \rho_1^+) \vee (\rho^+ \wedge \theta_1^+) = (\rho^+ \wedge \rho_1^+) \vee 0 = \rho^+ \wedge \rho_1^+$. That is, $\rho^+ \leq \rho_1^+$. By symmetry $\rho_1^+ \leq \rho^+$. That is, $\rho^+ = \rho_1^+$. Similarly $\tau^- = \rho^- \vee \theta^- = \rho_1^- \vee \theta_1^-$ yields $\rho^- = \rho_1^-$. So $\rho = \rho_1$. Finally $\theta = \tau - \rho = \tau - \rho_1 = \theta_1$. So ρ and θ are unique. $\square$

Exercises (§3.3).

1. Let $\sigma = [(\Delta v)^{\frac{1}{2}}]$ on a cell K where v is a strictly increasing, continuous function on K. Show that σ is continuous but not weakly archimedean.

2. Let σ be weakly archimedean and ρ be tag-finite on a cell K. Show that if ρ is continuous σ-everywhere then $\rho\sigma = 0$.

3. In §3.1 we proved that for $\mathbb{E}$ the set of all tag-finite differentials on a cell K: (P_1) $\mathbb{E}$ is a Riesz subspace of the Riesz space $\mathbb{D}$ of all differentials on K, (P_2) If σ belongs to $\mathbb{E}$ then so does $f\sigma$ for every function f on K, (P_3) If σ belongs to $\mathbb{E}$ then so does $\rho\sigma$ for every tag-finite ρ in $\mathbb{D}$, (P_4) $\mathbb{E}$ is closed in $\mathbb{D}$, hence complete. Show that $(P_1), \cdots, (P_4)$ hold for $\mathbb{E}$ the set of all: (i) damper-summable differentials on K, (ii) archimedean differentials on K, (iii) weakly archimedean differentials on K.

4. Let F be a function on a cell K in $\mathbb{R}$ such that dF is weakly archimedean. Let E be the set of all t in K where the derivative $F'(t) = \infty$. Show that E is dx-null.

5. Show the equivalence of the following conditions for σ a differential on a cell K:

(i) σ is dampable,

(ii) There exist functions G and v on K such that G is of bounded variation, v is a damper, and $\sigma = vdG$,

(iii) There exist functions G and h on K such that G is of bounded variation, (sgn h)dG is integrable, and $\sigma = hdG$.

6. (An extension of Theorem 5 (§2.8).) Let dg be dampable on a cell K, $dF_n = f_n dg$ for $n = 1, 2, \cdots$, $f_n \to f$, and $dF_n \to dF = fdg$. We contend $(*)$ $f_1\Delta g, f_2\Delta g, \cdots$ are uniformly integrable. To prove this let $H_n = F_n - F$, $h_n = \frac{1}{v}(f_n - f)$ where v is a damper such that $vdg = dG$ with G of bounded variation. Then verify the following:

(i) $h_1 \Delta G, h_2 \Delta G, \cdots$ are uniformly integrable. (Apply Theorem 5 (§2.8).)

(ii) There exists a function w on K such that $|h_n| \leq w$ for all n.

(iii) $|\Delta F_n - f_n \Delta g| \leq 2|\Delta H_n| + |\Delta F - f\Delta g| + w|v\Delta g - \Delta G| + |h_n \Delta G - \Delta H_n|$.

(iv) $(*)$ holds. $(dH_n \to 0, dF - fdg = w(vdg - dG) = 0$ in (iii). Apply (i) to the last term in (iii).)

7. *(Open Questions!)* Do there exist differentials σ, τ on a cell K such that:

(i) σ is archimedean but not damper-summable?

(ii) τ is weakly archimedean but not archimedean?

8. Let $\sigma \geq 0$ be an archimedean differential on a cell K. Prove:

(i) If $v_n \searrow 0$ σ-everywhere on K then $\bigwedge_{n=1}^{\infty}(v_n \sigma) = 0$.

(ii) If $w_n \geq 0$ on K and $\sup_{n=1,2,\cdots} w_n = w < \infty$ σ-everywhere then $\bigvee_{n=1}^{\infty}(w_n \sigma) = w\sigma$.

(See Exercise 6 in §2.7.)

9. Consider the following property of a set $\mathbb{E}$ of differentials on a cell K: If $E_1 \cup \cdots \cup E_n \cup \cdots = K$ and $1_{E_n}\sigma \in \mathbb{E}$ for all n then $\sigma \in \mathbb{E}$. Show that this holds for $\mathbb{E}$ the set of all: (i) continuous, (ii) tag finite, (iii) damper-summable, (iv) archimedean, (v) weakly archimedean, differentials on K.

10. Show that a differential σ on a cell K is tag-finite if and only if $1_E\sigma$ is damper-summable for every countable subset E of K.

11. Show that for a differential σ on a cell K the following conditions are equivalent:

(i) $1_p\sigma$ is integrable for every point p on K,

(ii) $1_p\sigma$ is absolutely integrable for every point p in K,

(iii) $1_E\sigma$ is dampable for every countable subset E of K,

(iv) $1_p\sigma = 1_p(fQ^- + gQ^+)\omega$ for some functions f, g on K and all p in K.

(See Exercise 11 in §3.1.)

12. For E a subset of a cell K prove the equivalence of the following three conditions:

(i) $1_E\omega$ is weakly archimedean on K,

(ii) E is σ-null for every continuous differential σ on K,

(iii) $1_E\omega$ is archimedean on K.

13. Prove: If u is a damper on an uncountable set E then there exist $\varepsilon > 0$ in $\mathbb{R}$ and an uncountable subset A of E such that $u > \varepsilon$ on A.

14. For E a subset of a cell K prove the equivalence of the following three conditions:

(i) $1_E\omega$ is damper-summable,

(ii) E is countable,

(iii) $1_E\omega$ is dampable.

(Apply Exercise 13 to prove (i) implies (ii).)

Note that if there exists an uncountable E satisfying any (hence, all) of the three conditions in Exercise 12 then $1_E\omega$ provides an affirmative answer to (i) in Exercise 7. But the existence of such E is conjectural.

15. For σ a differential on a cell K prove the equivalence of the following:

(i) σ is damper-summable,

(ii) $u|\sigma| \leq \tau$ for some damper u and some integrable τ on K,

(iii) $\sigma = [S]$ for some tame summant S on K.

(To prove (i) implies (ii) use Exercise 5 in §2.3.)

16. Let ρ, σ be continuous, dampable differentials on a cell K such that $\overline{\int_K} w\rho = \overline{\int_K} w\sigma$ for all functions $w \geq 0$ on K. Prove $\rho = \sigma$.

(Apply Theorem 2 (§2.5) and Theorem 5 (§2.4) appropriately to prove $\rho \leq \sigma$ and $\sigma \leq \rho$.)

17. Let $f_n \to f$ on a cell K. Let τ be a tag-finite differential on K such that for some summant T representing τ the summants $f_1 T, f_2 T, \cdots$ are uniformly integrable on K. By Theorem 4 (§1.6) $f\tau$ is integrable and $\int_J f_n \tau \to \int_J f\tau$ for every cell J in K. Prove that this sequential convergence is uniform for all cells J in K.

§3.4 Differentials on Open-ended Intervals.

Let L be a nondegenerate interval in $[-\infty, \infty]$. That is, $(a, b) \subseteq L \subseteq [a, b]$ where $-\infty \leq a < b \leq \infty$. So $\bar{L} = K = [a, b]$. In §1.9 we defined summants on L as functions on the set of all tagged cells in L. Such a summant S_L has upper and lower integrals given according to (4) in §1.9 by

$$(1) \quad \overline{\int_L} S_L = \overline{\int_K} 1_L S = \overline{\int_K} 1_L S, \quad \underline{\int_L} S_L = \underline{\int_K} 1_L S = \underline{\int_K} 1_L S$$

where S is any extension of S_L to a summant on K, $\mathbf{1}_L$ is the cell summant on K indicating the set of all cells contained in L, and 1_L is the function on K indicating the subset L of K. So we could define a differential on L to be an equivalence class σ_L of summants S_L on L with equivalence of S_L and S'_L defined by

$$(2) \quad \overline{\int_L} |S_L - S'_L| = \overline{\int_K} \mathbf{1}_L |S - S'| = \overline{\int_K} 1_L |S - S'| = 0$$

for any extensions of S_L, S'_L to summants S, S' on K. We could then define the upper and lower integrals of σ_L over L by

$$(3) \quad \overline{\int_L} \sigma_L = \overline{\int_L} S_L \quad \text{and} \quad \underline{\int_L} \sigma_L = \underline{\int_L} S_L$$

under (1) with concommitant definitions of $\int_L \sigma_L$ and integrability of σ_L over L, consistent with our definitions for the case $L = K$.

But all of this is redundant since the Riesz space of differentials on L corresponds isomorphically to the Riesz space of all differentials σ on K such that $1_L\sigma = \sigma$. These are just the differentials σ on K for which $K - L$ is σ-null, that is, $1_p\sigma = 0$ for each endpoint p of L that does not belong to L. These form a complete Riesz space under differential convergence as a closed subspace of the complete Riesz space of all differentials on K. Let us trace the correspondence between differentials σ_L on L and differentials σ on K such that $1_L\sigma = \sigma$.

Given a differential σ_L on L we have $\sigma = [1_L S]$ on K where S is any summant on K whose restriction S_L to the tagged cells in L belongs to σ_L. In the other direction, given $\sigma = 1_L\sigma$ on K let σ_L be the differential on L represented by the restriction S_L of any representative S of σ to the tagged cells in L. The reader can easily verify the biuniqueness of the correspondence $\sigma_L \leftrightarrow \sigma$ and the induced isomorphism of the Riesz properties. In particular, note the following: $\sigma_L = \sigma$ on every cell J in L since both differentials are represented on J by the same summants, $\sigma_L = 0$ on L if and only if $\sigma_L = 0$ on every cell J contained in L, and for *any* differential ρ on K, $1_L\rho = 0$ on K if and only if $\rho = 0$ on every cell J in L. Thus, if $\sigma = 1_L\sigma$ and $\tau = 1_L\tau$ on K, and $\sigma = \tau$ on every cell J in L then $\sigma = \tau$ on K.

Theorem 4 (§1.9) which shows that the Cauchy extension is redundant has the following obvious reformulation in terms of differentials.

THEOREM 1. *A differential σ on a nondegenerate interval L in $[-\infty, \infty]$ is integrable over L if and only if*

(i) σ is integrable over every cell J contained in L, and for the set of these cells directed upward by containment,

(ii) $\lim\limits_{J \to L} \int_J \sigma$ *exists and is finite.*

Moreover, for σ integrable over L

(iii) $\int_L \sigma = \lim\limits_{J \to L} \int_J \sigma.$

Finally, if $\int_I \sigma = 0$ for every cell I contained in L then $\sigma = 0$ on L.

The next three results show the redundancy of the Harnack extension They get integrability from appropriate forms of local integrability.

THEOREM 2. *Let $J_1, J_2, \cdots$ be a sequence of nonoverlapping cells in $K = [a, b]$. Let D be the complement in K of $J_1 \cup J_2 \cup \cdots$. Let $G_1, G_2, \cdots$ be a sequence of continuous functions on K such that*

$$(4) \qquad\qquad 1_{J_n} dG_n = dG_n \quad \text{for all } n$$

and

$$(5) \qquad\qquad \sum_{n=1}^{\infty} \|G_n\| < \infty \quad \text{where } \|G\| = \sup_{t \in K} |G(t)|.$$

Then there is a unique differential ϕ on K such that

$$(6) \qquad\qquad 1_{J_n} \phi = dG_n \quad \text{for all } n$$

and

$$(7) \qquad\qquad\qquad D \quad \text{is } \phi\text{-null.}$$

Moreover, ϕ is integrable. Specfically $\phi = dF$ for F the continuous function defined by

$$(8) \qquad\qquad F(x) = \sum_{n=1}^{\infty} G_n(x) \quad \text{for all } x \text{ in } K.$$

Thus

$$(9) \qquad \Delta F(I) = \sum_{n=1}^{\infty} \Delta G_n(I) \quad \textit{for every cell } I \textit{ in } K$$

and

$$(10) \qquad \int_K \phi = \sum_{n=1}^{\infty} \int_{J_n} \phi.$$

PROOF. Define the continuous function F by (8) where the series converges absolutely and uniformly by (5). (9) follows directly from (8). By (4) and the continuity of G_n

$$(11) \qquad \Delta G_n(I) = \Delta G_n(IJ_n) \quad \text{for every cell } I \text{ in } K,$$

with value 0 if I fails to overlap J_n.

If $I \subseteq J_n$ then, since $J_1, J_2, \cdots$ are nonoverlapping, (9) and (11) give $\Delta F(I) = \Delta G_n(I)$. So $dF = dG_n$ on J_n. Thus, since F and G_n are continuous, $1_{J_n} dF = 1_{J_n} dG_n = dG_n$ by (4). That is, (6) holds for $\phi = dF$.

Let $\varepsilon > 0$ be given. To prove D is dF-null we need a gauge δ on K such that

$$(12) \qquad (1_D |\Delta F|)^{(\delta)}(K) < \varepsilon.$$

To get such a gauge take N large enough so that by (5)

$$(13) \qquad \sum_{n>N} \|G_n\| < \varepsilon/4.$$

Then take δ on K so that for all t in D

$$(14) \qquad \delta(t) < |t - s| \quad \text{for all } s \text{ in } J_1 \cup \cdots \cup J_N.$$

Such a gauge δ exists because D is disjoint from the closed set $J_1 \cup \cdots \cup J_N$.

To prove (12) take any δ-division $\mathcal{K}$ of K. Let $(I_1, t_1), \cdots ,$ (I_m, t_m) be those members of $\mathcal{K}$ whose tags t_i belong to D. By (14) $I_1 \cup \cdots \cup I_m$ is disjoint from $J_1 \cup \cdots \cup J_N$. So by (9) and (11)

(15)
$$\left(\sum 1_D |\Delta F| \right)(\mathcal{K}) = \sum_{i=1}^{m} |\Delta F(I_i)| = \sum_{i=1}^{m} \left| \sum_{n>N} \Delta G_n(I_i) \right|$$
$$\leq \sum_{n>N} \sum_{i=1}^{m} |\Delta G_n(I_i)|.$$

None of the tags $t_1, \cdots , t_m$ belong to $J_1 \cup J_2 \cup \cdots$ since they all belong to the complement D. So no J_n can contain any of the cells $I_1, \cdots , I_m$. Hence, J_n meets at most two of these nonoverlapping cells. So by (11) the sum $\sum_{i=1}^{m} |\Delta G_n(I_i)|$ has at most two nonzero terms. Therefore since each such term is bounded by $2\|G_n\|$, $\sum_{i=1}^{m} |\Delta G_n(I_i)| \leq 4\|G_n\|$ which with (15) and (13) gives

(16) $$\left(\sum 1_D |\Delta F| \right)(\mathcal{K}) \leq 4 \sum_{n>N} \|G_n\| < \varepsilon.$$

Since (16) holds for every δ-division $\mathcal{K}$ of K it gives (12). So $1_D dF = 0$ which gives (7) for $\phi = dF$. ϕ is a continuous differential since F is continuous. To get (10) apply (9) with $I = K$ to get $\int_K \phi = \int_K dF = \Delta F(K) = \sum_{n=1}^{\infty} \Delta G_n(K) = \sum_{n=1}^{\infty} \int_K dG_n = \sum_{n=1}^{\infty} \int_K 1_{J_n} \phi = \sum_{n=1}^{\infty} \int_{J_n} \phi$ by (6) and the continuity of ϕ.

Now consider any differential ϕ on K satisfying (6) and (7). To show that ϕ is unique we must prove $\phi = dF$ for F defined by (8). Since both ϕ and dF satisfy (6), $1_{J_n} \phi = dG_n = 1_{J_n} dF$. So $1_{J_n} |\phi - dF| = 0$. That is, J_n is $(\phi - dF)$-null. So is D since it is both ϕ-null and dF-null by (7). Thus $K = D \cup J_1 \cup J_2 \cup \cdots$ is $(\phi - dF)$-null. That is, $\phi - dF = 0$. So $\phi = dF$. $\square$

THEOREM 3. *Let σ be a continuous differential on a cell K. Let $J_1, J_2, \cdots$ be a sequence of nonoverlapping cells in K such that (i) σ is integrable over each J_n, (ii) there exists $c < \infty$ such that $\sum_{n=1}^{\infty} |\int_{I_n} \sigma| < c$ for every sequence of cells $I_1, I_2, \cdots$ such that $I_n \subseteq J_n$ for all n, and (iii) for D the complement of $J_1 \cup J_2 \cup \cdots$ in K the differential $1_D\sigma$ is integrable over K. Then σ is integrable over K and*

$$(16) \qquad \int_K \sigma = \int_K 1_D\sigma + \sum_{n=1}^{\infty} \int_{J_n} \sigma.$$

PROOF. Since σ is integrable over J_n and continuous at the endpoints of J_n, $\int_{J_n} \sigma = \int_K 1_{J_n}\sigma$ by (i). So we can define G_n on $K = [a, b]$ by

$$(17) \qquad G_n(t) = \int_a^t 1_{J_n}\sigma.$$

The function G_n is continuous since σ is a continuous differential. By Theorem 1 (§2.5), (17) implies

$$(18) \qquad dG_n = 1_{J_n}\sigma \quad \text{for all } n.$$

By (18) and (ii), $\sum_{n=1}^{\infty} |\Delta G_n(I_n)| < c$ for all $I_n \subseteq J_n$. Hence,

$$(19) \qquad \sum_{n=1}^{\infty} \text{diam } G_n(K) \le c < \infty.$$

Now for the sup norm in (5) of Theorem 2,

$$(20) \qquad \|G_n\| \le \text{diam } G_n(K)$$

since $G_n(a) = 0$ by (17). The inequalities (19) and (20) give the convergence $\sum_{n=1}^{\infty} \|G_n\| < \infty$ in (5) of Theorem 2. (18)

gives (4). So we can apply Theorem 2 to get ϕ satisfying (6) and (7). Now, since $J_n \cap D = \emptyset$, (18) gives

$$(21) \qquad 1_{J_n}(\sigma - 1_D\sigma) = 1_{J_n}\sigma = dG_n.$$

Clearly

$$(22) \qquad 1_D(\sigma - 1_D\sigma) = 0.$$

Comparing (21), (22) with (6), (7) we get $\sigma - 1_D\sigma = \phi$ by uniqueness of ϕ. That is, $\sigma = 1_D\sigma + \phi$ which is integrable by (iii) and Theorem 2. Integration over K gives (16) by (10) since $1_{J_n}\phi = 1_{J_n}\sigma$ by (21).   □

For D σ-null Theorem 3 gives the following result as an immediate corollary.

THEOREM 4. *Let σ be a continuous differential on a cell K. Let $J_1, J_2, \cdots$ be a sequence of nonoverlapping cells in K covering σ-all of K such that:*

(i) σ is integrable over J_n for all n

and

(ii) there exists $c < \infty$ such that $\sum_{n=1}^{\infty} |\int_{I_n} \sigma| < c$ for every sequence of cells $I_1, I_2, \ldots$ such that $I_n \subseteq J_n$ for all n.

Then σ is integrable over K and

$$\int_K \sigma = \sum_{n=1}^{\infty} \int_{J_n} \sigma \quad \text{where the series converges absolutely.}$$

Theorem 4 in turn yields the following result on integration over an interval.

THEOREM 5. *Let F be a continuous function on an interval L. Let $J_1, J_2, \cdots$ be a sequence of nonoverlapping cells in L covering dF-all of L such that*

$$\sum_{n=1}^{\infty} \operatorname{diam} F(J_n) < \infty.$$

Then $\int_L dF = \sum_{n=1}^{\infty} \Delta F(J_n)$ and the series converges absolutely.

PROOF. Let K be the cell $\bar{L}$ in $[-\infty, \infty]$. Let $\sigma = [S]$ on K where S is the cell summant on K defined by $S(I) = \Delta F(I)$ if $I \subseteq L$, and $S(I) = 0$ otherwise. Since F is continuous on L the differential σ is continuous on K. For every cell I in L, $\int_I \sigma = \int_I dF = \Delta F(I)$. So (i) in Theorem 4 holds. Moreover, $|\int_{I_n} \sigma| = |\Delta F(I_n)| \leq \operatorname{diam} F(J_n)$ for $I_n \subseteq J_n$. Hence, (ii) holds with $c = 1 + \sum_{n=1}^{\infty} \operatorname{diam} F(J_n) < \infty$. Finally, $\int_K \sigma = \int_L dF$ in the conclusion of Theorem 4 which gives Theorem 5 since $\int_{J_n} \sigma = \Delta F(J_n)$.   □

Although $\int_L dg$ need not exist on a noncompact interval L we always have the existence of $\int_L |dg| \leq \infty$. This is our next result.

THEOREM 6. *For g a function on a nondegenerate interval L*

(i)
$$\int_L |dg| = \sup_{J \subseteq L} \int_J |dg| \leq \infty$$

where the supremum is taken over all cells J in L.

PROOF. Let c be the supremum in (i). If $c < \infty$ then (i) follows from Theorem 1. If $c = \infty$ then clearly $\int_L |dg| = \infty$ since $c \leq \int_L |dg|$.   □

THEOREM 7. *Let g be a function on a nondegenerate interval L such that $\int_L |dg| < \infty$. Then dg is integrable over the interval L.*

PROOF. Take a sequence of nonoverlapping cells $J_1, J_2, \cdots$ such that $J_1 \cup \cdots \cup J_n$ is a cell K_n and $K_n \nearrow L$. For every cell J in L, diam $g(J) \le \int_J |dg|$. So $\sum_{n=1}^{\infty}$ diam $g(J_n) \le \sum_{n=1}^{\infty} \int_{J_n} |dg| = \int_L |dg| < \infty$ by Theorem 6. Therefore, $\int_L dg = \sum_{n=1}^{\infty} \Delta g(J_n)$ by Theorem 5.   □

We extend Theorem 6 to give the total variation of functions on open sets.

THEOREM 8. *Given a function g on an open subset D of $\mathbb{R}$ extend g arbitrarily to $K = [\infty, \infty]$. Then*

(i)
$$\int_K 1_D |dg| = \sup_{A \subset D} \int_A |dg| \le \infty$$

where the supremum is taken over all figures A contained in D.

For $D_1, D_2, \cdots$ the components of D

(ii)
$$\int_K 1_D |dg| = \sum_i \int_{D_i} |dg|.$$

PROOF. For each component D_i of D Theorem 6 gives the existence of $\int_{D_i} |dg|$ which equals $\int_K 1_{D_i} |dg|$ since D_i is an open interval. So (ii) follows from Theorem 1 (§1.7) applied to $S = |\Delta g|$ with $1_D = \sum_{i=1}^{\infty} 1_{D_i}$. (For notational convenience we set $D_i = \emptyset$ for $i \ge n$ if D has fewer than n components.) (i) then follows from (i) in Theorem 6 since for cells $J_i \subset L_i$ we can consider the figures $A_n = J_1 \cup \cdots \cup J_n$.   □

For the case of bounded variation on an open set we have the following result.

THEOREM 9. *Let g be a function on a cell K. Let D be an open subset of K such that $1_D dg$ is summable. Then $1_D dg$ is absolutely integrable over K. Moreover,*

$$\text{(i)} \qquad \int_K 1_D dg = \sum_i \int_{D_i} dg$$

where $D_1, D_2, \cdots$ are the components of D.

PROOF. For $1_D dg$ summable the integrals in (ii) of Theorem 8 are finite. So integrability of dg over D_i follows from Theorem 7. Thus, both $(dg)^+$ and $(dg)^-$ are integrable over D_i since dg is absolutely integrable over D_i. By Theorem 1 (§1.7) applied to $S = (\Delta g)^+$ and to $S = (\Delta g)^-$ we get $\sum_i \int_{D_i} (dg)^+ = \int_K 1_D (dg)^+ < \infty$ and $\sum_i \int_{D_i} (dg)^- = \int_K 1_D (dg)^- < \infty$. Taking the difference between these two we get (i). $\square$

Theorem 3 (§1.8) yields the following formulation in terms of differentials.

THEOREM 10. *Let σ be a differential on a nondegenerate interval L. Then for all cells J in L directed upward by containment*

$$\overline{\int_J} |\sigma| \nearrow \overline{\int_L} |\sigma| \quad \text{as } J \nearrow L.$$

In Theorem 4 (and in its corollary, Theorem 5) we can relax the demand that the series be absolutely convergent if the sequence $J_1, , J_2, \cdots$ is a chain of abutting cells. We can also drop the requirement that σ be continuous. Here is one such result.

THEOREM 11. *In $L = [a, b)$ let $J_n = [a_{n-1}, a_n]$ for $n = 1, 2, \cdots$ where $a = a_0 < a_1 < \cdots$ and $a_n \nearrow b$. Let σ be a differential on L such that:*

(i) σ is integrable on each J_n,

(ii) $\int_{I_n} \sigma \rightrightarrows 0$ *uniformly for all sequences* $I_1, I_2, \cdots$ *of cells such that* $I_n \subseteq J_n$ *for all* n,

(iii) *The series* $\sum_{n=1}^{\infty} \int_{J_n} \sigma$ *is convergent.*

Then σ *is integrable on* L *and* $\int_L \sigma = \sum_{n=1}^{\infty} \int_{J_n} \sigma$.

PROOF. Given x in L $a_n > x$ ultimately as $n \to \infty$ since $a_n \to b$ and $x < b$. Let $n(x)$ be the smallest integer n such that $a_n > x$. For $n = n(x)$ we have $[a, x] \subseteq [a, a_n] = J_1 \cup \cdots \cup J_n$. So (i) implies that σ is integrable on $[a, x]$ for all x in L. That is, σ is integrable on every cell in L. So we can define F on L by

$$F(x) = \int_a^x \sigma \quad \text{for all } x \text{ in } L.$$

Then $dF = \sigma$ on L and $\Delta F(I) = \int_I \sigma$ for all cells I in L. By the uniformity of the convergence in (ii), diam $F(J_n) < \infty$ ultimately as $n \to \infty$ and

(23) $$\text{diam } F(J_n) \to 0.$$

Since $F(a_n) = \int_a^{a_n} \sigma = \sum_{j=1}^{n} \int_{J_n} \sigma$ (iii) yields c in $\mathbb{R}$ such that

(24) $$F(a_n) \to c.$$

To deduce Theorem 11 from Theorem 1 we need only prove that $F(b-) = c$.

Clearly $x \in J_{n(x)}$ for all x in L. So

(25) $\qquad |F(x) - F(a_n)| \le \text{ diam } F(J_n) \quad \text{for } n = n(x).$

Also

(26) $$n(x) \to \infty \quad \text{as } x \to b-.$$

Thus

(27) $\qquad |F(x) - F(a_{n(x)})| \to 0 \quad$ as $x \to b-$

by (23), (25), (26). Finally

$$|F(x) - c| \le |F(x) - F(a_{n(x)})| + |F(a_{n(x)}) - c| \to 0$$

as $x \to b-$ by (24), (26), 27). So $F(b-) = c$. $\qquad \square$

Exercises (§3.4).

1. Prove: A continuous function g on an interval L is of bounded variation if and only if $1_D dg$ is integrable for every open
subset D of L.

2. Let f be defined on $(0, 1]$ by $f(t) = (-1)^n n$ for $\frac{1}{n+1} < t \le \frac{1}{n}$
for $n = 1, 2, \cdots$. Prove:

(i) $\qquad \displaystyle\int_{\frac{1}{n+1}}^{1} f(t)dt = \sum_{k=1}^{n} \frac{(-1)^k}{k+1} \quad$ for $n = 1, 2, \cdots$.

(ii) $\qquad \displaystyle\int_{(0,1]} f dx = \sum_{k=1}^{\infty} \frac{(-1)^k}{k+1}.$

(iii) $\qquad \displaystyle\int_{(0,1]} |f| dx = \infty.$

3. Let σ be a differential on $\mathbb{R}$. Prove:

(i) If σ is dampable on every cell in $\mathbb{R}$ then σ is dampable
on $\mathbb{R}$.

(ii) If σ is damper-summable on every cell in $\mathbb{R}$ then σ is damper-summable on $\mathbb{R}$.

(iii) If σ is archimedean on every cell in $\mathbb{R}$ then σ is archimedean on $\mathbb{R}$.

(iv) If σ is weakly archimedean on every cell in $\mathbb{R}$ then σ is weakly archimedean on $\mathbb{R}$.

4. Let $a_n \nearrow b$ in $L = [a, b)$. Let σ be a differential on L such that

(i) σ is integrable on every cell in L,

(ii) $(-1)^n \sigma \geq 0$ on $[a_n, a_{n+1}]$ for $n = 1, 2, \cdots$,

(iii) $|\int_{a_n}^{a_{n+1}} \sigma| \searrow 0$ as $n \nearrow \infty$.

Prove that σ is integrable on L.

(Apply Theorem 11 by invoking a well-known convergence theorem for alternating series.)

5. Define f on $[\pi, \infty) = L$ by $f(x) = \frac{\sin x}{x}$.

(i) Show that $f dx$ is (absolutely) integrable on every cell in L.

(ii) Show that $f dx$ is integrable on L by applying Exercise 4 with $a_n = n\pi$, $\sigma = f dx$.

6. Show that a differential σ is summable on an interval L if and only if $\overline{\int_I} |\sigma|$ is uniformly bounded for all cells I in L. (See Theorem 10.)

7. Prove: If $u\sigma$ is summable on $L = [a, , b)$ for every function u on L such that $du \leq 0$ and $u(b-) = 0$ then σ is summable on L. (Give an indirect proof.)

8. (The Converse of Theorem 5.) Given a continuous function F on an interval L with dF integrable over L show that there is some sequence $J_1, J_2, \cdots$ of cells satisfying the hypothesis of Theorem 5. (One can get such a sequence that covers *all* of L.)

§3.5 σ-Nullity of the Union of All σ-Null Cells.

The result presented here is needed for the proof of Theorem 1 in §5.1.

THEOREM 1. *For any differential σ on a cell K the union Z of all σ-null cells in K is a union of countably many such cells. So Z is σ-null.*

PROOF. (We dismiss the trivial case $Z = \emptyset$.)

Given z in Z let L_z be the union of all σ-null cells I in K which contain z. L_z is a nondegenerate interval by the definition of Z. If J is any σ-null cell which meets L_z it must meet some σ-null cell I containing z. So $I \cup J$ is a σ-null cell containing z, hence must be contained in L_z. This implies that L_z is a component of Z. Since the components of Z are disjoint and nondegenerate there are only countably many of them. Thus we need only prove that for each z in Z

(1) L_z is a countable union of σ-null cells.

Now a nondegenerate interval is a union of countably many cells. So to prove (1) it suffices to show that

(2)   every cell J in L_z is σ-null.

Each point y in L_z lies in some σ-null cell I_y containing z. For a cell $J = [x, y]$ in L_z the cell $I_x \cup I_y$ is σ-null and contains J. So J is σ-null, which proves (2).   □

Exercises (§3.5).

1. Show that if the differential σ in Theorem 1 is continuous then Z is a union of disjoint cells.

2. Show that if A, B are complementary figures in a bounded cell K and $\sigma = 1_A dx$ on K then B is the union of all σ-null cells in K.

3. Verify that the only properties of σ-nullity that are required for the validity of Theorem 1 are:

   (i) If $D \subseteq E$ and E is σ-null then so is D,

   (ii) A countable union of σ-null sets is σ-null.

§3.6 Mappings of Differentials Induced by Lipschitz Functions.

Let F be a Lipschitz function on $\mathbb{R}^n$. That is, there exists a constant $c \geq 0$ such that

$$(1) \qquad |F(\mathbf{u}) - F(\mathbf{v})| \leq c\|\mathbf{u} - \mathbf{v}\| \quad \text{for all } \mathbf{u}, \mathbf{v} \text{ in } \mathbb{R}^n.$$

With appropriate adjustment of the value of the Lipschitz constant c the norm in (1) can be any norm on $\mathbb{R}^n$ since all such norms are equivalent. That is, given norms L, M on $\mathbb{R}^n$ there exists $c > 0$ such that $L(\mathbf{u}) \leq cM(\mathbf{u})$ for all $\mathbf{u}$ in $\mathbb{R}^n$. The familiar norms on $\mathbb{R}^n$ are of the form

$$(2) \qquad \|\mathbf{u}\|_p = (|u_1|^p + \cdots + |u_n|)^{1/p}$$

for $\mathbf{u} = (u_1, \cdots, u_n)$ where the parameter p satisfies $1 \leq p < \infty$. For $p = \infty$ we have the norm

$$(3) \qquad \|\mathbf{u}\|_\infty = |u_1| \vee \cdots \vee |u_n|.$$

The Euclidean norm is given by $p = 2$ in (2). For convenience we often use (2) with $p = 1$ since it has the simple expression

$$(4) \qquad \|\mathbf{u}\|_1 = |u_1| + \cdots + |u_n|.$$

For the space $\mathbb{D}$ of all differentials on a cell K each Lipschitz function F on $\mathbb{R}^n$ induces a mapping $\tilde{F}$ of $\mathbb{D}^n$ into $\mathbb{D}$ defined by

$$(5) \qquad \tilde{F}(\sigma_1, \cdots, \sigma_n) = [F(S_1, \cdots, S_n)]$$

for $\sigma_i = [S_i]$ on K with $i = 1, \cdots, n$. The value of the summant $F(S_1, \cdots, S_n)$ at (I, t) is $F(S_1(I, t), \cdots, S_n(I, t))$. To show that (5) is an effective definition we must verify that

(6) $F(S_1, \cdots, S_n) \sim F(S_1', \cdots, S_n')$ if $S_i \sim S_i'$ for all i.

To do so set $\mathbf{u} = (S_1, \cdots, S_n)$ and $\mathbf{v} = (S_1', \cdots, S_n')$ in (1) with the norm (4) to get the summant inequality

$$|F(S_1, \cdots, S_n) - F(S_1', \cdots, S_n')| \le c(|S_1 - S_1'| + \cdots + |S_n - S_n'|)$$

which immediately gives (6).

A norm L on $\mathbb{R}^n$ is a Lipschitz function with Lipschitz constant $c = 1$ by the triangle inequality. So L defines a differential $\widetilde{L}(\sigma_1, \cdots, \sigma_n)$ on K for $\sigma_1, \cdots, \sigma_n$ differentials on K. For $L(\mathbf{u}) = \|\mathbf{u}\|_p$ in (2) or (3) we shall use the notation

(7) $\|(\sigma_1, \cdots, \sigma_n)\|_p = (|\sigma_1|^p + \cdots + |\sigma_n|^p)^{1/p}$

for $1 \le p < \infty$ and

(8) $\|(\sigma_1, \cdots, \sigma_n)\|_\infty = |\sigma_1| \vee \cdots \vee |\sigma_n|$.

In general we may use "F" in place of "$\widetilde{F}$" in (5). This is convenient. But care must be taken since a Lipschitz function may be composed of functions some of which are not Lipschitzean. For example if for $i = 1, \cdots, n$ σ_i is continuous and weakly archimedean then $\sigma_i^2 = 0$. So one might draw the false conclusion that

$$\|(\sigma_1, \cdots, \sigma_n)\|_2 = (\sigma_1^2 + \cdots + \sigma_n^2)^{1/2} = (0 + \cdots + 0)^{1/2} = 0.$$

But the square root $|\tau|^{1/2}$ is not defined for differentials τ since $F(u) = |u|^{1/2}$ fails to be Lipschitzean on $\mathbb{R}$ near 0.

However, there are many legitimate calculations that can be made. For example given $\sigma_i = f_i \tau$ on K for $i = 1, \cdots, n$ and $1 \leq p < \infty$ it is valid to conclude that

$$\|(\sigma_1, \cdots, \sigma_n)\|_p = (|f_1\tau|^p + \cdots + |f_n\tau|^p)^{1/p} =$$

$$(|f_1|^p + \cdots + |f_n|^p)^{1/p}|\tau| = \|(f_1, \cdots, f_n)\|_p|\tau|.$$

We also have $\|(\sigma_1, \cdots, \sigma_n)\|_\infty = \|(f_1, \cdots, f_n)\|_\infty|\tau|$. In these formulas $\|(f_1, \cdots, f_n)\|_p$ is a function on K for $f_1, \cdots, f_n$ functions on K; $\|(\sigma_1, \cdots, \sigma_n)\|_p$ is a differential on K.

For continuous differentials $\sigma_1, \cdots, \sigma_n$ on a cell K a Lipschitz function F need only be defined on a neighborhood of $\mathbf{0}$ in $\mathbb{R}^n$ to define the differential $\widetilde{F}(\sigma_1, \cdots, \sigma_n)$ by (5) since $(S_1(I, t), \cdots, S_n(I, t))$ ultimately lies in any given neighborhood of $\mathbf{0}$ as $(I, t) \to t$ in K.

For tag-finite differentials definition (5) is valid under a condition weaker than (1), namely that F be Lipschitzean on every bounded subset of $\mathbb{R}^n$. Explicitly, given a norm $\|\cdot\|$ on $\mathbb{R}^n$ there exists a function $C \geq 0$ on $(0, \infty)$ such that given $r > 0$ in $\mathbb{R}$

$$(9) \qquad |F(\mathbf{u}) - F(\mathbf{v})| \leq C(r)\|\mathbf{u} - \mathbf{v}\|$$

for all $\mathbf{u}, \mathbf{v}$ in $\mathbb{R}^n$ such that $\|\mathbf{u}\| \leq r$, $\|\mathbf{v}\| \leq r$.

THEOREM 1. *Let F be Lipschitzean on every bounded subset of $\mathbb{R}^n$. Then $\widetilde{F}(\sigma_1, \cdots, \sigma_n)$ is effectively defined by (5) if $\sigma_1, \cdots, \sigma_n$ are tag-finite on a cell K.*

PROOF. Given tag-finite $\sigma_1, \cdots, \sigma_n$ on K there exists a damper w on K such that $\nu(1_p\sigma_i) < w(p)$ for all p in K and $i = 1, \cdots, n$. Thus, given summants S_i, S_i' representing σ_i for $i = 1, \cdots, n$ there is a gauge δ on K such that

$$(10) \qquad |S_i(I, t)| < w(t) \quad \text{and} \quad |S_i'(I, t)| < w(t)$$

for all δ-fine (I, t) in K and $i = 1, \cdots, n$. For the ∞-norm (3) the inequality (9) on the ball of radius r about $\mathbf{0}$ in $\mathbb{R}^n$ applies to (10) with $r = w(t)$ giving the summant inequality

$$
(11) \quad \begin{aligned} &|F(S_1, \cdots, S_n) - F(S'_1, \cdots, S'_n)| \leq \\ &C(w)(|S_1 - S'_1| \vee \cdots \vee |S_n - S'_n|) \end{aligned}
$$

on the set of all δ-fine tagged cells in K. Since $C(w)$ is a function on K (11) yields (6). $\qquad \square$

In the next section we shall examine the Riesz space $\mathbb{D}^n$ that we have implicitly introduced here.

Exercises (§3.6). Prove the following:

1. If F is a Lipschitz function on $\mathbb{R}$ with Lipschitz constant c then $|\widetilde{F}(\sigma) - \widetilde{F}(\tau)| \leq c|\sigma - \tau|$ for all σ, τ in $\mathbb{D}$.

2. For every Lipschitz function F on $\mathbb{R}$ the mapping $\widetilde{F}$ is continuous on $\mathbb{D}$: If $\sigma_j \to \sigma$ on K then $\widetilde{F}(\sigma_j) \to \widetilde{F}(\sigma)$.

3. $\widetilde{F} \circ \widetilde{G} = \widetilde{F \circ G}$ for all Lipschitz functions F, G on $\mathbb{R}$.

4. If $G(x) = c$ for all x in $\mathbb{R}$ then $\widetilde{G}(\sigma) = c\omega$ and $(\widetilde{F} \circ \widetilde{G})(\sigma) = F(c)\omega$ for every Lipschitz function F on $\mathbb{R}$ and all σ in $\mathbb{D}$.

5. Theorem 1 with $n = 1$ applies to every differentiable function F on $\mathbb{R}$ with a continuous derivative.

6. If $F(x, y) = xy$ for all (x, y) in $\mathbb{R}^2$ then F is Lipschitzean on every bounded subset of $\mathbb{R}^2$. So Theorem 1 confirms the existence of the product $\sigma\tau$ for all tag-finite σ, τ on a cell K.

7. For $n = 1$ (7) and (8) give $\|\sigma\|_p = |\sigma|$ for all differentials σ on a cell K and all p in $[1, \infty]$.

8. If F is a twice differentiable function on a neighborhood of 0 in $\mathbb{R}$ and the second derivative $F^{(2)}$ is continuous at 0 then $\widetilde{F}(\sigma) = F(0)\omega + F'(0)\sigma$ for all continuous, weakly

archimedean differentials σ on a cell K. For all such σ, $\sin \sigma = \sigma, \cos \sigma = \omega$, and $\tan \sigma = \sigma$. So $e^{\sigma} = \cos \sigma + \sin \sigma$.

9. If $\mathbf{f} = (f_1 \cdots, f_n)$ is a mapping of a cell K into $\mathbb{R}^n$ and F is a Lipschitz function on $\mathbb{R}^n$ then $\tilde{F}(f_1\omega, \cdots, f_n\omega) = F(\mathbf{f})\omega$ on K.

10. For all Lipschitz functions F, G on $\mathbb{R}^n$:

    (i) $F + G$ is Lipschitzean and $\widetilde{F + G} = \tilde{F} + \tilde{G}$ on $\mathbb{D}$.

    (ii) cF is Lipschitzean and $\widetilde{cF} = c\tilde{F}$ for all real c.

    (iii) $|F|$ is Lipschitzean and $\widetilde{|F|} = |\tilde{F}|$ where $|\tilde{F}|(\sigma) = |\tilde{F}(\sigma)|$ in $\mathbb{D}$ for all
    $\sigma = (\sigma_1 \cdots \sigma_n)$ in $\mathbb{D}^n$.

    (iv) $\tilde{F}(\sigma) \geq 0$ for all σ in $\mathbb{D}^n$ if and only if $F(\mathbf{x}) \geq 0$ for all $\mathbf{x}$ in $\mathbb{R}^n$.

    (v) $\tilde{F} = 0$ if and only if $F = 0$.

    (vi) $\tilde{F}(\sigma) \geq 0$ for all $\sigma \geq \mathbf{0}$ in $\mathbb{D}^n$ if and only if $F(\mathbf{x}) \geq 0$ for all $\mathbf{x} \geq \mathbf{0}$ in $\mathbb{R}^n$.

11. For any Riesz norm L on $\mathbb{R}^n$ the mapping $\tilde{L}$ inherits the formal properties of a Riesz norm. That is, for all σ, τ in the product Riesz space $\mathbb{D}^n$:

    (i) $\tilde{L}(\sigma) \geq 0$ in $\mathbb{D}$ with equality only for $\sigma = \mathbf{0}$,

    (ii) $\tilde{L}(c\sigma) = |c|\tilde{L}(\sigma)$ for all c in $\mathbb{R}$,

    (iii) $\tilde{L}(\sigma + \tau) \leq \tilde{L}(\sigma) + \tilde{L}(\tau)$,

    (iv) $\tilde{L}(\sigma) \leq \tilde{L}(\tau)$ if $|\sigma| \leq |\tau|$.

12. If $\sigma = (f_1\tau, \cdots, f_n\tau)$ where τ is a differential on a cell K and $\mathbf{f} = (f_1 \cdots, f_n)$ is a mapping of K into $\mathbb{R}^n$ then for any norm L on $\mathbb{R}^n$, $\tilde{L}(\sigma) = L(\mathbf{f})|\tau|$ where $L(\mathbf{f})$ is a function on the cell K.

§3.7 n-differentials on a cell K.

An **n-summant S** on a cell K is a mapping on the set of all tagged cells in K into $\mathbb{R}^n$. That is,

$$(1) \qquad\qquad \mathbf{S} = (S_1, \cdots, S_n)$$

where each S_i is a summant (1-summant) on K. Under coordinatewise operations the n-summants on K form the Riesz space $\mathbb{Y}^n$, the cartesian product of n copies of the Riesz space $\mathbb{Y}$ of all 1-summants on K.

The cartesian product $\mathbb{Z}^n$ which consists of all $\mathbf{S}$ in (1) whose coordinates S_i all belong to $\mathbb{Z}$ is a Riesz ideal in $\mathbb{Y}^n$. An **n-differential** $\sigma = [\mathbf{S}]$ consists of all n-summants $\mathbf{S}' = (S_1', \cdots, S_n')$ on K such that $\int_K |S_i - S_i'| = 0$ for all of the coordinates. Clearly by (1)

$$(2) \qquad\qquad \sigma = (\sigma_1, \cdots, \sigma_n)$$

where $\sigma_i = [S_i]$ for all i. That is, $\mathbb{Y}^n / \mathbb{Z}^n = \mathbb{D}^n$.

We define the lower and upper integrals

$$(3) \qquad \begin{aligned} \underline{\int_K} \sigma &= (\underline{\int_K} \sigma_1, \cdots, \underline{\int_K} \sigma_n) \quad \text{and} \\ \overline{\int_K} \sigma &= (\overline{\int_K} \sigma_1, \cdots, \overline{\int_K} \sigma_n) \end{aligned}$$

with concommitant definitions for existence of the integral $\int_K \sigma$ and integrability of σ.

According to §3.6 a norm $\| \cdot \|$ on $\mathbb{R}^n$ assigns to each n-differential $\sigma = [\mathbf{S}]$ on K a 1-differential $\|\sigma\| = [\|\mathbf{S}\|] \geq 0$ on K. Its upper integral is a norm $\nu(\|\sigma\|) \leq \infty$ on $\mathbb{D}^n$. All such norms on $\mathbb{D}^n$ are equivalent; the convergence $\sigma \to \mathbf{0}$ in

$\mathbb{D}^n$ defined by $\|\sigma\| \to 0$ in $\mathbb{D}$ is just convergence to 0 of the 1-differentials σ_i in (2) for each coordinate i.

For f a function on K we define

$$(4) \qquad\qquad f\sigma = (f\sigma_1, \dots, f\sigma_n)$$

in terms of (2). A subset E of K is σ-**null** if $1_E\sigma = \mathbf{0}$, that is, if E is σ_i-null for $i = 1, \cdots, n$. So we have the concept of σ-everywhere. We thus have the validity of (4) for f defined and finite σ-everywhere on K.

Any property of 1-differentials is defined for n-differentials by endowing σ with the property whenever all its coordinates σ_i have it. Thus we gain the concepts of summability, tag finiteness, and continuity for n-differentials. By (3) integrability is just integrability of the coordinates in (2).

For σ tag-finite in $\mathbb{D}^n$ and ρ tag-fnite in $\mathbb{D}$ we can define in terms of (2)

$$(5) \qquad\qquad \rho\sigma = (\rho\sigma_1, \dots \rho\sigma_n)$$

in $\mathbb{D}^n$.

Given $\mathbf{G} = (G_1, \cdots, G_n)$ mapping K into $\mathbb{R}^n$ we define the additive n-summant $\Delta\mathbf{G} = (\Delta G_1, \cdots, \Delta G_n)$ which yields the n-differential $d\mathbf{G} = [\Delta\mathbf{G}] = (dG_1, \cdots, dG_n)$. $d\mathbf{G}$ is integrable on K and $\int_K d\mathbf{G} = \Delta\mathbf{G}(K)$. Moreover, every integrable n-differential is of this form.

For any norm $\|\cdot\|$ on $\mathbb{R}^n$ we contend $\int_K \|d\mathbf{G}\|$ exists. It gives the distance (in norm $\|\cdot\|$) traveled by a particle in $\mathbb{R}^n$ over a time interval K with $\mathbf{G}(t)$ the position of the particle at time t. This distance includes the jumps made by the particle wherever $\mathbf{G}$ is discontinuous. In the 1-dimensional case it is just a positive scalar multiple of total variation.

THEOREM 1. *Let* $\mathbf{G} = (G_1, \cdots, G_n)$ *map a cell K into $\mathbb{R}^n$. Let $\|\cdot\|$ be a norm on $\mathbb{R}^n$. Then the integral of $\|d\mathbf{G}\|$ exists and*

(6) $$0 \le \int_K \|d\mathbf{G}\| \le \infty.$$

Moreover, $\int_K \|d\mathbf{G}\| < \infty$ if and only if $G_1, \cdots, G_n$ are of bounded variation.

PROOF. Since $\Delta\mathbf{G}$ is additive the triangle inequality for norms makes $\|\Delta\mathbf{G}\|$ a subadditive 1-summant on K. So (6) follows from (i) in Theorem 1 (§1.4). Since all norms on $\mathbb{R}^n$ are equivalent the final statement in the theorem follows from the case

$$\int_K \|d\mathbf{G}\|_1 = \int_K |dG_1| + \cdots + \int_K |dG_n|$$

of the norm (4) in §3.6.   □

Exercises (§3.7).

1. Let τ be a differential on a cell K. Let $\mathbf{f}$ map K into $\mathbb{R}^n$ with $\mathbf{f} = (f_1, \cdots, f_n)$. Given any norm $\|\cdot\|$ on $\mathbb{R}^n$ let $\|\mathbf{f}\|$ be the function on K whose value at t is $\|\mathbf{f}(t)\|$. Prove $\|\mathbf{f}\tau\| = \|\mathbf{f}\|\|\tau\|$ for the n-differential $\mathbf{f}\tau = (f_1\tau, \cdots, f_n\tau)$.

2. For any norm $\|\cdot\|$ on $\mathbb{R}^n$ and n-differential $\boldsymbol{\sigma} = (\sigma_1, \cdots, \sigma_n)$ on a cell K verify the following for the 1-differential $\|\boldsymbol{\sigma}\|$:

   (i) $\|\boldsymbol{\sigma}\|$ is summable if and only if $\boldsymbol{\sigma}$ is summable.

   (ii) $\|\boldsymbol{\sigma}\|$ is tag-finite if and only if $\boldsymbol{\sigma}$ is tag-finite.

   (iii) $\|\boldsymbol{\sigma}\| = 0$ if and only if $\boldsymbol{\sigma} = \mathbf{0}$.

   (iv) $\|\boldsymbol{\sigma}\|$ is continuous if and only if $\boldsymbol{\sigma}$ is continuous.

   (v) A subset E of K is $\|\boldsymbol{\sigma}\|$-null if and only if E is $\boldsymbol{\sigma}$-null.

3. Given tag-finite n-differentials $\boldsymbol{\sigma}, \boldsymbol{\tau}$ on a cell K and ρ a tag-finite 1-differential on K verify:

(i) The inner product $\sigma \cdot \tau = \sum_{i=1}^{n} \sigma_i \tau_i$ is a well defined 1-differential on K.

(ii) The inner product is commutative and distributive.

(iii) $(f\sigma) \cdot \tau = f(\sigma \cdot \tau)$ for every function f on K.

(iv) $(\rho\sigma) \cdot \tau = \rho(\sigma \cdot \tau)$.

(v) $\|\rho\sigma\| = |\rho| \, \|\sigma\|$ for any norm on $\mathbb{R}^n$.

(vi) For the Euclidean norm $\| \cdot \|$, $|\sigma \cdot \tau| \leq \|\sigma\| \, \|\tau\|$.

(vii) If $\tau = \rho\sigma$ then equality holds in (vi).

(viii) If $\mathbf{f}$ maps K into $\mathbb{R}^n$ then $\sigma \cdot (\mathbf{f}\rho) = (\mathbf{f} \cdot \sigma)\rho = \mathbf{f} \cdot (\rho\sigma)$.

4. Let $\sigma = (\sigma_1, \cdots, \sigma_n)$ be an n-differential on a cell K with disjoint coordinates, that is $|\sigma_i| \wedge |\sigma_j| = 0$ for $i \neq j$. Prove that $\|\sigma\|_2 = \|\sigma\|_1$ for the norms defined by (7) in §3.6. (Apply Exercise 4 of §2.2.)

5. Let ρ, σ be tag-finite n-differentials on a cell K such that $|\rho| \wedge |\sigma| = 0$. Show that $\rho \cdot \sigma = 0$. (Apply the special case $n = 1$ given by Theorem 9 (§3.1).)

6. (Line segments are geodesics in $\mathbb{R}^n$ for all norms.) Given distinct points $\mathbf{a}, \mathbf{b}$ in $\mathbb{R}^n$ define the mapping $\mathbf{G}$ on $K = [0, 1]$ into $\mathbb{R}^n$ by $\mathbf{G}(t) = (1 - t)\mathbf{a} + t\mathbf{b}$ for all t in K. Show that $\int_K \|d\mathbf{G}\| = \|\mathbf{b} - \mathbf{a}\|$ for all norms on $\mathbb{R}^n$.

7. Let $\mathbf{a}, \mathbf{b}$ be distinct points in $\mathbb{R}^n$. Let $\mathbf{G}$ be a continuous mapping of $K = [0, 1]$ into $\mathbb{R}^n$ such that $d\mathbf{G} \geq \mathbf{0}, \mathbf{G}(0) = \mathbf{a} \wedge \mathbf{b}$, and $\mathbf{G}(1) = \mathbf{a} \vee \mathbf{b}$. Show that $\int_K \|d\mathbf{G}\|_1 = \|\mathbf{b} - \mathbf{a}\|_1$.

CHAPTER 4

MEASURABLE SETS AND FUNCTIONS

§4.1 Measurable Sets.

Given an integrable differential $\sigma \geq 0$ on a cell K we define a subset E of K to be σ-**measurable** if $1_E \sigma$ is integrable. (We remind the reader that $\sigma = dv$ for some monotone function v on K, $v(s) \leq v(t)$ for $s \leq t$.)

By (7) in Theorem 4 (§2.7) every point in K is σ-measurable. So is every open set in K by Theorem 9 (§3.4). The complement D in K of a σ-measurable subset E of K is σ-measurable since $1_D \sigma = \sigma - 1_E \sigma$. If A and B are σ-measurable then so is their union $A \cup B$ since $1_{A \cup B} \sigma = (1_A \sigma) \vee (1_B \sigma)$ for $\sigma \geq 0$, and the absolutely integrable differentials on K form a Banach lattice. (See the remarks following Theorem 3 (§2.4).) Finally the union E of a monotone sequence $E_1 \subseteq E_2 \subseteq \cdots$ of σ-measurable sets is σ-measurable since $1_{E_n} \nearrow 1_E$ which implies $1_{E_n} \sigma \nearrow 1_E \sigma$ with $1_E \sigma$ integrable by Theorem 3 (§2.7) on monotone convergence.

In summary the set $\mathcal{M}_\sigma$ of σ-measurable subsets of K is a **sigma-algebra**:
(i) $\emptyset \in \mathcal{M}_\sigma$, (ii) If $E \in \mathcal{M}_\sigma$ then $K - E \in \mathcal{M}_\sigma$, and (iii) If $E_1, E_2, \cdots \in \mathcal{M}_\sigma$ then $E_1 \cup E_2 \cup \cdots \in \mathcal{M}_\sigma$. Moreover, every open subset of K belongs to $\mathcal{M}_\sigma$.

As a consequence of (i) and (ii) $K \in \mathcal{M}_\sigma$. By (ii) and (iii) $E_1 \cap E_2 \cap \cdots \in \mathcal{M}_\sigma$ if $E_n \in \mathcal{M}_\sigma$ for $n = 1, 2, \cdots$. Therefore for every sequence $E_1, E_2, \cdots$ in $\mathcal{M}_\sigma$ both $\overline{\lim} E_n$ and $\underline{\lim} E_n$ belong to $\mathcal{M}_\sigma$ where

129

$$\overline{\lim} E_n = \bigcap_{n=1}^{\infty} (E_n \cup E_{n+1} \cup \cdots) \text{ and}$$

(1)

$$\underline{\lim} E_n = \bigcup_{n=1}^{\infty} (E_n \cap E_{n+1} \cap \cdots).$$

Their indicators are given by

$$(2) \quad 1_{\overline{\lim} E_n}(t) = \overline{\lim_{n \to \infty}} 1_{E_n}(t) \text{ and } 1_{\underline{\lim} E_n}(t) = \underline{\lim_{n \to \infty}} 1_{E_n}(t)$$

since $\overline{\lim} E_n$ consists of those points that belong to E_n for infinitely many n, $\underline{\lim} E_n$ of those that belong to E_n for all but finitely many n. So $\underline{\lim} E_n \subseteq \overline{\lim} E_n$. If equality holds we define $\lim E_n = \underline{\lim} E_n = \overline{\lim} E_n$. This is just convergence of the indicators, $1_{E_n} \to 1_{\lim E_n}$.

Since every open interval in K is σ-measurable so is every interval since every interval is the intersection of a sequence of open (relative to K) intervals.

Now we must often deal with more than one differential σ. So we define a subset E of K to be **measurable** if E is σ-measurable for every integrable $\sigma \geq 0$ on K. The measurable subsets of K form a sigma-algebra $\mathcal{M}$ since $\mathcal{M}$ is the intersection of the sigma-algebras $\mathcal{M}_\sigma$ for all integrable $\sigma \geq 0$. Moreover, $\mathcal{M}$ contains all intervals and all open sets in K. Note that E is measurable if and only if $1_E \sigma$ is (absolutely) integrable for every absolutely integrable differential σ on K. The adverb in parentheses is redundant.

In any topological space Y a **Borel set** is a set that belongs to every sigma-algebra in Y to which every open subset of Y belongs. The intersection of all such sigma-algebras is the sigma-algebra of Borel sets in Y.

Since every open subset of K is measurable every Borel set in K is measurable.

A subset E of $\mathbb{R}$ is **Lebesgue-measurable** if $E \cap K$ is dx-measurable on K for every cell K in $\mathbb{R}$, that is, if $1_E dx$ is integrable on every bounded interval in $\mathbb{R}$.

Since an integrable differential $\sigma \geq 0$ acts as a measure on the sigma-algebra $\mathcal{M}_\sigma$ (and thereby on the sigma-algebra $\mathcal{M}$ contained in $\mathcal{M}_\sigma$) we get some standard results of measure theory.

THEOREM 1. *Let $\sigma \geq 0$ be an integrable differential on a cell K. Let $E_1, E_2, \cdots$ be a sequence of σ-measurable subsets of K. Then*

(a) *If $E_n \searrow \emptyset$ (that is, $E_1 \supseteq E_2 \supseteq \cdots$ and $E_1 \cap E_2 \cap \cdots = \emptyset$) then $1_{E_n}\sigma \searrow 0$.*

(b) $-\infty < \int_K 1_{\underline{\lim} E_n}\sigma \leq \underline{\lim}_{n\to\infty} \int_K 1_{E_n}\sigma \leq \overline{\lim}_{n\to\infty} \int_K 1_{E_n}\sigma \leq \int_K 1_{\overline{\lim} E_n}\sigma < \infty.$

(c) *If $1_{E_n} \to 1_E$ then $1_{E_n}\sigma \to 1_E\sigma$.*

PROOF.

(a). Apply Theorem 3 (§2.7) on monotone convergence with the functions $f_0 = 0, f_n = 1 - 1_{E_n}$ for $n > 0$, and $f = 1$.

(b). Let $B = \overline{\lim} E_n$ and $B_n = E_n \cup E_{n+1} \cup \cdots$. Then $B_n \searrow B$. Apply continuity condition (a) to the monotone convergence $B_n - B \searrow \emptyset$ to conclude that $1_{B_n}\sigma \searrow 1_B\sigma$. So $\int_K 1_{B_n}\sigma \searrow \int_K 1_B\sigma$. Since $E_n \subseteq B_n, \int_K 1_{E_n}\sigma \leq \int_K 1_{B_n}\sigma$. Therefore $\overline{\lim} \int_K 1_{E_n}\sigma \leq \int_K 1_B\sigma$ which gives the upper limit inequalities in (b). The dual inequalities for the lower limit follow similarly from (a) applied to $A - A_n \searrow \emptyset$ for $A = \underline{\lim} E_n$ and $A_n = E_n \cap E_{n+1} \cap \cdots$.

(c). Let $D_n = E \cup E_n - E \cap E_n$. This symmetric difference $D_n \to \emptyset$ since $E_n \to E$. Apply (b) to D_n to conclude that $1_{D_n}\sigma \to 0$. That is, $|1_{E_n} - 1_E|\sigma \to 0$. $\square$

As the next result shows, the study of measurability is not a futile exercise.

THEOREM 2.. *Every cell K in $\mathbb{R}$ contains sets which are not dx-measurable.*

PROOF. Lebesgue measure M is defined by

$$M(E) = \int_{\mathbb{R}} 1_E dx \text{ for every } dx\text{-measurable set } E \text{ in } \mathbb{R}.$$

Call s, t in $\mathbb{R}$ equivalent if their difference $s - t$ is rational. This is easily seen to be an equivalence relation on $\mathbb{R}$. Given a cell K in $\mathbb{R}$ every equivalence class D meets K since D is dense in $\mathbb{R}$. The axiom of choice yields a subset E_0 of K consisting of exactly one member from each equivalence class. We give an indirect proof that E_0 is not dx-measurable.

Suppose E_0 were dx-measurable. Let A be the countable set of all rationals r such that $|r| \leq M(K)$. The translates $E_r = \{t + r : t \varepsilon E_0\}$ of E_0 for r in A are disjoint, cover K, and their union E is bounded. Also M is translation invariant, $M(E_r) = M(E_0)$. Thus

$$M(K) \leq M(E) = \sum_{r \varepsilon A} M(E_r) = \infty \cdot M(E_0).$$

Since $M(E) < \infty$ this implies $M(E_0) = 0$, hence $M(K) = 0$ which is false. $\square$

Exercises (§4.1).

1. Let $\mathcal{A}$ be a sigma-algebra in $\mathbb{R}$ such that $(-\infty, t]$ belongs to $\mathcal{A}$ for all t in $\mathbb{R}$. Show that every open subset (hence, every Borel subset) of $\mathbb{R}$ belongs to $\mathcal{A}$.

2. Explain how the existence of a nonconstant, monotone, continuous function v on a cell K implies that K is uncountable.

3. Prove:

   (i) If $A_n \subseteq B_n$ for all but finitely many n then $\underline{\lim} A_n \subseteq \underline{\lim} B_n$ and $\overline{\lim} A_n \subseteq \overline{\lim} B_n$.

(ii) If $A_n \subseteq B_n$ for infinitely many n then $\underline{\lim}A_n \subseteq \overline{\lim}B_n$.

(iii) $\underline{\lim}(K-E_n) = K-\overline{\lim}E_n$ and $\overline{\lim}(K-E_n) = K-\underline{\lim}E_n$ if $E_n \subseteq K$ for all n.

(iv) $\underline{\lim}(A_n \cap B_n) = \underline{\lim}A_n \cap \underline{\lim}B_n$.

(v) $\overline{\lim}(A_n \cup B_n) = \overline{\lim}A_n \cup \overline{\lim}B_n$.

4. Let $\langle t_n \rangle$ be a sequence in $\mathbb{R}$ with $\underline{\lim}t_n = p$ and $\overline{\lim}t_n = q$ (See example (i) in §11.1.) Prove:

(i) $[-\infty, p) \subseteq \underline{\lim}[-\infty, t_n)$, (ii) $\overline{\lim}[-\infty, t_n] \subseteq [-\infty, q]$,

(iii) $(p, \infty] \subseteq \overline{\lim}(t_n, \infty]$, (iv) $\underline{\lim}[t_n, \infty] \subseteq [q, \infty]$.

Which of these inclusions become equality if $\langle t_n \rangle$ is:

(a) increasing? (b) decreasing?

§4.2 The Hahn Decomposition for Differentials.

In §4.1 we defined a subset E of a cell K to be measurable if $1_E\sigma$ is integrable for every integrable $\sigma \geq 0$ on K. Equivalently, E is measurable of $1_E\sigma$ is absolutely integrable for every absolutely integrable σ on K.

For some differentials σ on K we can get a **Hahn decomposition** of K into complementary measurable subsets A, B such that $1_A\sigma \geq 0$ and $1_B\sigma \leq 0$. This pair of inequalities for complementary A, B is equivalent to $1_A\sigma = \sigma^+, 1_B\sigma = -\sigma^-$. (See Exercise 3.)

THEOREM 1. *Let σ be a continuous, dampable differential on a cell K. Then there exist complementary measurable subsets A, B of K such that $1_A\sigma = \sigma^+$ and $1_B\sigma = -\sigma^-$. So $\sigma = w|\sigma|$ and $|\sigma| = w\sigma$ for $w = 1_A - 1_B$.*

PROOF. $\sigma = u\rho$ for some damper u and some continuous, absolutely integrable ρ. So we need only get a Hahn decomposition for ρ since $1_A\sigma = u1_A\rho$ and $1_B\sigma = u1_B\rho$ with $u > 0$.

So $1_A\sigma \geq 0$ and $1_B\sigma \leq 0$ if $1_A\rho \geq 0$ and $1_B\rho \leq 0$. In effect we need only consider the case where σ is continuous and absolutely integrable.

By Theorem 6 (§2.4)

$$(1) \qquad \int_K \sigma^+ = \sup_{A \in \mathcal{F}} \int_A \sigma < \infty$$

where $\mathcal{F}$ is the set of all figures in K. Since σ is continuous every finite set is σ-null. In particular the boundary of any figure is σ-null. So $1_A\sigma = 1_{A^\circ}\sigma$ for the interior A° of a figure A. Hence

$$(2) \qquad \int_A \sigma = \int_A 1_{A^\circ}\sigma = \int_K 1_{A^\circ}\sigma = \int_K 1_A\sigma.$$

By (1) and (2) we can for each positive integer n choose a figure A_n such that

$$(3) \qquad \int_K \sigma^+ < \int_K 1_{A_n}\sigma + \frac{1}{2^n}.$$

Let $A_n^* = A_n \cup A_{n+1} \cup \cdots$. Then $A_n^* \searrow A$ where $A = \overline{\lim} A_n$. Clearly A_n, A_n^*, A are all measurable. By (3)

$$(4) \qquad \int_K \sigma^+ < \int_K 1_{A_n^*}\sigma^+ + \frac{1}{2^n}$$

since $1_{A_n}\sigma \leq 1_{A_n}\sigma^+ \leq 1_{A_n^*}\sigma^+$. (3) also implies

$$(5) \qquad \int_K 1_{A_n}\sigma^- < \frac{1}{2^n}$$

since $1_{A_n}\sigma^- \leq 1_{A_n}\sigma^- + (1 - 1_{A_n})\sigma^+ = \sigma^+ - 1_{A_n}\sigma.$

Now $1_A \leq 1_{A_n^*} \leq \sum_{i=n}^{\infty} 1_{A_i}$. So Theorem 2 (§2.7) gives

$$(6) \qquad \int_K 1_A \sigma^- \leq \sum_{i=n}^{\infty} \int_K 1_{A_i} \sigma^- < \sum_{i=n}^{\infty} \frac{1}{2^i} = \frac{1}{2^{n-1}}$$

by (5). Since (6) holds for all n,

$$(7) \qquad\qquad\qquad 1_A \sigma^- = 0.$$

Since $A_n^* \searrow A, f_n \nearrow f$ for $f_n = 1 - 1_{A_n^*}$ and $f = 1 - 1_A$. So Theorem 3 (§2.7) gives $1_{A_n^*} \sigma^+ \searrow 1_A \sigma^+$ which implies by (4) that $\int_K \sigma^+ \leq \int_K 1_A \sigma^+$. That is, for B the complement of A in K

$$(8) \qquad\qquad\qquad 1_B \sigma^+ = 0.$$

By (7) and (8) $1_A \sigma = 1_A \sigma^+ - 1_A \sigma^- = 1_A \sigma^+ = 1_A \sigma^+ + 1_B \sigma^+ = \sigma^+$ and $1_B \sigma = (1 - 1_A)\sigma = \sigma - \sigma^+ = -\sigma^-$. $\square$

We can extend Theorem 1 to allow discontinuities as long as σ^+ and σ^- have none in common.

THEOREM 2. *Let σ be a dampable differential on a cell K such that σ^+ and σ^- have no common point of discontinuity. Then there exist complementary measurable subsets A, B of K such that $1_A \sigma = \sigma^+$ and $1_B \sigma = -\sigma^-$.*

PROOF. Let D be the set of points at which σ is discontinuous. Since D is countable for σ dampable, D is measurable. So its complement C, the set of points in K at which σ is continuous, is measurable. By hypothesis D is the union of disjoint sets D^+, D^- which are respectively the discontinuities of σ^+, σ^-. Also $1_{D^+} \sigma^- = 1_{D^-} \sigma^+ = 0$ since σ^- is continuous at each point in the countable set D^+, and σ^+ is continuous at each point in the countable set D^-.

Since σ is dampable so is $1_C \sigma$. Indeed, for $\sigma = u\rho$ with u a damper and ρ absolutely integrable $1_C \rho$ is absolutely integrable and $1_C\sigma = u1_C\rho$. So Theorem 1 gives a Hahn decomposition A, B for $1_C\sigma$. Then $(A \cap C) \cup D^+, (B \cap C) \cup D^-$ is a Hahn decomposition for $\sigma = 1_C\sigma + 1_D\sigma$. $\square$

Exercises (§4.2).

1. Let g be the indicator 1_0 of the point 0 in $[-\infty, \infty]$.

   (a) Prove:

   (i) $g(dg)^+ = (dg)^+ > 0$ and $g(dg)^- = (dg)^- > 0$.

   (ii) There is no Hahn decomposition of $[-\infty, \infty]$ for dg.

   (b) Find differentials σ, τ on $[-\infty, \infty]$ such that $\sigma + \tau = w, \sigma \wedge \tau = 0, \sigma dg = (dg)^+$, and $\tau dg = -(dg)^-$.

2. Let h be a continuous function of bounded variation on a cell K. For f a function on K prove the equivalence of the following three conditions:

   (i) fdh is absolutely integrable.

   (ii) $f(dh)^+$ and $f(dh)^-$ are absolutely integrable.

   (iii) $f|dh|$ is absolutely integrable.

   (Apply Theorem 1 to $\sigma = dh$.)

3. For σ a differential on a cell K and A, B complementary subsets of K prove the equivalence of the following six conditions:

   (i) $1_A\sigma \geq 0$ and $1_B\sigma \leq 0$.

   (ii) $1_A\sigma^- = 1_B\sigma^+ = 0$.

   (iii) $1_A\sigma = 1_A\sigma^+$ and $1_B\sigma = -1_B\sigma^-$.

   (iv) $\sigma = 1_A\sigma^+ - 1_B\sigma^-$.

   (v) $1_A\sigma = \sigma^+$ and $1_B\sigma = -\sigma^-$.

   (vi) $1_A\sigma^+ = \sigma^+$ and $1_B\sigma^- = \sigma^-$.

4. For w in Theorem 1 verify that $w^+ = 1_A, w^- = 1_B$, and $w^2 = |w| = 1$.

§4.3 Measurable Functions.

Given any sigma-algebra $\mathcal{A}$ of subsets of K we shall define $\mathcal{A}$-measurability of functions in the usual way. In our applications $\mathcal{A}$ will be either $\mathcal{M}_\sigma$ or $\mathcal{M}$ which were introduced in §4.1.

For f a function on K let $(f > c) = f^{-1}(c, \infty)$ for each c in $\mathbb{R}$, $(f \geq c) = f^{-1}[c, \infty)$, $(f < c) = f^{-1}(-\infty, c)$, $(f \leq c) = f^{-1}(-\infty, c]$, and $(f = c) = f^{-1}(c)$. f is $\mathcal{A}$-**measurable** if $(f > c)$ belongs to $\mathcal{A}$ for every c in $\mathbb{R}$. For an equivalent definition $(f > c)$ may be replaced by any of the sets $(f \geq c), (f < c)$, or $(f \leq c)$. If f is $\mathcal{A}$-measurable then $f^{-1}(D)$ belongs to $\mathcal{A}$ for D any interval, open set, or closed set in $\mathbb{R}$.

For $f = 1_E$ with E a subset of K the set $(f > c)$ is K for $c < 0, E$ for $0 \leq c < 1$, and $\emptyset$ for $c \geq 1$. So 1_E is $\mathcal{A}$-measurable if and only if E belongs to $\mathcal{A}$. In particular, 0 and 1 are $\mathcal{A}$-measurable.

If f is $\mathcal{A}$-measurable then so is af for every real constant a. Indeed, for the nontrivial case $a \neq 0, (af > c)$ is $(f > \frac{c}{a})$ if $a > 0$ and $(f < \frac{c}{a})$ if $a < 0$.

If f and g are $\mathcal{A}$-measurable then so are $f \wedge g$ and $f \vee g$ since both $(f \wedge g > c) = (f > c) \cap (g > c)$ and $(f \vee g > c) = (f > c) \cup (g > c)$. Applying this to each of the two cases $g = 0$ and $g = -f$ we infer that f^+, f^-, and $|f|$ are $\mathcal{A}$-measurable if f is $\mathcal{A}$-measurable.

If f and g are $\mathcal{A}$-measurable then so is $f + g$. This follows from

$$(1) \qquad (f + g > c) = \bigcup_{r \in \mathbb{Q}} (f > r) \cap (g > c - r)$$

where $\mathbb{Q}$ is the countable set of all rationals r in $\mathbb{R}$. Clearly $f(t) + g(t) > c$ if $f(t) > r$ and $g(t) > c - r$ for some r. So the left side of (1) contains the right side. In the other direction $f(t) + g(t) > c$ implies $c - g(t) < f(t)$, hence $c - g(t) < r < f(t)$ for some r in $\mathbb{Q}$ since $\mathbb{Q}$ is dense in $\mathbb{R}$. That is $t \in (f > r) \cap (g > c - r)$ for some r in $\mathbb{Q}$. So (1) holds.

An analogous proof applied to the positive and negative parts of f, g gives the $\mathcal{A}$-measurability of the product fg if both f and g are $\mathcal{A}$-measurable.

We have concluded so far that the $\mathcal{A}$-measurable functions form a Riesz space under pointwise operations and ordering. Moreover this function space is an algebra containing the indicators of all members of $\mathcal{A}$. Now these indicators generate all $\mathcal{A}$-measurable functions according to the following theorem.

THEOREM 1. *Every $\mathcal{A}$-measurable function f is a uniform limit of countably valued, $\mathcal{A}$-measurable functions.*

PROOF. Let f be $\mathcal{A}$-measurable. We may assume $f \geq 0$ since this case applies to f^+ and f^-. Given $\varepsilon > 0$ we need to find a countably valued, $\mathcal{A}$-measurable function g such that $|f - g| \leq \varepsilon$.

Let $E_n = ((n-1)\varepsilon \leq f < n\varepsilon)$ for $n = 1, 2, \cdots$. Then $E_1, E_2, \cdots$ are disjoint members of $\mathcal{A}$ whose union is K. Let g be the function on K with value $n\varepsilon$ on E_n for $n = 1, 2, \cdots$. Clearly $(g > c) = \bigcup_{n > \frac{c}{\varepsilon}} E_n$ which belongs to $\mathcal{A}$. So g is $\mathcal{A}$-measurable. Also $g(K) \subseteq \{\varepsilon, 2\varepsilon, 3\varepsilon, \cdots\}$ which is countable. Finally $g - \varepsilon \leq f < g$, so $0 < g - f \leq \varepsilon$. $\square$

The converse of Theorem 1 follows from the next theorem which shows that a limit of a sequence of $\mathcal{A}$-measurable functions is $\mathcal{A}$-measurable.

THEOREM 2. *Let $f_1, f_2, \cdots$ be a sequence of $\mathcal{A}$-measurable functions.*

(a) If $\underline{\lim}_{n \to \infty} f_n$ is finite everywhere then it is $\mathcal{A}$-measurable.

(b) If $\overline{\lim}_{n \to \infty} f_n$ is finite everywhere then it is $\mathcal{A}$-measurable.

(c) If $f_n \to f$ on K then f is $\mathcal{A}$-measurable.

PROOF. (a) follows from the identity

$$\left(\lim_{n \to \infty} f_n \geq c \right) = \bigcap_{k=1}^{\infty} \bigcup_{j=1}^{\infty} \bigcap_{i=j}^{\infty} (f_i > c - \frac{1}{k}) = \bigcap_{k=1}^{\infty} \lim_{i \to \infty} (f_i > c - \frac{1}{k})$$

which implies that this set belongs to $\mathcal{A}$. To prove (b) apply (a) to $-f_n$ in place of f_n. Finally, (c) follows from either (a) or (b). □

Hereafter we shall call the $\mathcal{M}_\sigma$-measurable functions **σ-measurable** and the $\mathcal{M}$-measurable functions **measurable** for the special cases $\mathcal{A} = \mathcal{M}_\sigma$ and $\mathcal{A} = \mathcal{M}$. Since $\mathcal{M} \subseteq \mathcal{M}_\sigma$ for every integrable $\sigma \geq 0$ on K every measurable function is σ-measurable for all such σ. Every continuous function f on K is measurable since $(f > c)$ is open, hence measurable. Clearly, every monotone function on K is measurable. So every function f of bounded variation on K is measurable since f is the difference of two monotone functions.

The main role of measurable functions in integration is revealed by the following result.

THEOREM 3. *Let $\sigma \geq 0$ be an integrable differential on a cell K. Let f be a function on K. Then the following two conditions are equivalent:*

(i) $f\sigma$ is absolutely integrable.

(ii) f is σ-measurable and $f\sigma$ is summable.

PROOF. **(i) ⇒ (ii)**. We need only prove f is σ-measurable, that is, $(f > c)$ is a σ-measurable set for all c in $\mathbb{R}$. Now $(f-c)\sigma$ is absolutely integrable since both $c\sigma$ and $f\sigma$ are absolutely integrable by (i). Hence $u\sigma$ is integrable for $u = (f - c)^+$ Since $(u > 0) = (f > c)$ we must prove $(u > 0)$ is σ-measurable. In the Banach lattice of absolutely integrable differentials on K we have $[(nu) \wedge 1]\sigma = (nu\sigma) \wedge \sigma$ for $n = 1, 2, \cdots$ with $(nu) \wedge 1 \nearrow 1_{(u>0)}$. So $1_{(u>0)}\sigma$ is integrable by Theorem 3 (§2.7). That is, $(u > 0)$ is σ-measurable.

(ii) ⇒ (i). We may assume $f \geq 0$ since this case can be applied to f^+ and f^- for f satisfying (ii).

Consider first the case where $f(K)$ is countable. Then $f = \sum_{k=1}^{\infty} f_k$ with $f_k = c_k 1_{E_k} \geq 0$ for $E_1, E_2, \cdots, E_k, \cdots$ disjoint, σ-measurable subsets of K. $f_k\sigma$ is integrable since

E_k is σ-measurable. So Theorem 2(§2.7) gives $\sum_{k=1}^{\infty} \int_K f_k \sigma = \int_K f\sigma < \infty$ for $f\sigma$ summable. So $f\sigma$ is integrable.

For the more general case of σ-measurable $f \geq 0$ with $f\sigma$ summable let $\varepsilon > 0$ be given. Apply Theorem 1 to get a σ-measurable function $g \geq 0$ with countable range $g(K)$ such that $|f - g| < \varepsilon$. Then $0 \leq g\sigma \leq f\sigma + \varepsilon\sigma$. So $g\sigma$ is summable since both $f\sigma$ and $\varepsilon\sigma$ are summable. By the preceding case $g\sigma$ is integrable. Hence, since $g - \varepsilon < f < g + \varepsilon, -\infty < \int_K g\sigma - \varepsilon \int_K \sigma \leq \underline{\int}_K f\sigma \leq \overline{\int}_K f\sigma \leq \int_K g\sigma + \varepsilon \int_K \sigma < \infty$. Thus $0 \leq \overline{\int}_K f\sigma - \underline{\int}_K f\sigma \leq 2\varepsilon \int_K \sigma$. Since this holds for all $\varepsilon > 0$ it implies integrability of $f\sigma$. $\square$

The measurability of continuous functions is the special case $C = K$ of the next theorem.

THEOREM 4. *Given a function f on a cell K let C be the set of all points at which f is continuous. Then C is a measurable set and $1_C f$ is a measurable function.*

PROOF. Let C_n be the union of all intervals L in K such that L is open relative to K and diam $f(L) < \frac{1}{n}$. For $n = 1, 2, \cdots$ C_n is open in $K, C_n \supseteq C_{n+1}$, and $C_n \searrow C$. So C is measurable.

To prove $1_C f$ is measurable it suffices to prove $1_C f^+$ and $1_C f^-$ are measurable since their difference is just $1_C f$. Clearly $(1_C f^+ > c) = (1_C f^- > c) = K$ for $c < 0$. For $c \geq 0$ the reader can easily verify that $(1_C f^+ > c) = C \cap (f > c) = C \cap (f > c)^\circ$ and $(1_C f^- > c) = C \cap (f < -c) = C \cap (f < -c)^\circ$ where the interiors are taken relative to K. Since K, C, and open subsets of K are measurable, $1_C f^+$ and $1_C f^-$ are measurable functions. $\square$

There are sets other than C for which the conclusion of Theorem 4 holds.

THEOREM 5. *Let f be a function on a cell K and E be a countable subset of K. Then $1_E f$ is a measurable function. If* *f has only countably many points of discontinuity then f is measurable.*

PROOF. Since E is countable every subset of E is measurable. Hence $K - E$ is measurable. Given c in $\mathbb{R}$ $(1_E f > c)$ is a subset of E if $c \geq 0$, the union of a subset of E with $K - E$ if $c < 0$. In either case $(1_E f > c)$ is measurable set. That is, $1_E f$ is a measurable function.

Consequently if the set D of points at which f is discontinuous is countable then $1_D f$ is measurable. $1_C f$ is measurable by Theorem 4. So $1_C f + 1_D f = f$ is measurable. $\square$

Lebesgue proved $f dx$ is Riemann-integrable on a bounded interval if and only if f is continuous dx- everywhere and bounded. The implication of integrability is a special case of the next result.

THEOREM 6. *Let σ be an integrable differential on a cell K. Let f be a function on K such that f is continuous σ-everywhere and $f\sigma$ is summable. Then $f\sigma$ is absolutely integrable.*

PROOF. The hypothetical conditions on f hold as well for f^+ and f^-. Also, absolute integrability of $f^+\sigma$ and $f^-\sigma$ implies absolute integrability of $f\sigma$. So we need only consider the case $f \geq 0$.

Let C be the set of points at which f is continuous. Since f is continuous σ-everywhere,

(2) $$\sigma = 1_C \sigma \text{ and } f\sigma = 1_C f\sigma.$$

For $n = 1, 2, \cdots$ let $B_n = (f > \frac{1}{n})$. Then $C \cap B_n = C \cap B_n^\circ$ for the interior relative to K. So by (2) $B_n - B_n^\circ$ is σ-null. Hence,

(3) $$1_{B_n} \sigma = 1_{B_n^\circ} \sigma$$

which is summable since $|1_{B_n}\sigma| \leq n|f\sigma|$ with $f\sigma$ summable. Apply Theorem 9 (§3.4) with the integrable $\sigma = dg$ and $B_n^\circ = D$ in (3) to conclude that

$$(4) \qquad \sigma_n = 1_{B_n}\sigma \text{ is absolutely integrable.}$$

Since $|\sigma_n| \leq |\sigma|, |f\sigma_n| \leq |f\sigma|$ which implies summability of $f\sigma_n$. By (2) and (4) $f\sigma_n = 1_C f\sigma_n$ which is absolutely integrable by Theorems 3 and 4. That is, $1_{B_n} f\sigma$ is absolutely integrable by (4). Apply the Monotone Convergence Theorem, Theorem 3 (§2.7), to σ^+ and σ^- with $f_n = 1_{B_n}f$ to conclude since $f_n \nearrow f$ with $f\sigma$ summable that $f_n\sigma \to f\sigma$. Therefore, since $f_n\sigma$ is absolutely integrable, so is $f\sigma$. $\square$

For measurable functions we can prove a dominated convergence theorem. The special case with integrable $\sigma \geq 0$ and $f_1 = 0$ is the Lebesgue Dominated Convergence Theorem.

THEOREM 7. *Let σ be a differential on a cell K. Let*

$$(5) \qquad f_n \to f$$

be a convergent sequence of measurable functions on K for which there is a function g on K such that $g\sigma$ is summable and

$$(6) \qquad |f_n - f_1| \leq g \text{ for } n = 1, 2, \cdots.$$

Then $f_n\sigma \to f\sigma$.

PROOF. Clearly we may assume $\sigma \geq 0$. Let $g_n = |f_n - f|$. Then g_n is measurable and $g_n \to 0$ by (5). We contend $g_n\sigma \to 0$ which proves the theorem.

By (5) and (6) $|f - f_1| \leq g$. So $0 \leq g_n \leq 2g$ since $g_n \leq |f_n - f_1| + |f_1 - f| \leq 2g$. Define the measurable function $G = g_1 \vee g_2 \vee \cdots \leq 2g$. Then define G_n on K by $G_n(t) = g_n(t)/G(t)$ if $G(t) > 0, G_n(t) = 0$ if $G(t) = 0$. Then

$$(7) \qquad G_n \text{ is measurable, } 0 \leq G_n \leq 1, \text{ and } G_n \to 0.$$

$G\sigma$ is summable since $0 \le G \le 2g$ and $g\sigma$ is summable. So Theorem 2 (§2.5) yields a function v on K such that $0 \le G\sigma \le dv$. Thus $0 \le g_n\sigma = G_nG\sigma \le G_ndv$. By (7) and Theorem 3 G_ndv is integrable. So Theorem 4 (§2.8) under (7) yields the conclusion $G_ndv \to 0$ which implies $g_n\sigma \to 0$.   □

For σ summable the norm $\nu(f\sigma)$ of differentials $f\sigma$ with f measurable behaves as it does in the case of integrable $|\sigma|$. Such behavior follows from our next theorem.

THEOREM 8. *Given σ summable on $K = [a, b]$ define w on K by $w(t) = \overline{\int}_a^t |\sigma|$. Then $dw \ge |\sigma|$ and*

$$(8) \quad \nu(f\sigma) = \int_K |f| dw \le \infty \text{ for every measurable } f \text{ on } K.$$

PROOF. Theorem 2 (§2.5) gives $dw \ge |\sigma|$ and implies

$$(9) \quad \underline{\int}_K udw \le \nu(u\sigma) \le \overline{\int}_K udw$$

for all bounded functions $u \ge 0$ on K. (See Exercise 7.) If such a function u is measurable then udw is integrable since it is clearly summable. (9) then becomes

$$(10) \quad \nu(u\sigma) = \int_K udw$$

for all bounded, measurable $u \ge 0$. Apply (10) to $u_k = |f| \wedge k$ for $k = 1, 2, \cdots$ and invoke the monotone convergence $u_k \nearrow |f|$ to conclude from Theorem 3 (§2.7) that

$$(11) \quad \nu(u_k\sigma) = \int_K u_k dw \nearrow \int_K |f| dw \le \infty.$$

Now $u_k|\sigma| \le |f\sigma| \le |f| dw$ since $u_k \le |f|$ and $|\sigma| \le dw$. So $\nu(u_k\sigma) \le \nu(f\sigma) \le \int_K |f| dw$ which together with (11) gives (8).   □

THEOREM 9. *Given σ summable on a cell K and measurable functions $v_i \geq 0$ on K for $i = 1, 2, \cdots$ let $v = v_1 + v_2 + \cdots \leq \infty$. Then*

$$\nu(v\sigma) = \sum_{i=1}^{\infty} \nu(v_i\sigma) \leq \infty.$$

PROOF. By Theorem 8 and Theorem 2 (§2.7)

$$\sum_i \nu(v_i\sigma) = \sum_i \int_K v_i dw = \int_K v dw = \nu(v\sigma).$$

☐

Similarly Theorem 8 yields the following extension of Theorem 3 (§2.7) on monotone convergence.

THEOREM 10. *Given σ summable on a cell K and measurable functions $0 \leq u_1 \leq u_2 \leq \cdots$ let $u = \lim_{n \to \infty} u_n \leq \infty$. Then $\nu(u_n\sigma) \nearrow \nu(u\sigma)$ as $n \nearrow \infty$.*

In Theorem 10 we can drop the demand that the u_n's be measurable if $|\sigma|$ is integrable. (See Theorem 3 in §4.6.)

Exercises (§4.3).

1. Given a function G on a cell K such that dG is damper-summable show that G is measurable. (Apply Theorem 5.)

2. Let $\rho_1, \rho_2, \cdots$ be a sequence of differentials on a cell K such that $\rho_k \to 0$ and $|\rho_k| \leq \sigma$ for all k. Let f be a measurable function on K such that $f\sigma$ is summable. Show that $f\rho_k \to 0$. (Apply Theorem 7 with $f_n = f1_{(|f|<n)}, g = |f|$, and $h = 0$. Verify and use the inequality $|f\rho_k| \leq |(f - f_n)\sigma| + n|\rho_k|$.)

3. Given σ summable on a cell K and a sequence $D_1, D_2, \cdots$ of disjoint, measurable subsets of K show that

$$\sum_{i=1}^{\infty} \nu(1_{D_i}\sigma) = \nu(1_D\sigma) < \infty \text{ for } D = D_1 \cup D_2 \cdots.$$

(So σ defines a countably additive measure $\nu(1_E\sigma)$ on the measurable subsets E of K.)

4. Show that a continuous function g on a cell K is of bounded variation if and only if $f\,dg$ is integrable for every bounded, measurable function f on K. (Use Exercise 2 in §2.6.)

5. Given a dampable differential σ and measurable function f on a cell K show that $f\sigma$ is dampable.

6. Let $\sigma = 1_E dx$ on $K = [0, 1]$ where the subset E of K is not dx-measurable. Show that σ is summable and continuous but not dampable. (Apply Theorem 3.)

7. Verify the first statement in the proof of Theorem 8.

8. Prove that a differential σ on a cell K is dampable by a measurable damper if and only if $K = \bigcup_{n=1}^{\infty} E_n$ where E_n is measurable and $1_{E_n}\sigma$ is absolutely integrable for all n.

9. Let $\sigma \geq 0$ be an integrable differential on a cell K. Let τ be a differential on K such that every σ-null subset of K is τ-null, and given a measurable subset A of K such that $1_A\sigma \neq 0$ there exists a measurable subset B of A such that $1_B\sigma \neq 0$ and $1_B\tau$ is absolutely integrable. Prove that τ is dampable by a measurable damper.

§4.4 Step Functions and Regulated Functions.

Bounded measurable functions g on a cell K have the useful property that $g\,dh$ is absolutely integrable for every function h of bounded variation on K. We examine here two special cases of such functions g, step functions and regulated functions. These have concise definitions in terms of their differentials.

A function g on a cell K is a **step function** if

(1) $dg = 1_E dg$ for some finite subset E of K.

g is a **regulated function** on K if

(2) $1_p dg$ is integrable over K for every point p in K.

There are many characterizations of step functions and regulated functions. We shall present only a few here. Others are relegated to the exercises at the end of this section.

THEOREM 1. *(Step Functions.) For g a function on a cell K the following conditions are equivalent:*
(i) g is a a step function. (That is, (1) holds.)

(ii) K is a finite union of disjoint intervals $I_1, \cdots, I_n$ (some of which may be degenerate) such that g is constant on each I_i.

(iii) The range $g(K)$ of g and the set D of points at which g is discontinuous are finite sets.

PROOF. (i) $\Rightarrow$ (ii). Given (1) we may assume E contains the end points a, b of $K = [a, b]$. Then

(3)
$$E = \{e_0, e_1, \cdots, e_m\} \text{ with } a = e_0 < e_1 < \cdots < e_m = b$$
$$\text{and } K - E = L_1 \cup \cdots \cup L_m \text{ for } L_j = (e_{j-1}, e_j).$$

By (1) $dg = 0$ on L_j, so g is constant on L_j for $j = 1, \cdots, m$. Obviously, g is constant on each degenerate interval $J_j = [e_j, e_j]$ for $j = 0, \cdots, m$. So (ii) holds for $\{I_1, \cdots, I_n\} = \{L_1, \cdots, L_m, J_0, \cdots, J_m\}$.

(ii) $\Rightarrow$ (iii). By (ii) $g(K)$ has at most n points. So $g(K)$ is finite. Since g is constant on each I_i it is continuous on each I_i. So each point in D must be an endpoint of some I_i. Hence, D has at most $n + 1$ points.

(iii) $\Rightarrow$ (i). Take E in the form (3) consisting of the finite set D and the endpoints of K. Then g is continuous on each L_j in (3). Also , $g(L_j)$ is finite since $g(K)$ is finite. So $g(Lj)$ is connected and finite, hence consists of a single point. That is,

g is constant on each L_j. So each L_j is dg-null. Hence, their union $K - E$ in (3) is dg-null. So $1_E dg = dg$. □

By (1) it is obvious that the step functions form a linear space. That this is a Riesz space follows from (ii) since $|g|$ is constant on each interval on which g is constant. So $|g|$ is a step function if g is a step function. From (iii) it is obvious that fg is a step function if f and g are step functions.

According to Theorem 4 (§2.7) our definition (2) of regulated function on $K = [a, b]$ is equivalent to

(4) $\quad \begin{cases} g(p+) \text{ exists and is finite for } a \le p < b, \text{ and} \\ g(p-) \text{ exists and is finite for } a < p \le b. \end{cases}$

Indeed, with the notational convention $a- = a$ and $b+ = b$ Theorem 4 (§2.7) gives

$$(5) \qquad \int_a^b 1_p dg = g(p+) - g(p-)$$

where existence of either side implies existence of the other. Moreover, by Theorem 5 (§2.7), if $1_p dg$ is integrable then it is absolutely integrable and

$$(6) \qquad \int_a^b 1_p |dg| = |g(p) - g(p-)| + |g(p+) - g(p)|$$

where $a- = a$ and $b+ = b$.

A condition equivalent to (4) is given by the Cauchy criterion for the existence of the limits in (4),

(7) $\quad$ Given $\varepsilon > 0$ there exists a gauge δ on K such that diam $g(I°) < \varepsilon$ for every δ-fine tagged cell (I, p) in K.

So (2), (4), and (7) are equivalent characterizations of regulated functions. Since finite sets are measurable every function of bounded variation on K is regulated according to (2). (2) also implies that continuous functions are regulated since $1_p dg = 0$ for g continuous.

For g a step function on K the set E in (1) is finite. So each point p in E is interior (relative to K) to some cell I in K containing no other points in E. For such p and I, $\Delta g(I) = \int_I dg = \int_I 1_E dg = \int_I 1_p dg = \int_K 1_p dg$. So $1_p dg$ is integrable for all p in E. At all other p in K (1) implies $1_p dg = 0$ which is integrable. So (2) holds. That is, every step function is regulated. Furthermore, since integrability of $1_p dg$ is absolute integrability and $dg = \sum_{p \in E} 1_p dg$ over the finite set E in (1), dg is absolutely integrable. That is, every step function is of bounded variation.

Under definition (2) it is clear that the regulated functions on K form a linear space. Also, since integrability of $1_p \, dg$ is absolute integrability, $|g|$ is regulated if g is regulated. So the regulated functions on K form a Riesz space. The uniform norm

$$(8) \qquad \|g\| = \sup_{t \in K} |g(t)|$$

is a Riesz norm on this space since (7) assures us that every regulated function on K is bounded. Indeed, g is bounded on I for every (I, t) belonging to a δ-division of K according to (7).

We contend that the Riesz space of regulated functions on K is a Banach lattice under (8). We need only prove completeness which we do by showing that the regulated functions on K form a *closed* subspace of the Banach lattice of all bounded functions on K under (8).

THEOREM 2. *A uniform limit g of regulated functions on a cell K is regulated.*

PROOF. We need only show that g satisfies (4).

Given $a \leq p < b$ in $K = [a, b]$ we contend $g(p+)$ exists and is finite. By the Cauchy criterion (7) this is equivalent to

$$(9) \qquad g(s) - g(t) \to 0 \text{ as } s, t \to p+.$$

To prove (9) let $\varepsilon > 0$ be given. By hypothesis $|g - h| < \varepsilon$ for some regulated function h. So

$$(10) \quad \begin{aligned} |g(s) - g(t)| &\leq \\ |g(s) - h(s)| + |h(s) - h(t)| &+ |h(t) - g(t)| < \\ \varepsilon + |h(s) - h(t)| &+ \varepsilon. \end{aligned}$$

Since h is regulated it satisfies (9), $h(s) - h(t) \to 0$ as $s, t \to p+$. So (10) gives $\varlimsup_{s,t \to p+} |g(s) - g(t)| \leq 2\varepsilon$. Since ε is arbitrary this gives (9).

A similar proof for $p-$ gives (4). $\square$

The next result shows that step functions are dense in the Banach lattice of regulated functions on K.

THEOREM 3. *A function g on a cell K is regulated if and only if g is a uniform limit of step functions on K.*

PROOF. Since step functions are regulated, a uniform limit of step functions is a regulated function by Theorem 2.

To prove the converse let g be regulated . Consider any $\varepsilon > 0$. Apply (7) to get δ such that diam $g(I^\circ) < \varepsilon$ for every δ-fine (I, p). Take a δ-division $\{(I_1, p_1), \cdots, (I_n, p_n)\}$ of K and let $e_0, e_1, \cdots, e_n$ be the endpoints of $I_1, \cdots, I_n$. Take any b_i in I_i° for $i = 1, \cdots, n$. Define the step function h on K by setting $h = g$ on $e_0, \cdots, e_n$ and $h = g(b_i)$ on I_i° for $i = 1, \cdots, n$. For all t in I_i°, $|g(t) - h(t)| = |g(t) - g(b_i)| \leq$ diam $g(I_i^\circ) < \varepsilon$ by (7). Hence, since $g = h$ on the complement $\{e_0, \cdots, e_n\}$ of $I_1^\circ \cup \cdots \cup I_n^\circ$ in K, $|g - h| < \varepsilon$. $\square$

From Theorem 3 we can conclude that a regulated function g has only countably many discontinuities since a step function has only finitely many. So measurability of g follows from the last statement in Theorem 5 (§4.3).

Given a regulated function g on a cell $K = [a, b]$ define the functions g_- and g_+ on K by

(11) $$g_-(t) = g(t-) \text{ and } g_+(t) = g(t+)$$

with the notational convention

(12) $a- = a$ and $b+ = b$ at the endpoints a, b, of K.

It is easy to verify that g_- and g_+ are regulated functions on K with g_- left continuous on $(a, b]$ and g_+ right continuous on $[a, b)$. Indeed

(13) $$g_{--} = g_{+-} = g_- \text{ on } (a, b]$$

and

(14) $$g_{++} = g_{-+} = g_+ \text{ on } [a, b).$$

At the endpoints a, b of K

(15) $$1_a dg_- = 1_a dg, 1_b dg_- = 0$$

and

(16) $$1_b dg_+ = 1_b dg, 1_a dg_+ = 0.$$

For $a \leq s < b$

(17) $$\int_K 1_s dg_- = (g_+ - g_-)(s), \int_K 1_s |dg_-| = |g_+ - g_-|(s).$$

For $a < s \leq b$

(18) $$\int_K 1_s dg_+ = (g_+ - g_-)(s), \int_K 1_s |dg_+| = |g_+ - g_-|(s).$$

For all s in K (11) and (12) give

(19) $$\int_K 1_s dg = (g_+ - g_-)(s)$$

with

(20) $$\int_K 1_s |dg| = (|g - g_-| + |g_+ - g|)(s).$$

By the Monotone Convergence Theorem (Theorem 3, §2.7) summation of (20) over all s in the countable set D of discontinuities of g gives

(21) $$\int_K 1_D |dg| = \sum_{s \in D} (|g - g_-| + |g_+ - g|)(s) \le \infty$$

Similarly the second equations in (15), (16), (17), and (18) give

(22) $$\int_K 1_D |dg_-| = \sum_{s \in D \cap [a,b)} |g_+ - g_-|(s)$$

and

(23) $$\int_K 1_D |dg_+| = \sum_{s \in D \cap (a,b]} |g_+ - g_-|(s).$$

By (22), (23), (21), and the triangle inequality

(24) $$\int_K 1_D |dg_-| \le \int_K 1_D |dg|$$

and

(25) $$\int_K 1_D |dg_+| \le \int_K 1_D |dg|.$$

For g of bounded variation both g_- and g_+ are of bounded variation according to the following theorem.

THEOREM 4. *Let g be a regulated function on $K = [a, b]$ such that $\int_K 1_D |dg| < \infty$ for D the set of all points at which g is discontinuous. Then for the set C of all points at which g is continuous*

(i) $1_C dg_- = 1_C dg = 1_C dg_+.$

So

(ii) $\int_K |dg_-| \leq \int_K |dg|$ and $\int_K |dg_+| \leq \int_K |dg|.$

PROOF. Let $\varepsilon > 0$ be given. Since $1_D dg$ is summable (20) yields a finite subset E of D large enough so that

$$(26) \qquad \int_K 1_{D-E} |dg| < \varepsilon.$$

Take a gauge δ on K small enough to ensure that I is disjoint from E for every δ-fine tagged cell (I, t) with t in C. (Such gauges exist because E is closed and disjoint from C.) For such (I, t) the other endpoint s of I belongs to $K - E, g_-(t) = g(t)$, and

$$(27) \qquad \begin{aligned} |\Delta(g - g_-)(I)| &= |(g(t) - g_-(t)) - (g(s) - g_-(s))| = \\ |g - g_-|(s) &\leq \int_K 1_s |dg| \end{aligned}$$

by (20).

For any δ-division $\mathcal{K}$ of K summation of (27) over (I, t) in $\mathcal{K}$ with t in C gives

$$(28) \qquad \left(\sum 1_C |\Delta(g - g_-)| \right)(\mathcal{K}) \leq 2\varepsilon$$

by (26). The doubling of ε is needed in (28) because some s in (27) may be an endpoint of two abutting members of $\mathcal{K}$. Since (28) holds for all δ-divisions $\mathcal{K}$ of K we have the first equation in (i). The second follows by a similar proof.

From (i) and (24) we conclude that $\int_K |dg_-| = \int_K 1_C |dg_-| + \int_K 1_D |dg_-| \leq \int_K 1_C |dg| + \int_K 1_D |dg| = \int_K |dg|$ which gives the first inequality in (ii). The second follows similarly from (i) and (25). □

In Theorem 4 the demand that $1_D dg$ be summable cannot be dropped from the hypothesis without replacement by some other suitable condition. For the regulated function g in Exercise 5 below both $1_D dg$ and $1_C dg$ fail to be summable. (Nonsummability of $1_D dg$ is easy to prove. That of $1_C dg$ is much harder; a clever application of the Baire Category Theorem (§11.4) effects a proof.) Now $1_C dg_- = 1_C dg_+ = 0$ but the nonsummable $1_C dg \neq 0$. So the conclusion (i) of Theorem 4 is false for this function g.

Exercises (§4.4).

1. For g a function on a cell K prove the equivalence of the following conditions:

   (i) g is a step function. (i.e. (1) holds.)

   (ii) For every sequence $I_1, I_2, \cdots$ of nonoverlapping cells in K, $dg = 0$ on I_n ultimately as $n \to \infty$.

   (iii) Condition (ii) holds with "nonoverlapping" replaced by "disjoint".

   (iv) g is of bounded variation and its range $g(K)$ is finite.

   (v) g is regulated and its range $g(K)$ is finite.

   (vi) $f dg$ is absolutely integrable on K for every function f on K.

2. For g a function on a cell K prove the equivalence of the following conditions:

   (i) g is regulated.

   (ii) For all points t in K, diam $g(I - t) \to 0$ as $(I, t) \to t$.

(iii) $1_L dg$ is integrable for every open interval $L = (p, q)$ in K.

(iv) $1_L dg$ is integrable for every nondegenerate interval L in K.

(v) dg is integrable over every nondegenerate interval L in K.

(vi) $f dg$ is integrable on K for every step function f on K.

(vii) $\int_{I_n} dg \to 0$ for every sequence $I_1, I_2, \cdots$ of nonoverlapping cells in K.

(viii) Condition (vii) holds with "nonoverlapping" replaced by "disjoint". (See also Exercise 11 (§3.3). Another condition is given in Exercise 7 (§6.2).)

3. Prove that if g is regulated on a cell K then $1_{p_n} dg \to 0$ for every sequence $p_1, p_2, \cdots, p_n, \cdots$ of distinct points in K.

4. For g defined on $K = [0, 1]$ by

$$g(t) = \begin{cases} 0 \text{ if } t = 0 \\ \sin \frac{1}{t} \text{ if } t > 0 \end{cases}$$

show that $1_I dg$ is integrable on K for every cell I in K, but g is not regulated.

5. Define g on $[0,1]$ by

$$g(t) = \begin{cases} \frac{1}{n} \text{ if } t \text{ is rational and } n \text{ is the smallest positive} \\ \quad \text{integer such that } nt \text{ is an integer.} \\ 0 \text{ if } t \text{ is irrational.} \end{cases}$$

Show that g is regulated by finding a sequence of step functions converging uniformly to g on $[0,1]$.

6. (Saltus functions of bounded variation.) For g a function on a cell K prove the equivalence of the two conditions:

(i) $dg_n \to dg$ for some sequence of step functions $g_1, \cdots, g_n,$ $\cdots$ on K.

(ii) g is of bounded variation and $1_D dg = dg$ for some countable subset D of K.

7. Given $dF = f dg$ on a cell K prove:

   (i) If g is a step function then so is F,

   (ii) If g is regulated then so is F.

8. Let F be regulated on a cell K. Let $L_1, L_2, \cdots$ be a sequence of intervals in K converging to an interval L. Prove $\int_K 1_{L_n} dF \to \int_K 1_L dF$ but the conclusion that $\int_{L_n} dF \to \int_L dF$ may be false.

9. For F regulated on a cell K prove the equivalence of:

   (i) $1_D dF = dF$ for some countable subset D of K,

   (ii) $dF = [(F - F_-)Q^- + (F_+ - F)Q^+]\omega$, for Q defined by (20) in §1.4 and $\omega = [1]$,

   (iii) F is discontinuous dF-everywhere.

10. For F a function on a cell K prove the equivalence of (i) and (ii):

    (i) F is regulated, $1_D dF = dF$ for some countable subset D of K, and $F = \frac{1}{2}(F_- + F_+)$ on the interior of K.

    (ii) $dF = f\omega$ for some function f on K. (Use Exercise 9.)

11. For f a function on $K = [a, b]$ prove that $f\omega$ is integrable if and only if there exists a regulated function F on K such that $f^{-1}(0)$ is dF-null, its complement $K - f^{-1}(0)$ is countable, $F - F_- = f$ on $(a, b]$, and $F_+ - F = f$ on $[a, b)$. (Use Exercise 10.)

12. For g a function on $K = [a, b]$ prove the equivalence of the following three conditions:

(i) For each $\varepsilon > 0$ the set $(|g| > \varepsilon)$ is finite.

(ii) $g(t_n) \to 0$ for every sequence $t_1, \cdots, t_n, \cdots$ of distinct points in K.

(iii) g is regulated on K, $g_- = 0$ on $(a, b]$ and $g_+ = 0$ on $[a, b)$.

Under the additional restriction that $g(a) = 0$ show that each of the preceding conditions is equivalent to

(iv) $\int_K f dg = 0$ for every step function f on K. (Use (vi) in Exercise 2.)

13. For g of bounded variation on $K = [a, b]$ prove the equivalence of the following three conditions:

(i) $1_D g = g$ for some countable subset D of (a, b).

(ii) $g_- = g_+ = 0$ on K.

(iii) $g(a) = 0$ and $\int_K f dg = 0$ for every bounded function f on K.

14. Show that for every point p in $[a, b]$:

(i) $(1_p)_- = 1_a(p)1_a$ and (ii) $(1_p)_+ = 1_b(p)1_b$.

15. Given a cell $[r, s]$ in $K = [a, b]$ let $F = 1_{(r,b]}$ and $H = 1_{[s,b]}$ on K. Show that for every regulated function g on K:

(i) $\int_K (g + dg) dF = g(r+)$,

(ii) $\int_K (g - dg) dH = g(s-)$.

16. Let $g = g_-$ be a regulated, left continuous function on $[a, b]$. Given $\varepsilon > 0$ show that there exists a finite sequence $\{e_0, \cdots, e_n\}$ such that $a = e_0 < \cdots < e_n = b$ and $|g - h| < \varepsilon$ for the left continuous step function $h = g(a)1_a + \sum_{i=1}^n g(e_i)1_{(e_{i-1}, e_i]}$.

17. Given a continuous linear functional Φ on the Banach lattice of all regulated functions g on $K = [a, b]$ define the functions F and f on K by

(⋆) $F(s) = \Phi(1_{[a,s]})$ and $f(s) = \sum_{r \in [a,s]} \Phi(1_r)$.

(The definition of f is valid because the defining sum has only countably many nonzero terms, and these form an absolutely convergent sum.) Now F and f are of bounded variation, $f_+ = f$, and $(f - f_-)(t) = 1_{(a,b]}(t)\Phi(1_t)$ for all t in K. Moreover, it can be shown that for every regulated function g on K

(†) $\Phi(g)$ $=$ $(Fg_-)(b)$ $-$ $\int_K F dg_-$ $+$ $\int_K (g - g_-)df$

where

$(g - g_-)df = dgdf$.

Let p be any point in K.

(i) Let Φ be defined by $\Phi(g) = g(p)$ for every regulated function g on K. Show that (⋆) gives $F = f = 1_{[p,b]}$ and verify (†) for this case.

(ii) For $\Phi(g) = g(p-)$ show that (⋆) gives $F = 1_{[p,b]}$ and $f = 1_a$. Verify (†) for this case.

(iii) For $\Phi(g) = g(p+)$ show that (⋆) gives $F = 1_{(p,b]} + 1_b(p)1_b$ and $f = 1_b$. Verify (†) for this case.

18. Given a regulated function g on a cell K prove that g_- is of bounded variation if and only if g_+ is of bounded variation.

19. Given a regulated function g on $K = [a, b]$ define h on K by $h = g$ at the endpoints a, b of K and $h = \frac{1}{2}(g_+ + g_-)$ on the interior (a, b) of K. Prove:

(i) $h_- = g_-$ and $h_+ = g_+$ on K under (11), (12).

(ii) $h = \frac{1}{2}(h_+ + h_-)$ on (a, b).

(iii) At all points p in (a, b)

$$\int_a^p 1_p dh = \int_p^b 1_p dh = \frac{1}{2}(h_+ - h_-)(p).$$

(iv) For all p in K

$$\int_K 1_p|dh| = |\int_K 1_p dh| = |h_+ - h_-|(p).$$

20. Let g be regulated on $[a, b]$. Let C (alternatively C_-, C_+) be the set of points at which g (respectively g_-, g_+) is continuous.
Verify:

(i) $C = (g_- = g_+ = g)$.

(ii) $C_- = (g_- = g_+) \cup \{b\}$

(iii) $C_+ = (g_- = g_+) \cup \{a\}$

(iv) $C_- \cap (a, b) = C_+ \cap (a, b) = (g_- = g_+) \cap (a, b)$

(v) $C \subseteq C_- \cap C_+$.

21. Let g be a regulated function on a cell K with D the set of points at which g is discontinuous. Show that if the set E of limit points of D is finite then the conclusions of Theorem 4 hold. (To get (i) on $K - E$ apply Theorem 4 to an arbitrary cell in it. Since E is finite, $1_{C \cap E} dg_- = 1_{C \cap E} dg_+ = 1_{C \cap E} dg = 0$.)

22. (Integrable Idempotents.) For g a function on a cell K prove the equivalence of the three conditions:

(i) g is bounded and $(dg)^2 = dg$ on K,

(ii) g is a step function, $g - g_- = 1$ at every left discontinuity of g, and $g_+ - g = 1$ at every right discontinuity of g,

(iii) There exists a gauge δ on K such that $(\Delta g)^2 = \Delta g$ at every δ-fine tagged cell in K. (See Exercise 12 in §3.1.)

23. Let g be a function on a cell K and p be a point in K. Prove:

(i) $d1_p = -1_p Q\omega = 1_p d1_p$,

(ii) $d(g1_p) = g(p)d1_p = gd1_p$.

iii) If g is regulated then $1_p dg = (g - g_- Q^- - g_+ Q^+)d1_p$.

(iv) If g is regulated and $g_- = g_+ = 0$ then $1_p dg = gd1_p$ and the differential $dg + gQ\omega$ is continuous on K.

(v) If g is regulated and $g = 1_D g$ for some countable subset D of K then $1_D dg = -gQ\omega$ and $1_D |dg| = |g|\omega$. (Use (iv).)

(vi) If g is a step function then $dg = (g - g_- Q^- - g_+ Q^+)d1_E$ for some finite subset of E of K. (Use (iii).)

24. For g regulated on a cell K and $[p, q]$ a cell or point in K prove:

(i) $\int_K 1_{[p,q)} dg = \int_p^q dg_-$,

(ii) $\int_K 1_{(p,q]} dg = \int_p^q dg_+$.

25. Let g be regulated on a cell K such that dg is dampable and $(g - g_-)(g_+ - g) \geq 0$. Show that dg has a Hahn decomposition on K. (Use Theorem 2 in §4.2.)

26. Let f be of bounded variation and g be regulated on $K = [a, b]$. Prove:

(i) For all points p in K the differential $1_p df dg$ is absolutely integrable,

$$\int_K 1_p df dg = [(f_+ - f)(g_+ - g) + (f - f_-)(g - g_-)](p),$$

$$\int_K |1_p df dg| = [|f_+ - f||g_+ - g| + |f - f_-||g - g_-|](p).$$

(ii) $df dg$ is absolutely integrable on K with $\int_K df dg = \sum_{p \in D} \int_K 1_p df dg$ and $\int_K |df dg| = \sum_{p \in D} \int_K |1_p df dg|$ where D is the set of all points at which both f and g are discontinuous. (Apply Theorem 2 in §2.7.)

27. Show that if g is a function on a cell K with dg dampable then g is regulated. (See Exercise 5 in §3.3.)

28. Given a sequence $a_1 > a_2 > \cdots$ in $\mathbb{R}$ such that $a_n \searrow 0$ define the regulated function $g = 1_D x$ on $\mathbb{R}$ where $D = \{a_1, a_2, \cdots\}$. Let f be any function on $\mathbb{R}$. Prove:

    (i) $dg = 1_D dg = -gQ\omega$,

    (ii) $f dg = d(fg)$,

    (iii) $\int_{\mathbb{R}} f dg = 0$,

    (iv) $\int_{\mathbb{R}} |f dg| = 2 \sum_1^\infty a_n |f(a_n)| \leq \infty$,

    (v) g is of bounded variation if and only if $\sum_1^\infty a_n < \infty$.

§4.5 The Radon-Nikodym Theorem for Differentials.

THEOREM 1. *Let σ and τ be continuous, dampable differentials on a cell K. Then the following conditions are equivalent:*

(i) *Every measurable σ-null subset of K is τ-null.*

(ii) *$\tau = f\sigma$ for some function f on K.*

(iii) *Every σ-null subset of K is τ-null.*

PROOF. Theorem 1 (§4.2) yields a function w on K such that $w^2 = 1$ and $\sigma = w|\sigma|$. So (ii) is equivalent to

(1) $\tau = g|\sigma|$ for some function g on K.

Indeed, (ii) gives (1) for $g = fw$ while (1) gives (ii) for $f = gw$. Since σ is dampable and continuous there exist a damper v and a continuous, integrable differential $\rho \geq 0$ such that $|\sigma| = v\rho$. Thus (1) is equivalent to $\tau = gv\rho$, hence to

(2) $\tau^+ = g^+ v\rho, \tau^- = g^- v\rho$ for some function g on K.

Since τ is dampable there is a damper u such that $u\tau$ is absolutely integrable. In terms of (2) condition (ii) is equivalent to $u\tau^+ = (ug^+v)\rho$, $u\tau^- = (ug^-v)\rho$ for some g on K. Now

$u\tau^{+}, u\tau^{-}$, and ρ are continuous and integrable. The ρ-null sets are precisely the σ-null sets.

Consequently to prove that (i) implies (ii) we need only treat the special case of continuous, integrable σ, τ with $\sigma \geq 0, \tau \geq 0$. This case applies to the pairs $u\tau^{+}, \rho$ and $u\tau^{-}, \rho$ to give the general case.

So let (i) hold with $\sigma \geq 0, \tau \geq 0$ integrable and continuous. Let c be the supremum of $\int_K f\sigma$ taken over the set Φ of all functions $f \geq 0$ on K such that $f\sigma$ is integrable and $f\sigma \leq \tau$. Since 0 belongs to $\Phi, 0 \leq c \leq \int_K \tau$. We contend that the supremum c is actually a maximum.

If f, g belong to Φ then so does $f \vee g$. Indeed, since $\sigma \geq 0, (f \vee g)\sigma = (f\sigma) \vee (g\sigma)$ which belongs to the Banach lattice of absolutely integrable differentials on K. (See §2.4) So $(f \vee g)\sigma$ is integrable, $f \vee g \geq 0$, and $(f \vee g)\sigma \leq \tau$. That is, $f \vee g$ belongs to Φ.

Thus we can choose a monotone sequence $f_1 \leq f_2 \leq \cdots$ in Φ such that $\int_K f_n\sigma \nearrow c$ as $n \nearrow \infty$. By the Monotone Convergence Theorem (Theorem 3, §2.7) there is a function f on K such that $f_n \nearrow f$ σ-everywhere and $\int_K f_n\sigma \nearrow \int_K f\sigma$. Therefore

$$(3) \qquad \int_K f\sigma = c.$$

Moreover, since $f_n\sigma \nearrow f\sigma$ and $f_n\sigma \leq \tau$ for all n,

$$(4) \qquad f\sigma \leq \tau.$$

So f belongs to Φ. To get (ii) we must show that equality holds in (4).

Given $\varepsilon > 0$ in $\mathbb{R}$ apply Theorem 1 (§4.2) to get a Hahn decomposition for the continuous, absolutely integrable differential $\tau - (f+\varepsilon)\sigma$. That is, we have complementary, measurable subsets A, B of K such that

$$(5) \qquad 1_A[\tau - (f + \varepsilon)\sigma] \geq 0$$

and

(6) $$1_B[\tau - (f+\varepsilon)\sigma] \le 0.$$

By (5)

(7) $$1_A(f+\varepsilon)\sigma \le 1_A\tau.$$

By (4)

(8) $$1_B f\sigma \le 1_B\tau.$$

By addition of (7) and (8) we get $(f+\varepsilon 1_A)\sigma \le \tau$. So $f+\varepsilon 1_A$ belongs to Φ since $1_A\sigma$ is integrable. Hence, $\int_K(f+\varepsilon 1_A)\sigma \le c = \int_K f\sigma$ by (3). This implies $1_A\sigma = 0$. Therefore $1_A\tau = 0$ by (i). That is, $1_B\sigma = \sigma$ and $1_B\tau = \tau$ which imply by (4) and (6) that

(9) $$0 \le \tau - f\sigma \le \varepsilon\sigma.$$

Since (9) holds for all $\varepsilon > 0$ and σ is integrable, $\tau - f\sigma = 0$. That is, (ii) holds.

So (i) implies (ii). The implications (ii) $\Rightarrow$ (iii) $\Rightarrow$ (i) are obvious. $\square$

We remark that the demand that τ be continuous in the hypothesis of Theorem 1 is redundant since continuity of τ follows from continuity of σ under each of the conditions (i), (ii), (iii).

Exercises (§4.5).

1. Show that Theorem 1 remains valid if we adjoin the condition: (iv) $\tau = \rho\sigma$ for some tag-finite differential ρ on K.

2. Let g, h be functions on $\mathbb{R}$ such that $h = 0$ at dx-all points where $g = 0$. Let $\sigma = gdx$ and $\tau = hdx$ on $\mathbb{R}$. Show that Theorem 1 yields the conclusion that $f = h/g$ σ-everywhere.

§4.6 Minimal Measurable Dominators.

THEOREM 1. *Let $f dv$ be summable on a cell K where $f \geq 0$ and $dv \geq 0$. Then there exists a measurable function $\bar{f} \geq 0$ on K such that: (i) $\bar{f} dv$ is integrable on K, (ii) $\int_K \bar{f} dv = \overline{\int_K} f dv$, (iii) $\bar{f} \geq f$ dv-everywhere, and (iv) $g \geq \bar{f}$ dv-everywhere for every measurable function g on K such that $g \geq f$ dv-everywhere,*

PROOF. Define F on $K = [a, b]$ by

$$(1) \qquad F(t) = \overline{\int_a^t} f dv$$

which is finite since $f dv$ is summable. By Theorem 2 (§2.5)

$$(2) \qquad \int_I dF = \overline{\int_I} f dv \text{ for every cell } I \text{ in } K$$

and
(3)
dF is the smallest integrable differential such that $dF \geq f dv$.

Let D be the set of all points in K at which the monotone function v is discontinuous. Since D is countable $1_D f$ is measurable. Therefore, since $f dv$ is summable, $1_D f dv$ is integrable. So there exists a function F_D on K such that

$$(4) \qquad dF_D = 1_D f dv \text{ and } F_D(a) = 0.$$

Let C be the complement of D in K. That is, C is the set of all points in K at which v is continuous. Define F_C on K by

$$(5) \qquad F_C(t) = \overline{\int_a^t} 1_C f dv.$$

By the definition of C, $1_C dv$ is continuous. Since C is measurable, $1_C dv$ is integrable. By Theorem 2 (§2.5) dF_C is the smallest integrable differential such that $dF_C \geq 1_C f dv$. So $1_C dF_C = dF_C$ since $1_C dF_C$ is integrable by measurability of C, and $1_C dF_C \geq 1_C f dv$.

Now consider any measurable $1_C dv$-null set A. Its complement B in K is measurable and $1_B 1_C dv = 1_C dv$. So $1_B dF_C$ is integrable and $1_B dF_C \geq 1_C f dv$. Hence, $1_B dF_C = dF_C$ by the minimality of dF_C. That is, $1_A dF_C = 0$. This proves that every measurable $1_C dv$-null set A is dF_C-null. By the Radon-Nikodym Theorem (Theorem 1 (§4.5)) for $\sigma = 1_C dv$ and $\tau = dF_C$

(6) $dF_C = 1_C \tilde{f} dv$ for some measurable $\tilde{f}$.

Since $dF_C = (dF_C)^+ = 1_C \tilde{f}^+ dv$ we may assume $\tilde{f} \geq 0$, in effect replacing $\tilde{f}$ by $\tilde{f}^+$. Since 1_C and $\tilde{f}$ are measurable, so is $1_C \tilde{f}$. Hence, since $1_D f$ is measurable, $\bar{f}$ defined by

(7) $\bar{f} = 1_C \tilde{f} + 1_D f$

is measurable.

Now $\overline{\int_a^t} f dv = \overline{\int_a^t} (1_C f + 1_D f) dv = \overline{\int_a^t} 1_C f dv + \int_a^t 1_D f dv$ since $1_D f dv$ is integrable. That is, $F(t) = F_C(t) + F_D(t)$ by (1), (5), (4). So $dF = dF_C + dF_D = 1_C \tilde{f} dv + 1_D f dv = \bar{f} dv$ by (6), (4), (7). That is,

(8) $dF = \bar{f} dv$.

(i) follows from (8), (ii) from (8) and (2). Now $1_C \tilde{f} dv = dF_C \geq 1_C f dv$ by (6), (5), and Theorem 2 (§2.5). Adding $1_D f dv$ to both sides of this inequality and invoking (7) we get $\bar{f} dv \geq f dv$ which is equivalent to (iii).

Finally, let g be measurable and

(9) $g \geq f$ dv-everywhere on K.

Since both g and $\bar{f}$ are measurable so is $g \wedge \bar{f}$. Hence $(g \wedge \bar{f})dv$ is integrable since summability follows from $(g \wedge \bar{f})dv \leq \bar{f}dv$ and (i). Thus by (3), (8), (9) $dF \geq (g \wedge \bar{f})dv \geq fdv$. By (3) and (8) $\bar{f}dv = dF = (g \wedge \bar{f})dv$. That is, $(\bar{f} - g)^+ dv = 0$ which is just $(\bar{f} - g)^+ = 0$ dv-everywhere. Equivalently, $g \geq \bar{f}$ dv-everywhere which proves (iv). □

THEOREM 2. *Let σ and $f\sigma$ be absolutely integrable differentials on a cell K. Then $f\sigma = \bar{f}\sigma$ for some measurable function $\bar{f}$ on K. That is, $f = \bar{f}$ σ-everywhere.*

PROOF. We need only consider the case $f \geq 0, \sigma = dv \geq 0$ which applies to each of the four cases in which f^+ or f^- is paired with σ^+ or σ^-. Combining the results of these four cases yields the general result since $\sigma^+ \wedge \sigma^- = 0$.

For $f \geq 0$ and $\sigma = dv \geq 0$ Theorem 1 yields a measurable $\bar{f} \geq 0$ satisfying (i), $\cdots$, (iv). Since fdv is integrable (ii) and (iii) imply $v[(\bar{f} - f)\sigma] = \int_K (\bar{f} - f)dv = \int_K \bar{f}dv - \int_K fdv = 0$. So $(\bar{f} - f)\sigma = 0$. That is, $\bar{f}\sigma = f\sigma$. □

Theorem 1 enables us to extend Theorem 3 (§2.7) on monotone convergence to upper integrals.

THEOREM 3. *Let $dv \geq 0$ on a cell K. Let $0 \leq f_1 \leq f_2 \leq \cdots$ be an ascending sequence of nonnegative functions on K such that $f_n \nearrow f \leq \infty$. Then*

$$(10) \qquad \overline{\int_K} f_n dv \nearrow \overline{\int_K} f dv \text{ as } n \nearrow \infty.$$

PROOF. By monotoneity

$$(11) \qquad \overline{\int_K} f_n dv \nearrow c$$

where $c \leq \overline{\int_K} fdv \leq \infty$. So we need only treat the case $c < \infty$. For this case Theorem 1 yields for each n a minimal measurable

$\bar{f}_n \geq f_n \, dv$-everywhere such that

(12)
$$\int_K \bar{f}_n dv = \overline{\int_K f_n dv} \leq c.$$

Since the measurable function $\bar{f}_n \wedge \bar{f}_{n+1} \wedge \cdots \geq f_n \, dv$-everywhere the minimality of $\bar{f}_n$ implies $\bar{f}_n \leq \bar{f}_{n+1} \, dv$-everywhere. So $\bar{f}_n \nearrow \bar{f} \, dv$-everywhere for some $\bar{f} \leq \infty$. By Theorem 3 (§2.7)

(13)
$$\int_K \bar{f}_n dv \nearrow \int_K \bar{f} dv.$$

Now $f_n \nearrow f, \bar{f}_n \nearrow \bar{f}$, and $f_n \leq \bar{f}_n \leq \bar{f} \, dv$-everywhere. So $f_n \leq f \leq \bar{f} \, dv$-everywhere. Hence $\int_K f_n dv \leq \overline{\int_K f dv} \leq \int_K \bar{f} dv$ which by (11), (12,), and (13) implies (10). $\square$ (Note that (10) is just $\nu(f_n dv) \nearrow \nu(f dv)$.)

We can now prove Fatou's Theorem for upper integrals.

THEOREM 4. *Let* $dv \geq 0$ *and* $f_n \geq 0$ *on a cell* K *for* $n = 1, 2, \cdots$. *Then for* $\underline{f} = \varliminf_{n \to \infty} f_n$ *we have* $0 \leq \underline{f} \leq \infty$ *and*

(14)
$$\overline{\int_K \underline{f} dv} \leq \varliminf_{n \to \infty} \overline{\int_K f_n dv}.$$

PROOF. The case $\overline{\int_K f_n dv} \to \infty$ is trivial. So we may assume that the right-hand side of (14) is finite, $\varliminf_{n \to \infty} \overline{\int_K f_n dv} = c < \infty$. Take a subsequence such that

(15)
$$\overline{\int_K f_{n_i} dv} \to c \text{ as } i \to \infty.$$

Since $\underline{f} \leq \varliminf_{i \to \infty} f_{n_i}$ we need only consider the case in which this subsequence is the original sequence. Then (15) is just

(16)
$$\overline{\int_K f_n dv} \to c.$$

To prove (14) we must show that

$$(17) \qquad \overline{\int_K} \underline{f} \, dv \leq c.$$

Apply Theorem 3 to $\underline{f_n} \nearrow \underline{f}$ where $\underline{f_n} = f_n \wedge f_{n+1} \wedge \cdots$ to conclude from (10) that

$$(18) \qquad \overline{\int_K} \underline{f_n} \, dv \nearrow \overline{\int_K} \underline{f} \, dv.$$

Since $\underline{f_n} \leq f_n$,

$$(19) \qquad \overline{\int_K} \underline{f_n} \, dv \leq \overline{\int_K} f_n \, dv.$$

From (16), (18), and (19) we get (17). $\quad \square$

Theorem 4 yields the following dominated convergence theorem for which the special case $g = -h$ gives the Lebesgue Dominated Convergence Theorem.

THEOREM 5. *Let* $dv \geq 0$ *on a cell* K. *Let* $f_n \to f$ *on* K *with* $g \leq f_n \leq h$ *for all* n, *where* g, h, *and* f_n *are* dv-*integrable. Then*

$$(20) \qquad f \, dv \text{ is integrable and } \int_K f_n \, dv \to \int_K f \, dv.$$

PROOF. (All integrals here are over K.) Application of Theorem 4 to $f_n - g \geq 0$ gives $\int \underline{f} \, dv - \int g \, dv = \underline{\int}(f - g) \, dv \leq \underline{\lim} \int (f_n - g) \, dv = \underline{\lim} \int f_n \, dv - \int g \, dv$. Since the last integral is finite this implies

$$(21) \qquad \overline{\int} f \, dv \leq \underline{\lim} \int f_n \, dv \leq \int h \, dv < \infty.$$

In a dual manner application of Theorem 4 to $h - f_n \geq 0$ yields

$$(22) \qquad -\infty < \int g \, dv \leq \overline{\lim} \int f_n \, dv \leq \int \overline{f} \, dv.$$

Finally (20) follows from (21) and (22). $\square$

Exercises (§4.6).

1. Given differentials $\rho \leq \overline{\rho}$ on a cell K with $\overline{\rho}$ integrable and $\int_K \overline{\rho} = \overline{\int}_K \rho$ prove that $\int_A \overline{\rho} = \overline{\int}_A \rho$ for every figure A in K. (So in Theorem 1 $\int_A \overline{f} \, dv = \overline{\int}_A f \, dv$ for every figure A in K.)

2. Apply Theorem 1 with $f = 1_E$ to get $\overline{f}$ satisfying (i), $\cdots$, (iv). Show that $\overline{f} = 1_A$ dv-everywhere, where A is the measurable set $(\overline{f} = 1)$.

3. Let E be the union of an ascending sequence $E_1 \subseteq E_2 \subseteq \cdots$ of subsets of a cell K. Given $dv \geq 0$ on K show that $\overline{\int}_K 1_{E_n} \, dv \nearrow \overline{\int}_K 1_E \, dv$ as $n \nearrow \infty$.

4. Let $f_1, f_2, \cdots$ be a convergent sequence $f_n \to f$ of functions on a cell K. Let g be a function of bounded variation on K such that $f_n \, dg \to 0$. Show that $f = 0 \, dg$-everywhere. (Apply Theorem 4.)

5. Verify that Theorem 4 yields the inequality (22) in the proof of Theorem 5.

6. Show that Theorem 5 yields the conclusion that $f_n \, dv \to f \, dv$.

CHAPTER 5

THE VITALI COVERING THEOREM
APPLIED TO DIFFERENTIALS

§5.1 The Vitali Covering Theorem with Some Applications to Upper Integrals.

An **outer measure** on a set Y is a function M on the power set of Y such that $0 \leq M(E) \leq \infty$ for every subset E of Y, $M(\emptyset) = 0$, and $M(E) \leq \sum_{n=1}^{\infty} M(E_n)$ for $E \subseteq \bigcup_{n=1}^{\infty} E_n$. The last two conditions imply $M(A) \leq M(B)$ for $A \subseteq B$.

THEOREM 1. *Let M be an outer measure on a topological space Y which is a countable, disjoint union $Y = A \cup Y_1 \cup Y_2 \cup \cdots$ such that $M(A) = 0$ and each Y_i is open in Y with $M(Y_i) < \infty$. Let C be a set of closed subsets I of Y such that:*

(i) $M(I_1 \cup \cdots \cup I_n) = M(I_1) + \cdots + M(I_n)$ for every finite, disjoint subset $\{I_1, \cdots, I_n\}$ of C,

(ii) $M(Z) = 0$ for Z the union of all members I of C such that $M(I) = 0$,

(iii) there exists $c > 0$ in $\mathbb{R}$ such that given $r > 0$ in $\mathbb{R}$ and a member I of C with $M(I) \leq r$ the union L of all members J of C such that J meets I and $M(J) \leq r$ has $M(L) \leq cr$.

Let E be a subset of Y such that

(iv) given p in E and a neighborhood N of p in Y some member I of C contains p and is a subset of N.

169

Then there is a countable, disjoint subset $\mathcal{E}$ of C and a subset D of E such that $\mathcal{E}$ covers $E - D$ and $M(D) = 0$. So $M(E) \leq \sum_{I \in \mathcal{E}} M(I)$.

PROOF. The reader can verify that we need only prove the theorem on the space Y_i for $i = 1, 2, \cdots$ with the covering C_i of $E \cap Y_i$ consisting of all members I of C which are contained in Y_i. In effect we may assume that $M(Y) < \infty$.

By (ii) we may replace E by $E - E \cap Z$ and delete from C all members I contained in Z. In effect we may assume that $M(I) > 0$ for all members I of C.

If some finite, disjoint subset of C covers E we are done. So we need only treat the case where E has no such covering. In this case $E \neq \emptyset$, otherwise the empty subset of C would cover E. So C is nonvoid. Hence we can choose some member I_0 of C. Having chosen disjoint members $I_0, \cdots, I_{n-1}$ of C for some positive integer n, let C_n consist of all members I of C which are disjoint from $I_0, \cdots, I_{n-1}$. C_n is nonvoid by (iv) since the finite, disjoint subset $\{I_0, \cdots, I_{n-1}\}$ of C cannot cover E, and $Y - (I_0 \cup \cdots \cup I_{n-1})$ is open. So for $r_n = \sup_{I \in C_n} M(I)$, $0 < r_n < \infty$ since $0 < M(I) \leq M(Y)$ for all members I of C. We can therefore choose a member I_n of C_n such that

$$(1) \qquad\qquad M(I_n) > \frac{1}{2} r_n.$$

We contend that the subset $\mathcal{E}$ of C formed by the inductively chosen $I_0, \cdots, I_n, \cdots$ satisfies the conclusion of the theorem.

Clearly $r_n \geq r_{n+1} > 0$ since $C_n \supseteq C_{n+1}$. Moreover by (1) and (i)

$$\frac{1}{2} \sum_{n=1}^{k} r_n < \sum_{n=1}^{k} M(I_n) = M(I_1 \cup \cdots \cup I_k) \leq M(Y) < \infty$$

for all k. So

(2)
$$\sum_{n=1}^{\infty} r_n < \infty.$$

Let L_n be the union of all members I of C_n which meet I_n. That is, L_n is the union of all members of $C_n - C_{n+1}$. By (iii)

(3)
$$M(L_n) \leq cr_n.$$

Given $n \geq 1$ consider any point p in E that does not lie in the closed set $I_0 \cup \cdots \cup I_{n-1}$. By (iv) some member I of C_n contains p. Now I cannot belong to C_k for all k since $0 < M(I) \leq r_k$ for I in C_k, and $r_k \to 0$ by (2). So I belongs to $C_k - C_{k+1}$ for some $k \geq n$. Hence $I \subseteq L_k$ which implies $p \in L_k$. We can therefore conclude that

$$E \subseteq (I_0 \cup \cdots \cup I_{n-1}) \cup (L_n \cup L_{n+1} \cup \cdots)$$

for all $n \geq 1$. Hence for $D = E \cap \overline{\text{Lim}} \, L_n$

(4)
$$E \subseteq (I_0 \cup I_1 \cup \cdots) \cup D.$$

Since $D \subseteq L_n \cup L_{n+1} \cup \cdots$ for all n, (3) implies

$$0 \leq M(D) \leq \sum_{k=n}^{\infty} M(L_k) \leq c \sum_{k=n}^{\infty} r_k$$

for all n. Hence, $M(D) = 0$ by (2). With (4) the proof is complete. □

There are two ways in which we can apply Theorem 1 to a positive, integrable differential $dv \geq 0$ on a cell K. We can use the outer measure defined by $\overline{\int_K} 1_E dv$ for all subsets E of K, or the topologically induced outer measure M on K defined by

(5)
$$M(E) = \inf_{U \supseteq E} \int_K 1_U dv$$

where the infimum is taken over all subsets U of K that are open relative to K and contain E. Integrability of $1_U dv$ is given by Theorem 9 (§3.4). Clearly $\overline{\int_K} 1_E dv \leq M(E)$ which implies that every M-null set is dv-null. So the conclusion of Theorem 1 is stronger if we use M defined by (5). (Actually the two outer measures are equal as will be shown by Theorem 3. But we must apply Theorem 1 to M in (5) to prove this.)

The reader can verify that M defined by (5) for a given $dv \geq 0$ on K is a finite outer measure on K. Moreover, if E is open in K then

(6)
$$M(E) = \int_K 1_E dv.$$

Furthermore (6) holds for E any interval in K, degenerate or nondegenerate. To verify this recall that, with the notational convention $a- = a$ and $b+ = b$ for $K = [a, b]$, Theorem 4 (§2.7) gives $\int_K 1_p dv = v(p+) - v(p-)$ which is the infimum of $\int_K 1_U dv$ over all intervals U in K which contain p and are open relative to K. So (6) holds for $E = p$. Now any interval L in K is the union of its interior with none, one, or two of its endpoints. Therefore

(7) $M(L) = \int_K 1_L dv$ for every interval L in K.

THEOREM 2. *Let E be a subset of a cell K. Let C be a set of cells in K such that given p in E and a neighborhood N of p some member I of C contains p and is a subset of N. Given*

$dv \geq 0$ on K define M by (5). Then there exists a countable set $\mathcal{E}$ of disjoint members of C and a subset D of E such that $\mathcal{E}$ covers $E - D$ and $M(D) = 0$. So $M(E) \leq \sum_{I \in \mathcal{E}} \int_K 1_I dv$.

PROOF. To apply Theorem 1 we have (iv) but must verify (i), (ii), (iii). (i) follows from (7) and additivity of the integral. (ii) follows from (7) and Theorem 1 (§3.5) for $\sigma = dv$. We contend (iii) holds for $c = 3$.

Let C_r consist of all members I of C such that $M(I) \leq r$. Given a member I of C_r let L be the union of all members of C_r that meet I. Then L is an interval such that $L_n \nearrow L$ for some sequence of cells $L_1 \subseteq L_2 \subseteq \cdots$ of the form $L_n = I_n \cup I \cup J_n$ where I_n and J_n are members of C_r that meet I. So $M(L_n) \nearrow M(L)$ by (7) and the Monotone Convergence Theorem (Theorem 3, (§2.7)). Therefore $M(L) \leq 3r$ since $M(L_n) \leq M(I_n) + M(I) + M(J_n) \leq 3r$. So Theorem 2 follows from Theorem 1 and (7). $\square$

THEOREM 3. *Given a function v on a cell K such that $dv \geq 0$ let M be the outer measure defined by (5). Then $M(E) = \overline{\int_K} 1_E dv$ for every subset E of K.*

PROOF. We need only prove

$$(8) \qquad M(E) \leq \overline{\int}_K 1_E dv$$

since the reverse inequality is obvious under (5). Since v has only countably many discontinuities the set C of all points at which v is continuous is dense in K. C is the set of all dv-null points. Consider any subset E of K.

Given a gauge δ on K let C consist of all cells H in $K = [a, b]$ such that either of the following conditions holds for some point t in E:

(i) (H, t) is a δ-fine tagged cell with t an endpoint of K and the other endpoint of H in C,

(ii) $H = I \cup J$ where I and J are cells abutting at t in (a, b), (I, t) and (J, t) are δ-fine, and both endpoints of H belong to C.

Now v is continuous at endpoints of H that lie in (a, b). So $\int_K 1_H dv = \int_H dv = \Delta v(H)$ for every member H of C. Thus Theorem 2 yields a countable set $\mathcal{E}$ of disjoint members of C such that $M(E) \leq \sum_{H \in \mathcal{E}} \Delta v(H) \leq (1_E \Delta v)^{(\delta)}(K)$ since $\Delta v(H) = 1_E(t) \Delta v(H)$ under (i), and $\Delta v(H) = 1_E(t) \Delta v(I) + 1_E(t) \Delta v(J)$ under (ii). Since $M(E) \leq (1_E \Delta v)^{(\delta)}(K)$ for every gauge δ on K we get (8).   □

We can now extend the result of Exercise 2 ($\S4.6$) which gave $1_A \geq 1_E$ dv-everywhere, a weaker conclusion than $A \supseteq E$.

THEOREM 4. *Let $dv \geq 0$ on a cell K. Given a subset E of K there exists a measurable subset A of K such that $A \supseteq E$ and*

$$(9) \qquad \int_K 1_A dv = \overline{\int_K} 1_E dv.$$

PROOF. Using the definition (5) of $M(E)$ take a nested sequence of open subsets U_n of K such that $U_n \supseteq U_{n+1} \supseteq E$ for $n = 1, 2, \cdots$ and

$$\int_K 1_{U_n} dv \searrow M(E) \quad \text{as } n \nearrow \infty.$$

The intersection $A = U_1 \cap U_2 \cap \cdots$ is measurable, contains E, and $\int_K 1_A dv = M(E)$ by the Monotone Convergence Theorem (Theorem 3, ($\S2.7$)) since $1_{U_n} \searrow 1_A$. (9) then follows from Theorem 3.   □

Exercises ($\S5.1$).

1. Show that Theorem 2 holds on *any* interval K (where the monotone function v may be unbounded at an open end of

K) by invoking the initial hypothesis in Theorem 1. Specifically, show that K is a disjoint union $A \cup Y_1 \cup Y_2 \cup \cdots$ where each Y_i is an interval open relative to K with $\Delta v(Y_i) < \infty$, and A is the dv-null set of all endpoints of $Y_1, Y_2, \cdots$ not belonging to any Y_i.

2. Let $dv \geq 0$ on a cell K and E be a subset of K.
   (i) Show that there is a measurable subset B of E such that $\int_K 1_B dv = \underline{\int_K} 1_E dv$. (Apply Theorem 4 to $K - E$.)
   (ii) Show that there is a measurable subset D of K such that both $E \cup D$ and $E - E \cap D$ are measurable and

$$\int_K 1_D dv = \overline{\int_K} 1_E dv - \underline{\int_K} 1_E dv.$$

3. Show that in Theorem 4 (9) implies

$$\int_K 1_{A \cap C} \, dv = \overline{\int_K} 1_{E \cap C} \, dv$$

   for every measurable subset C of K.

4. For A in Theorem 4 prove:
   (i) $\int_I 1_A dv = \overline{\int_I} 1_E dv$ for every cell I in K,
   (ii) A is unique modulo measurable, dv-null sets. (That is, if A and A' are measurable supersets of E satisfying (9) then $1_A dv = 1_{A'} dv$.)

§5.2 $\nu(1_E df)$ and Lebesgue Outer Measure of $f(E)$.

Lebesgue measure enters here because the diameter of a subset A of $\mathbb{R}$ is the Lebesgue measure (length) of its convex closure $\widehat{A} = [\inf A, \sup A]$.

THEOREM 1. *Let f be a function on a cell K and E be a subset of K. Then*

$$(1) \qquad \overline{\int_{\mathbb{R}} 1_{fE} dx} \leq 2 \overline{\int_K 1_E |df|}.$$

PROOF. Let $\mathcal{Z}$ be the set of all nondegenerate intervals L on which f is constant, that is, on which diam $f(L) = 0$. Let Z be the union of all members of $\mathcal{Z}$. The component L of Z that contains a point t is the union of all members of $\mathcal{Z}$ that contain t. Hence $L \in \mathcal{Z}$ since $f(L) = f(t)$. Now Z has only countably many components since they are disjoint and nondegenerate. Therefore $f(Z)$ is countable, hence of Lebesgue measure zero. So we need only prove (1) with E replaced by $E - E \cap Z$. In effect we assume

$$(2) \qquad \text{diam } f(J) > 0 \quad \text{for every cell } J \text{ in } K \text{ that meets } E.$$

If $1_E df$ is not summable then the right-hand side of (1) is infinite and there is nothing to prove. So we need only consider the case in which $1_E df$ is summable. This implies that the set D of all points at which $1_E df$ is discontinuous is countable. Thus $f(D)$ is countable, hence of Lebesgue measure zero. So we can delete from E the subset D. Replacing E by $E - D$ we assume in effect that $1_E df$ is a continuous differential. That is,

$$(3) \qquad f \text{ is continuous at every point in } E.$$

We need the following lemma: Given a tagged cell (J, t) in K with t in E, $c > 1$, and $\varepsilon > 0$ there exists a cell I in J with t as an endpoint such that

$$(4) \qquad 0 < \text{ diam } f(I) \leq 2c|\Delta f(I)|$$

and

$$(5) \qquad \text{diam } f(I) < \varepsilon.$$

To get such a cell I we may assume by (3) that J is small enough to ensure that diam $f(J) < \varepsilon$. Since $c > 1$ there is a point s in J such that

(6) $$0 < \sup_{r \in J} |f(r) - f(t)| < c|f(s) - f(t)|$$

where the sharpness of the inequalities is made possible by (2).

Let $I = \widehat{s,t}$. Clearly $I \subseteq J$ since both s and t belong to J. So $f(I) \subseteq f(J)$ which implies (5) since diam $f(J) < \varepsilon$. Finally (4) follows from (6) since the triangle inequality implies that the diameter of a set is at most twice its radius about any given point. So the lemma holds.

Given $c > 1$, a positive integer n, and a gauge δ on K let $\mathcal{C}$ be the set of all cells H in $[-n, n]$ of the form

(7)
$$H = \widehat{f(I)} \quad \text{for some } \delta\text{-fine } (I, t)$$
$$\text{such that } t \in E \text{ and (4) holds.}$$

By (5) and (2) each point y in $E_n = (-n, n) \cap f(E)$ is contained in members H of $\mathcal{C}$ of arbitrarily small, positive diameter. Apply Theorem 2 (§5.1) to the subset E_n of the cell $K_n = [-n, n]$ with $v = x$ to get a countable set $\mathcal{E} = \{H_i\}$ of disjoint members of $\mathcal{C}$ covering dx-all of E_n. For each H_i choose (I_i, t_i) generating H_i according to (7). The I_i's are disjoint since the H_i's are disjoint and $f(I_i) \subseteq H_i$. Therefore

(8)
$$\overline{\int_{-n}^{n}} 1_{f(E)} dx \leq \sum_i \Delta x(H_i) = \sum_i \text{diam } f(I_i) \leq$$
$$2c \sum_i |\Delta f(I_i)| \leq 2c|1_E \Delta f|^{(\delta)}(K)$$

since $H_i = \widehat{f(I_i)}$ and (4) holds for I_i by (7).

Since the gauge δ is arbitrary, (8) implies

$$(9) \qquad \overline{\int_{-n}^{n}} 1_{f(E)} dx \leq 2c \overline{\int_{K}} 1_E |df|.$$

Finally, since (9) holds for all positive integers n and all $c > 1$, (9) gives (1) by Theorem 10 (§3.4). $\square$

An immediate corollary of Theorem 1 is the following important result.

THEOREM 2. *Let f be a function on a cell K. Then $f(E)$ is dx-null for every df-null subset E of K.*

Under appropriate conditions on f the coefficient 2 in (1) can be halved according to the next theorem.

THEOREM 3. *Let f be a function on a cell K, and E be a subset of K such that for df-all t in E*

$$(10) \qquad \begin{array}{l} f(t) \text{ is either the maximum or minimum of } f(J) \\ \text{for some cell } J \text{ in } K \text{ having } t \text{ as an endpoint.} \end{array}$$

Then

$$(11) \qquad \overline{\int_{\mathbb{R}}} 1_{f(E)} dx \leq \overline{\int_{K}} 1_E |df|.$$

PROOF. By hypothesis the set W of all t in E where (10) fails to hold is df-null. So $f(W)$ is dx-null by Theorem 2. So we may delete W from E, replacing E by $E - W$, with no change in value for either side of (11). In effect we may assume that (10) holds for *all t* in E. Under (10) the radius $\sup_{r \in J} |f(r) - f(t)|$ of $f(J)$ equals the diameter of $f(J)$. For such (J, t) (6) in the proof of Theorem 1 gives (4) with $2c$ replaced by c. With this modification the proof of Theorem 1 gives (11). $\square$

By Theorem 3 (11) holds for monotone functions. Indeed, for such functions we can reach a stronger conclusion.

THEOREM 4. *Let v be a function on a cell K such that $dv \geq 0$. Let C be the set of all points in K at which v is continuous. Then for every subset E of K*

(i)
$$\overline{\int_{\mathbb{R}} 1_{v(E)} dx} = \overline{\int_K 1_{E \cap C} dv}.$$

If, moreover, the monotone function v is continuous at every point in E then

(ii)
$$\overline{\int_{\mathbb{R}} 1_{v(E)} dx} = \overline{\int_K 1_E dv}.$$

PROOF. Since v is monotone the set $D = K - C$ of its discontinuities is countable. So $v(D)$ is countable, hence dx-null since x is continuous. Therefore $1_{v(E)} = 1_{v(E \cap C)}$ dx-everywhere which gives

(12)
$$\overline{\int_{\mathbb{R}} 1_{v(E)} dx} \leq \overline{\int_K 1_{E \cap C} dv}$$

by application of Theorem 3 to $E \cap C$. To reverse the inequality (12) we can replace E by $E \cap C$ and then, in view of Theorem 3 (§5.1), we need only prove that if v is continuous at every point in E and U is any open subset of $\mathbb{R}$ containing $v(E)$ then

(13)
$$\overline{\int_K 1_E dv} \leq \int_{\mathbb{R}} 1_U dx.$$

Since v is continuous at every t in E there is a gauge δ on K such that $\widehat{v(I)} \subseteq U$ for every δ-fine tagged cell (I, t) in K with t in E. Thus, for every δ-division $\mathcal{K}$ of K

(14)
$$\left(\sum 1_E \Delta v\right)(\mathcal{K}) \leq \int_{\mathbb{R}} 1_U dx$$

since the nonzero terms of the sum are of the form $\Delta v(I) =$ diam $v(I) = \int_{\underset{v(I)}{\frown}} dx$ where the $\widehat{v(I)}$'s are nonoverlapping cells in U. (14) implies

$$\overline{\int_K} 1_E dv \le (1_E \Delta v)^{(\delta)}(K) \le \int_{\mathbb{R}} 1_U dx$$

which gives (13). This proves (i).

For $C \supseteq E$ (i) gives (ii).   □

(Equation (ii) in Theorem 4 gives the well-known result in probability theory that if a random variable has a continuous distribution function then the distribution function is a uniformly distributed random variable.)

Exercises (§5.2). Prove:

1. If $dF = f\sigma$ on a cell K then $F(E)$ is dx-null for every σ-null subset E of K.

2. Let E be a dx-null subset of $\mathbb{R}$. Let F be a function on $\mathbb{R}$ with all four of its Dini derivates finite at every point in E. Then $F(E)$ is dx-null. (*Hint:* $1_E|dF| \le g dx$ for some function g on $\mathbb{R}$.)

3. Let $dv \ge 0$ on a cell $K = [a, b]$. Let D be the set of all points in K at which v is discontinuous. Given a subset E of K let $D(E)$ be the open subset of $\mathbb{R}$ whose components are all the nonempty open intervals of the form $(v(t-), v(t+))$ with t in E. (We use the notational convention $a- = a$ and $b+ = b$.) Then

$$\int_{\mathbb{R}} 1_{D(E)} dx = \int_K 1_{E \cap D} dv$$

and

$$\overline{\int_{\mathbb{R}}} 1_{v(E) \cup D(E)} dx = \overline{\int_K} 1_E dv.$$

4. Let F be a function on a cell K such that diam $F(K) = \int_K |dF| < \infty$. Then F is monotone.

5. If a function f on a cell K is differentiable at df-all points of a subset E of K then (11) in Theorem 3 holds. If furthermore $|f'| \le c$ df-everywhere in E then

$$\overline{\int_{\mathbb{R}}} 1_{f(E)} dx \le c \overline{\int_K} 1_E dx.$$

6. (*Open Question*) Is the inequality (1) in Theorem 1 sharp? That is, given $c < 2$ do there exist f and E such that

$$\nu(1_{f(E)} dx) > c\nu(1_E df)?$$

§5.3 Continuity σ-Everywhere of ρ Given $\rho\sigma = 0$.

To explore the connection between differentials and derivatives we shall need a continuity theorem for differentials. From Theorem 4 (§3.3) it is easy to conclude that for σ weakly archimedean and ρ tag-finite on a cell continuity σ-everywhere of ρ implies $\rho\sigma = 0$. With stronger conditions on σ we can prove the converse. We need some definitions.

Given a differential σ on $K = [a, b]$ and $a < p \le b$ σ is **left continuous** at p if the restriction of σ to $[a, p]$ is continuous at p. That is, $\int_a^p |1_p\sigma| = 0$. For $a \le p < b$ σ is **right continuous** at p if the restriction of σ to $[p, b]$ is continuous at p. Explicitly, $\int_p^b |1_p\sigma| = 0$.

THEOREM 1. *Let σ be a differential on a cell K such that σ is continuous at every point where it is either left or right continuous, and $|\sigma|$ is dampable. Let ρ be a tag-finite differential on K such that $\rho\sigma = 0$. Then ρ is continuous σ-everywhere.*

PROOF. Since $|\sigma|$ is dampable there exist a damper u on K and a monotone function v on K such that $dv = u|\sigma| \ge 0$.

Equivalently, $|\sigma| = \frac{1}{u}dv$. So σ and dv have the same null sets, hence the same points of continuity. These are just the points at which the monotone function v is continuous. The hypothetical continuity condition on σ is equivalent to continuity of v wherever v is either left or right continuous. The hypothetical condition $\rho\sigma = 0$ is equivalent to $\rho dv = 0$ since $|\rho dv| = u|\rho\sigma|$ and $|\rho\sigma| = \frac{1}{u}|\rho dv|$. Therefore, we need only consider the case $\sigma = dv \geq 0$ and $\rho \geq 0$, the latter condition attained by replacing ρ by $|\rho|$.

Take $R \geq 0$ representing ρ. For each positive integer n let D_n consist of all points p in K at which $\overline{\lim}\, R(I,p) > 1/n$ as $(I,p) \to p$. We need only prove that each D_n is dv-null to conclude from dv-nullity of $D_1 \cup D_2 \cup \cdots$ that $R(I,p) \to 0$ at dv-all p as $(I,p) \to p$. This conclusion is just continuity dv-everywhere of ρ.

Consider any fixed n and let $D = D_n$. Consider any point p in $K = [a,b]$ at which v is discontinuous. By hypothesis $v(p-) < v(p)$ if $p > a$ and $v(p) < v(p+)$ if $p < b$. So there exists $\lambda > 0$ in $\mathbb{R}$ such that $\Delta v(I) > \lambda$ for every cell I in K which contains p. Now ρdv is obviously continuous at p since $\rho dv = 0$. So $R(I,p)\Delta v(I) \to 0$ as $(I,p) \to p$ in K. This implies $R(I,p) \to 0$ since $\Delta v(I) > \lambda$. That is, ρ is continuous at every point where v is discontinuous. Contrapositively

(1) v is continuous at every point in D.

Let $\varepsilon > 0$ be given in $\mathbb{R}$. Since $\rho dv = 0$ there is a gauge δ on K such that

(2) $$(R\Delta v)^{(\delta)}(K) < \frac{\varepsilon}{n}.$$

Let C be the set of all cells I in K such that for some endpoint p of I

(3) $p \in D, (I,p)$ is δ-fine, and $R(I,p) > \dfrac{1}{n}$.

Theorem 2 (§5.1) yields a countable set $\mathcal{E}$ of disjoint cells belonging to $\mathcal{C}$ such that $\mathcal{E}$ covers dv-all points in D. We contend that the interiors I° of the members I of $\mathcal{E}$ cover dv-all points in D.

Clearly we can delete any endpoint of I which does not belong to D. An endpoint of I that does belong to D is dv-null by (1). Since $\mathcal{E}$ is countable there are only countably many such endpoints, so they form a dv-null set.

Thus there exists a dv-null subset E of K such that $1_D \leq 1_E + \sum_{I \in \mathcal{E}} 1_{I^\circ}$. Hence

$$\overline{\int_K} 1_D dv \leq \int_K \left(\sum_{I \in \mathcal{E}} 1_{I^\circ} \right) dv =$$

$$\sum_{I \in \mathcal{E}} \int_K 1_{I^\circ} dv \leq \sum_{I \in \mathcal{E}} \int_I dv$$

by Theorem 2 (§2.7). That is,

$$(4) \qquad \overline{\int_K} 1_D dv \leq \sum_{I \in \mathcal{E}} \Delta v(I).$$

Assign to each member I of $\mathcal{E}$ an endpoint p such that (3) holds. Then since $1 < nR(I, p)$

$$(5) \qquad \sum_{I \in \mathcal{E}} \Delta v(I) \leq n \sum_{I \in \mathcal{E}} R(I, p)\Delta v(I) \leq n(R\Delta v)^{(\delta)}(K).$$

Thus by (4), (5), and (2)

$$(6) \qquad \overline{\int_K} 1_D dv \leq \varepsilon.$$

Finally, since (6) holds for all $\varepsilon > 0$, $1_D dv = 0$. $\square$

As an indication of how Theorem 1 applies to derivatives we present the following result.

THEOREM 2. *Let f be a continuous function on a cell K in $\mathbb{R}$ such that $|df|$ is dampable. Then the set E of all points in K where f has either a left or right derivative equal to zero is df-null. So $f(E)$ is dx-null.*

PROOF. We need consider only the case where E is the set where f has a vanishing right derivative, the case for the left derivative being similar. So let

(7) $\Delta f(I) = o(\Delta x(I))$ as $(I, t) \to t+$ for all t in E.

For Q defined in (20) of §1.4 $Q^+(I, t)$ indicates that t is the left endpoint of I. By (7) $Q^+ 1_E \Delta f = o(\Delta x)$. So $Q^+ 1_E df = 0$ on K. By Theorem 1 the differential $[Q^+ 1_E]$ is continuous df-everywhere. That is, for df-all t in E

(8) $Q^+(I, t) = 0$ ultimately as $(I, t) \to t$.

Clearly (8) can occur only at the right endpoint of K, a point that cannot belong to E. So the empty set is df-all of E. That is, E is df-null. Hence $f(E)$ is dx-null by Theorem 2 (§5.2). □

Exercises (§5.3).

1. Apply Theorem 1 with $\sigma = dx$ and $\rho = [R]$ where $R(I, t) = \left| \frac{\Delta F}{\Delta x}(I) - f(t) \right| \wedge 1$ to prove that if $dF = f dx$ on an interval L in $\mathbb{R}$ then $F'(t) = f(t)$ at almost all t in L. (We shall study such results in §6.3.)

2. Let $dF = f dx$ on an interval L in $\mathbb{R}$. Show that the set E of all points in L at which F has either a local minimum or a local maximum is dF-null, hence $F' = 0$ at almost every point in E and $F(E)$ is dx-null. (Apply Exercise 1.)

3. Show that a bounded function f on a cell K is continuous if and only if $df dg = 0$ for every step function g on K.

4. Show that a locally bounded function f on $\mathbb{R}$ is continuous almost everywhere if and only if $df\,dx = 0$ on $\mathbb{R}$.

5. Show that a locally bounded function f on $\mathbb{R}$ is continuous almost everywhere if and only if $d(xf) = xdf + fdx$. (See Exercise 12 in §1.4.)

6. Let σ be a weakly archimedean differential on a cell K such that for ρ tag-finite on K, $\rho\sigma = 0$ implies ρ is continuous σ-everywhere. Prove that σ is archimedean.

   (*Outline of a proof:* $\sigma \geq 0$ say. Let $0 \leq \tau \leq \varepsilon\sigma$ for all $\varepsilon > 0$ in $\mathbb{R}$. To prove $\tau = 0$ take T, S representing τ, σ with $0 \leq T \leq S$ and let $\rho = [R]$ where $R = \frac{T}{S}$ if $S > 0$, 0 if $S = 0$. Then $(\rho - \varepsilon\omega)^+\sigma = 0$, so $(\rho - \varepsilon\omega)^+$ is continuous σ-everywhere. Since this holds for all $\varepsilon > 0$, ρ is continuous σ-everywhere. Hence $\tau = \rho\sigma = 0$ since σ is weakly archimedean.)

7. Let $\tau = [T]$ be an idempotent differential on $\mathbb{R}$ with $T^2 = T$. Show that $\tau dx = dx$ on $\mathbb{R}$ if and only if for almost all t in $\mathbb{R}$, $T(I, t) = 1$ ultimately as $(I, t) \to t$. (See §3.1 for idempotent differentials.)

CHAPTER 6

DERIVATIVES AND DIFFERENTIALS

§6.1 Differential Coefficients from the Gradient.

For $\mathbf{x} = (x_1, \cdots, x_n)$ and $\mathbf{y} = (y_1, \cdots, y_n)$ in $\mathbb{R}^n$ the **inner product** is $\mathbf{x} \cdot \mathbf{y} = x_1 y_1 + \cdots + x_n y_n$. (See §11.3.)

A function F on a neighborhood of a point $\mathbf{u}$ in $\mathbb{R}^n$ is **differentiable** at $\mathbf{u}$ if there exist a neighborhood $\mathbb{V}$ of $\mathbf{0}$ in $\mathbb{R}^n$ and a mapping $\mathbf{P}$ on $\mathbb{V}$ into $\mathbb{R}^n$ such that

(1) $\mathbf{P}$ is continuous at $\mathbf{0}$

and

(2) $F(\mathbf{u} + \mathbf{h}) - F(\mathbf{u}) = \mathbf{P}(\mathbf{h}) \cdot \mathbf{h}$ for all $\mathbf{h}$ in $\mathbb{V}$.

The **n-dimensional derivative** (or **gradient**) $\nabla F(\mathbf{u})$ at $\mathbf{u}$ is defined by

(3) $\nabla F(\mathbf{u}) = \mathbf{P}(\mathbf{0})$.

The reader may verify that $\mathbf{P}(\mathbf{0})$ does not depend on the choice of $\mathbf{P}$ and $\mathbb{V}$ in (1) and (2). Of course $\mathbf{P}$ does depend on $\mathbf{u}$. (Exercises 1 and 2 show the equivalence of the Carathéodory definition (3) with the usual definition of ∇F.)

Under the restriction that $\mathbf{h} = (h_1, \cdots, h_n)$ in (2) have all but a particular one of its coordinates equal to 0 we get the partial derivatives $\frac{\partial F}{\partial x_i}$ as the coordinates of the gradient

$$(4) \qquad \nabla F(\mathbf{u}) = \left(\frac{\partial F}{\partial x_1}(\mathbf{u}), \cdots, \frac{\partial F}{\partial x_n}(\mathbf{u}) \right).$$

If $n = 1$ then the gradient is just the classical derivative $\frac{dF}{dx}(u)$.

Using the gradient we can formulate Theorem 1 which concerns the differential of a function F on a curve $\mathbf{G}$ in $\mathbb{R}^n$.

THEOREM 1. *Let $\mathbf{G}$ be a continuous mapping of a cell K into $\mathbb{R}^n$ where $\mathbf{G}(t) = (G_1(t), \cdots, G_n(t))$ with each dG_i weakly archimedean. Let F be a function on a neighborhood of $\mathbf{G}(K)$ in $\mathbb{R}^n$ such that F is differentiable at every point $\mathbf{u}$ in $\mathbf{G}(D)$ for some subset D of K which contains $d(F \circ \mathbf{G})$-all points t in K. For $i = 1, \cdots, n$ define f_i on K by*

$$(5) \qquad f_i(t) = \begin{cases} \left(\frac{\partial F}{\partial x_i} \circ \mathbf{G} \right)(t) & \text{if } t \in D \\ 0 & \text{if } t \in K - D. \end{cases}$$

Then

$$(6) \qquad d(F \circ \mathbf{G}) = \sum_{i=1}^{n} f_i dG_i \text{ on } K.$$

PROOF. Take a gauge δ on K small enough so that if (I, t) is δ-fine in K with its tag t in D then (2) holds for $\mathbf{u} = \mathbf{G}(t)$ and $\mathbf{h} = \mathbf{G}(t + r) - \mathbf{G}(t) = (sgn\, r)(\Delta G_1(I), \cdots, \Delta G_n(I))$ where $t + r$ is the endpoint of I opposed to the endpoint t. For these data (2) takes the form

$$(7) \qquad \Delta(F \circ \mathbf{G})(I) = (sgn\, r)\mathbf{P}(\mathbf{h}) \cdot \mathbf{h}$$

while (3), (4), (5) give

(8)
$$\sum_{i=1}^{n} f_i(t)\Delta G_i(I) = (sgn\,r)\mathbf{P}(0) \cdot \mathbf{h}$$

with the mapping $\mathbf{P}$ dependent on t in D. Subtraction of (8) from (7) gives

(9)
$$\Delta(F \circ \mathbf{G})(I) - \sum_{i=1}^{n} f_i(t)\Delta G_i(I) =$$
$$(sgn\,r)(\mathbf{P}(\mathbf{h}) - \mathbf{P}(0)) \cdot \mathbf{h} = \sum_{i=1}^{n} R_i(I,t)\Delta G_i(I)$$

where $R_i(I,t)$ is the i-th coordinate of $\mathbf{P}(\mathbf{h}) - \mathbf{P}(0)$. By continuity of $\mathbf{P}$ at $\mathbf{0}$, $R_i(I,t) \to 0$ as $(I,t) \to t$ (that is, as $r \to 0$). For $i = 1, \cdots, n$ extend R_i to a summant on K by setting $R_i = 0$ at all (I,t) where either t fails to lie in D or (I,t) fails to be δ-fine. So R_i is a continuous summant on K.

At all δ-fine tagged cells in K (9) gives the summant equation

(10)
$$1_D\Delta(F \circ \mathbf{G}) - \sum_{i=1}^{n} f_i\Delta G_i = \sum_{i=1}^{n} R_i\Delta G_i.$$

By hypothesis, $1_D d(F \circ \mathbf{G}) = d(F \circ \mathbf{G})$. Also dG_i is weakly archimedean. So $R_i dG_i = 0$ since R_i is a continuous summant. Therefore (10) gives the differential equation

$$d(F \circ \mathbf{G}) - \sum_{i=1}^{n} f_i dG_i = \sum_{i=1}^{n} R_i dG_i = 0$$

which gives (6). $\square$

There are three remarks on Theorem 1.

(i) Since $\mathbf{G}$ is continuous on K so is $F \circ \mathbf{G}$. To verify this statement note that if $1_p dG_i = 0$ for p in K and $i = 1, \cdots, n$ then (6) implies $1_p d(F \circ \mathbf{G}) = 0$.

(ii) If D contains dG_i-all of K for $i = 1, \cdots, n$ then $\frac{\partial F}{\partial x_i} \circ \mathbf{G}$ exists and is finite dG_i-everywhere on K. So f_i in (6) can in this case be replaced by $\frac{\partial F}{\partial x_i} \circ \mathbf{G}$ for $i = 1, \cdots, n$. So (6) can be expressed as $d(F \circ \mathbf{G}) = ((\nabla F) \circ \mathbf{G}) \cdot d\mathbf{G}$ under (4).

(iii) The case in (ii) holds under the following conditions on F and D: The function F on a neighborhood of $\mathbf{G}(K)$ is continuous at every point in $\mathbf{G}(K)$ and differentiable at every point in $\mathbf{G}(D)$ for some subset D of K with countable complement $K - D$.

For the case $n = 1$ in Theorem 1 we need not require that F be defined on a neighborhood of $G(K)$, but only on the bounded cell $G(K)$ itself. Specifically we have the following theorem.

THEOREM 2. *Let G be a nonconstant, continuous function on a cell K with dG weakly archimedean. Let F be a function on the bounded cell $G(K)$ such that F is differentiable at all points u in $G(D)$ for some subset D of K whose complement $K - D$ is $d(F \circ G)$-null. Let*

$$f(t) = \begin{cases} (F' \circ G)(t) & \text{if } t \in D \\ 0 & \text{if } t \in K - D. \end{cases}$$

Then

$$d(F \circ G) = f dG \text{ on } K.$$

PROOF. At each endpoint p of $G(K)$ extend F to the open halfline L abutting $G(K)$ at p by defining

$$F(t) = F(p) + \phi(p)(t - p) \text{ for all } t \text{ in } L$$

where $\phi(p) = F'(p)$ if F on $G(K)$ is differentiable at p, and $\phi(p) = 0$ otherwise. Apply Theorem 1 with $n = 1$. $\square$

For G the identity function x on a bounded cell K Theorem 2 gives as a corollary the following version of the Fundamental Theorem of Calculus. Note that the hypothesis implies continuity of F.

THEOREM 3. *Let F be a function on a bounded cell K such that F is differentiable dF-everywhere. Let $f(t) = F'(t)$ if F is differentiable at t, and $f(t) = 0$ otherwise. Then*

$$dF = f dx \text{ on } K.$$

Theorem 3 has the following obvious corollary which is suggested by Remark (iii) above. It is a version of the Fundamental Theorem of Calculus that is of great utility in elementary calculus.

THEOREM 4. *Let F be a continuous function on a bounded cell K such that F is differentiable at all but countably many points in K. Then $dF = F' dx$ on K.*

Theorems 3 and 4 actually apply to functions F on any nondegenerate subinterval L of $\mathbb{R}$. We refer the reader back to the initial part of §3.4.

Theorem 3 is a special case of the following theorem.

THEOREM 5. *Let $\sigma = [S]$ and $\tau = [T]$ be differentials on a cell K such that σ is weakly archimedean and for τ-all t in K*

(a) $S(I,t) \neq 0$ ultimately as $(I,t) \to t$ and $\lim\limits_{(I,t) \to t} \frac{T(I,t)}{S(I,t)}$ exists and is finite.

Define the function f on K by

(b) $f(t) = \lim\limits_{(I,t) \to t} \frac{T(I,t)}{S(I,t)}$ if (a) holds, 0 otherwise.
Then $\tau = f\sigma$.

PROOF. Let E be the set of all t at which (a) holds. Then $1_E \tau = \tau$ and $1_E f = f$. Let $\rho = [R]$ where

$$R = \begin{cases} 1_E \left| \frac{T}{S} - f \right| & \text{if } S \neq 0, \\ 0 & \text{if } S = 0. \end{cases}$$

By the definition of f and $E, R(I, t) \to 0$ as $(I, t) \to t$ for all t in K. So ρ is continuous. Hence, $\rho\sigma = 0$ since σ is weakly archimedean. Now $\rho\sigma = [RS]$ and $|RS| = 1_E|T - fS|$ ultimately. So $0 = |\rho\sigma| = 1_E|\tau - f\sigma| = |1_E\tau - 1_E f\sigma| = |\tau - f\sigma|$. Hence, $\tau = f\sigma$. $\square$

Theorem 4 yields the classical algorithm

$$(11) \qquad\qquad dF = F'dx$$

for F differentiable on an interval L in $\mathbb{R}$. Attempting to make some sense of the symbols in this formula, but lacking the appropriate concepts, the early analysts introduced a bogus definition of (11). By force of tradition and the lack of a viable alternative it has persistently survived in calculus textbooks. According to this putative definition the differential of a differentiable function F is

$$(12) \qquad\qquad dF = F'\Delta x$$

where Δx is merely an independent variable. In particular, $dx = \Delta x$ since $x' = 1$. So (12) effortlessly yields (11). By this subterfuge the valid approximation formula $F'\Delta x \doteq \Delta F$ for summants is presented as

$$(13) \qquad\qquad dF \doteq \Delta F$$

which is given by (12). This obfuscation has perpetuated the myth that differentials are approximations. The specious formula (13) should not be confused with our definition $dF = [\Delta F]$ which is firmly grounded in the integration process and applies to all functions F on L, differentiable or not.

The modern theory of differential forms is of no help in understanding (11). It gives us the formula $dF = F'$ in place of (11), making no distinction between differential and derivative in dimension 1.

Exercises (§6.1).

1. Prove that if $\mathbf{P}$ satisfies (1) and (2), and $\mathbf{P}(0) = \mathbf{p}$ then

$$(14) \qquad \frac{F(\mathbf{u} + \mathbf{h}) - F(\mathbf{u}) - \mathbf{p} \cdot \mathbf{h}}{\|\mathbf{h}\|} \longrightarrow 0 \text{ as } \mathbf{h} \to 0, (\mathbf{h} \neq 0).$$

(This is the usual definition of $\nabla f(\mathbf{u}) = \mathbf{p}$.)

2. Given $\mathbf{p}$ in $\mathbb{R}^n$ satisfying (14) prove that (1) and (2) hold for

$$\mathbf{P}(\mathbf{h}) = \begin{cases} \frac{F(\mathbf{u}+\mathbf{h}) - F(\mathbf{u}) - \mathbf{p} \cdot \mathbf{h}}{\|\mathbf{h}\|^2} \mathbf{h} + \mathbf{p} & \text{if } \mathbf{h} \neq 0 \\ \mathbf{p} & \text{if } \mathbf{h} = 0. \end{cases}$$

3. Define the function F on $\mathbb{R}^2$ by $F(u, v) = uv$. Given (u, v) in $\mathbb{R}^2$ define the mapping $\mathbf{P}$ of $\mathbb{R}^2$ into itself by

$$\mathbf{P}(\mathbf{h}) = \begin{cases} \left(v + \frac{h^2 k}{h^2 + k^2}, u + \frac{hk^2}{h^2 + k^2} \right) & \text{for } \mathbf{h} = (h, k) \neq 0 \\ (v, u) & \text{for } \mathbf{h} = 0. \end{cases}$$

Prove that (1) and (2) hold, and $\nabla F(u, v) = (v, u)$.

4. Apply Theorem 1 with $n = 2$, $\mathbf{G}(t) = (u(t), v(t))$, and $F(u, v) = uv$ to prove: If u, v are continuous functions on a cell K with du, dv weakly archimedean then $d(uv) = v\,du + u\,dv$.

(We shall investigate the product rule more thoroughly in the next section.)

5. Apply Theorem 4 to verify that $\int_0^1 x^{-1/2}\,dx = 2$.

6. Let G be a nonconstant, continuous function on $K = [a, b]$ in $\mathbb{R}$ and F a continuous function on $G(K)$ such that $G'(t) = g(t)$ at all but countably many t in K, and $F'(y) = f(y)$ at all y in $G(K)$ except for y in a set E with $G^{-1}(E)$ countable. Prove:

(i) $(F \circ G)'(t) = f(G(t))g(t)$ at all but countably many t,

(ii) $(f \circ G)g dx = d(F \circ G)$ on K,

(iii) $\int_a^b f(G(t))g(t)dt = F(G(b)) - F(G(a))$.

7. Apply Theorem 2 to prove that if G is continuous on a cell K with dG weakly archimedean then $d(G^j) = jG^{j-1}dG$ on K for $j = 1, 2, \cdots$.

8. Define F on $\mathbb{R}$ by $F(x) = x \sin \frac{1}{x}$ for $x \neq 0$ and $F(0) = 0$. Prove:

(i) $dF(x) = (\sin \frac{1}{x} - \frac{1}{x} \cos \frac{1}{x})dx$ on $\mathbb{R}$.

(ii) $\int_0^1 (\sin \frac{1}{x} - \frac{1}{x} \cos \frac{1}{x})dx = \sin 1$.

(iii) $(\frac{1}{x} \cos \frac{1}{x})dx$ is integrable on $[0, 1]$.
(*Hint:* $\sin \frac{1}{x}$ is bounded and continuous on $(0, 1]$. So $(\sin \frac{1}{x})dx$ is integrable on $[0, 1]$.)

9. Let f, G, H be functions on an interval L in $\mathbb{R}$ such that $\overline{D}G \leq f \leq \underline{D}H$ on L where $\overline{D}$ denotes the upper derivative and $\underline{D}$ the lower derivative. Show that $dG \leq f dx \leq dH$ on L. (This is relevant to Exercise 6 in §1.5.)

10. Given $dF \geq 0$ on an interval L in $\mathbb{R}$ show that the lower derivative $\underline{D}F$ is finite dx-everywhere on L and $\underline{D}F dx \leq dF$ on L. (See Exercise 10 in §6.3 for a well known stronger result.)

11. Let F be a function on $\mathbb{R}$ with finite left and right derivatives $F'_{(-)}(t)$ and $F'_{(+)}(t)$ at all t in $\mathbb{R}$. Show that $dF = [F'_{(-)}Q^- + F'_{(+)}Q^+]dx$ on $\mathbb{R}$ where Q is the summant defined by (20) (§1.4). (The next two exercises lead to a sharper conclusion.)

12. The set of all points where a function F on $\mathbb{R}$ has finite left and right derivatives with unequal values is countable [41]. To prove this let $E = (F'_{(-)} < F'_{(+)})$. (The case of the reverse inequality is similar.) Show that E is the union of

countably many sets each of which has at most one point. Specifically, for each ordered triple (p, q, r) of rational numbers let $E_{(p,q,r)}$ be the set of all t in E such that

(15) $$F'_{(-)}(t) < r < F_{(+)}(t),$$

$p < t < q$, and for all s in (p, q) such that $s \neq t$

(16) $$F(s) - F(t) > r(s - t).$$

Prove:

(i) Each t in E lies in some $E_{(p,q,r)}$. (First choose r so that (15) holds. Then use the first inequality in (15) to get p, the second to get q.)

(ii) If $s, t \in E_{(p,q,r)}$ and $s \neq t$ then the inequality (16) and its reversal both hold – a contradiction!

13. Show that in Exercise 11

$$dF = F'_{(-)}dx = F'_{(+)}dx = F'dx.$$

14. The conclusion of Theorem 3 can be validly expressed as $dF = F'(1_E dx)$ where E is the set of all points at which F is differentiable. But the expression $dF = 1_E(F'dx)$ may be invalid. Why?

15. Let f be a bounded measurable function on a neighborhood of c in $\mathbb{R}$. Prove:

(i) For all sufficiently small $\varepsilon > 0$ the function $F(\varepsilon) = \frac{3}{2\varepsilon}\int_{-1}^{1} f(c + \varepsilon t)t\,dt$ exists and is finite.

(ii) $F(\varepsilon) = \frac{3}{2\varepsilon}\int_{0}^{1}[f(c + \varepsilon t) - f(c - \varepsilon t)]t\,dt = \frac{1}{2}\int_{0}^{1}\left[\frac{f(c+\varepsilon t)-f(c)}{\varepsilon t} + \frac{f(c)-f(c-\varepsilon t)}{\varepsilon t}\right]d(t^3).$

(iii) If f has finite left and right derivatives at c then $F(0+)$
$= \frac{1}{2}\left[f'_{(+)}(c) + f'_{(-)}(c)\right]$.

(iv) If f is differentiable at c then $F(0+) = f'(c)$.

16. Let f be a differentiable function on $\mathbb{R}$ such that $f' \geq 0$ almost everywhere. Show that $f' \geq 0$ everywhere.

17. Show that $x^2 dg = dx$ on $\mathbb{R}$ where $g(x) = -x^{-1}$ for $x \neq 0$ and $g(0) = 0$. (Apply Theorem 5 with $S = \Delta x$ and $T = x^2 \Delta g$. Note that $dg \neq x^{-2} dx$ on $\mathbb{R}$ since 0 is dx-null but not dg-null.)

§6.2 Integration by Parts and Taylor's Formula.

Define the summant Q on $[-\infty, \infty]$ by letting $Q(I,t) = 1$ if $I = [t, s]$, -1 if $I = [s, t]$. Then for any function f on $I = \overset{\frown}{s,t}$

(1)
$$\Delta f(I) = Q(I,t)[f(s) - f(t)]$$

for either $t < s$ or $t > s$. (See (20) and Exercise 12 in §1.4.)

THEOREM 1. *For functions f, g on an interval L in the extended real line $[-\infty, \infty]$ the summant identity*

(2)
$$\Delta(fg) = f\Delta g + g\Delta f + Q\Delta f\Delta g$$

holds on the set of all tagged cells in L. If, moreover, both f and g are locally bounded on L then

(3)
$$d(fg) = fdg + gdf + Qdfdg$$

on L. Finally, if f is continuous and dg is weakly archimedean then

(4)
$$d(fg) = fdg + gdf.$$

PROOF. To get (2) at (I, t) for $I = \overset{\frown}{s, t}$ multiply the identity $f(s)g(s) - f(t)g(t) = f(t)[g(s) - g(t)] + g(t)[f(s) - f(t)] + [f(s) - f(t)][g(s) - g(t)]$ through by $Q(I, t)$ and apply (1).

If f and g are locally bounded then the differential $df\,dg$ is well defined , in which case (2) gives (3).

If f is continuous then the differential df is continuous. Thus $df\,dg = 0$ for dg weakly archimedean, in which case (3) gives (4). □

If f, g are sufficiently smooth we can extend (4) to get the formula ((6) below) for iterated integration by parts. (4) is the case $n = 0$ in (6) with dg weakly archimedean.

THEOREM 2. *Let f, g be functions on an interval L in $\mathbb{R}$ with continuous derivatives $f^{(j)}, g^{(j)}$ of order $j = 0, 1, \cdots, n$ for some $n \geq 1$. Let w be the continuous function on L defined by*

(5)
$$w = \sum_{j=0}^{n} (-1)^j f^{(j)} g^{(n-j)}.$$

Then

(6)
$$dw = f\,dg^{(n)} + (-1)^n g\,df^{(n)}$$

on L.

PROOF. By Theorem 4 (§6.1) $df^{(j)} = f^{(j+1)}dx$ and $dg^{(j)} = g^{(j+1)}dx$ for $j = 0, \cdots, n-1$. Thus since dx is weakly archimedean so are $df^{(j)}$ and $dg^{(j)}$ for $0 \leq j < n$. So (4) applies to each term of the sum in (5) giving $dw =$

$$\sum_{j=0}^{n} (-1)^j d(f^{(j)} g^{(n-j)}) = \sum_{j=0}^{n} (-1)^j [f^{(j)} dg^{(n-j)} + g^{(n-j)} df^{(j)}] =$$

$$f \, dg^{(n)} +$$

$$\left\{ \sum_{j=1}^{n} (-1)^j f^{(j)} g^{(n-j+1)} + \sum_{k=0}^{n-1} (-1)^k f^{(k+1)} g^{(n-k)} \right\} dx$$

$$+ (-1)^n g df^{(n)}.$$

The integrand in braces vanishes since term j in the first sum is the negative of term $k = j - 1$ in the second sum. Hence (6). $\square$

Iterated integration by parts is the key to Taylor's Formula which we consider next.

THEOREM 3. *Let f be a function on an interval L in $\mathbb{R}$ with continuous derivatives $f^{(j)}$ of order $j = 0, 1, \cdots, n$. Let a be a point in L. Then at all t in L*

$$(7) \qquad f(t) = \sum_{j=0}^{n} \frac{f^{(j)}(a)}{j!} (t - a)^j + R_n(t)$$

where

$$(8) \qquad R_n(t) = \int_a^t \frac{(t - s)^n}{n!} df^{(n)}(s).$$

PROOF. We shall apply Theorem 2 to get (7) and (8) at a given point $t = b$ in L. For $j = 0, 1, \cdots, n$ let g_j be the polynomial defined by

$$(9) \qquad g_j(s) = \frac{(s - b)^j}{j!}.$$

By repeated differentiation of g_n we get

$$(10) \qquad g_n^{(n-j)} = g_j \text{ for } j = 0, \cdots, n.$$

In particular at $j = 0$ (10) gives

(11) $$g_n^{(n)} = g_0 = 1.$$

By (9) and (10)

(12) $$g_n^{(n-j)}(b) = g_j(b) = 0 \text{ for } j = 1, \cdots, n.$$

By (11) $dg_n^{(n)} = 0$. So (6) in Theorem 2 applied with $g = g_n$ gives $dw = (-1)^n g_n df^{(n)}$. Integrating this from a to b we get

(13) $$w(b) - w(a) = \int_a^b (-1)^n g_n(s) df^{(n)}(s)$$

invoking the convention that $\int_b^a = -\int_a^b$. By (9) the integral in (13) is precisely the integral in (8) at $t = b$ which defines $R_n(b)$. That is, (13) is just $w(b) - w(a) = R_n(b)$. So

(14) $$w(b) = w(a) + R_n(b).$$

By (9) and (10) evaluation of (5) at a with $g = g_n$ gives

(15) $$w(a) = \sum_{j=0}^n \frac{f^{(j)}(a)}{j!}(b-a)^j.$$

By (11) and (12) evaluation of (5) at b with $g = g_n$ gives

(16) $$w(b) = (-1)^0 f^{(0)}(b) g_n^{(n)}(b) = f(b).$$

Thus (14), (15), and (16) give (7) at $t = b$. $\square$

In Theorem 3 existence of $f^{(n+1)}$ is not demanded. So Theorem 3 is more general than the usual formulations of Taylor's Theorem. Of course, if $f^{(n+1)}$ does exist it is useful in giving estimates for R_n since we can set $df^{(n)} = f^{(n+1)}dx$ in (8).

We turn now to some applications of Theorem 1. If f is continuous and g is of bounded variation on a cell K then gdf is integrable on K. This is a special case of the following definitive result.

THEOREM 4. *A function g on a cell K is of bounded variation if and only if gdf is integrable on K for every regulated function f on K.*

PROOF. Given g of bounded variation and f regulated on $K = [a, b]$ we contend that gdf is integrable. Since f is bounded and measurable fdg is integrable for g of bounded variation. (Apply Theorem 3 (§4.3) to $\sigma = (dg)^+$ and $\sigma = (dg)^-$.) By (3) $gdf = d(fg) - fdg - Qdfdg$. Since $d(fg)$ and fdg are integrable we need only prove that $Qdfdg$ is integrable to conclude that gdf is integrable.

Let C be the set of all points in K at which f is continuous, D the complementary set of all points at which f is discontinuous. Since $1_C df$ is a continuous differential and dg is absolutely integrable, $1_C df dg = 0$. So $df dg = 1_D df dg$. D is countable and for each point p in D,

$$\int_K 1_p Qdfdg =$$
$$(f(p+) - f(p))(g(p+) - g(p)) -$$
$$(f(p) - f(p-))(g(p) - g(p-))$$

and

$$\int_K |1_p Qdfdg| = \int_K 1_p |dfdg| =$$

$$|f(p+) - f(p)| \, |g(p+) - g(p)| + |f(p) - f(p-)| \, |g(p) - g(p-)|$$

where $a- = a$ and $b+ = b$. (See Exercise 26 (§4.4).) Application of Theorem 2(§2.7) to the positive and negative parts of $Qdfdg$ gives its (absolute) integrability with $\int_K Qdfdg = \int_K 1_D Qdfdg = \sum_{p \in D} \int_K 1_p Qdfdg$ and for the absolute value $\int_K |dfdg| = \sum_{p \in D} \int_K 1_p |dfdg|$.

To prove the converse let gdf be integrable for every regulated f on K. Suppose g is not of bounded variation. We shall show that this leads to a contradiction by constructing a regulated f for which gdf fails to be integrable.

For g of unbounded variation Exercise 2 (§2.6) yields a monotone sequence $a_1, a_2, \cdots$ in K such that

(17) $$\sum_{n=1}^{\infty} |g(a_{n+1}) - g(a_n)| = \infty.$$

We may assume that $a_n < a_{n+1}$ for all n since the case of a decreasing sequence reduces to this if we reflect K about 0. We may also assume that $a_1 = a$. Let $I_n = [a_n, a_{n+1}]$ and define the step function f_n on K by $f_n = -1_{(a_n, a_{n+1})}$. By (17) we can choose a sequence $0 = n_0 < n_1 < \cdots$ such that

(18) $$\sum_{n_{j-1} < n \leq n_j} |\Delta g(I_n)| > j$$

for $j = 1, 2, \cdots$. Define

(19) $$c_n = \frac{1}{j} sgn \Delta g(I_n) \text{ for } n_{j-1} < n \leq n_j.$$

Define f on K by $f = \sum_{n=1}^{\infty} c_n f_n$. Then $|f - \sum_{n=1}^{k} c_n f_n| < \frac{1}{j}$ for $k > n_j$. So f is a uniform limit of step functions. By Theorem 3 (§4.4) f is regulated. So gdf is integrable.

Now $a_n \nearrow p$ for some p in $(a, b]$, and by (19) $c_n \to 0$. Hence $f(p-) = 0$. Moreover $f = 0$ on $[p, b]$. So f is continuous at p. That is, $1_p df = 0$. Thus

(20) $$\int_K gdf = \lim_{n \to \infty} \int_a^{a_n} gdf = \sum_{n=1}^{\infty} \int_{I_n} c_n gdf_n$$

since $f = c_n f_n$ on I_n and $a_1 = a$. Now $\int_{I_n} gdf_n = \Delta g(I_n)$ by Exercise 5 (§2.7). So $\int_{I_n} c_n gdf_n = c_n \Delta g(I_n) = \frac{1}{j} |\Delta g(I_n)|$ for $n_{j-1} < n \leq n_j$ by (19). By (18) and (19) this implies

(21) $$\sum_{n_{j-1} < n \leq n_j} \int_{I_n} c_n gdf_n > 1$$

for $j = 1, 2, \cdots$. (20) and (21) imply $\int_K gdf = \infty$ contradicting the integrability of gdf. $\square$

(See Exercise 7 for the counterpart of Theorem 4.)

Although fdg and gdf are integrable on a cell K if f is regulated and g is of bounded variation, integrability may fail on an open-ended interval. On such intervals we invoke further conditions to ensure integrability. We treat the case $[a, b)$.

THEOREM 5. *Let f, g be functions on $L = [a, b)$ such that: (i) g is of bounded variation with $g(b-) = 0$, (ii) f is regulated and bounded, and (iii) f, g have no common left or right discontinuity. Then $dfdg = 0$ on L, both fdg and gdf are integrable on L, and*

$$\int_L fdg + \int_L gdf + f(a)g(a) = 0.$$

PROOF. From the proof of Theorem 4 we have $\int_L |dfdg| = \sum_{p \in D} \int_L 1_p |dfdg|$ where D is the countable set of discontinuities of f. Now

$$\int_L 1_p |dfdg| =$$
$$|f(p) - f(p-)| \, |g(p) - g(p-)| + |f(p+) - f(p)| \, |g(p+) - g(p)|.$$

In each term on the right-hand side at least one factor vanishes by (iii). So $dfdg = 0$. Hence $gdf = d(fg) - fdg$ on L by Theorem 1. So integrability of gdf will follow from that of $d(fg)$ and fdg.

Since $g(b-) = 0$ and f is bounded, $(fg)(b-) = 0$. So $d(fg)$ is integrable with $\int_L d(fg) = -(fg)(a)$.

By (i) and (ii) fdg is integrable on $[a, p]$ for all p in L. So $fdg = dh$ on L for $h(p) = \int_a^p fdg$. We contend that dh is integrable on L. By (ii) there exists a constant c such that $|f| \leq c$ on L. So $|dh| = |fdg| \leq c|dg|$ which implies $\int_L |dh| \leq$

$c \int_L |dg| < \infty$ by (i). Thus dh is integrable on L by Theorem 7 (§3.4). Integration of (4) over L completes the proof. □

We can relax the demand that f be bounded in (ii) if we impose other conditions. The next theorem is useful in probability theory where it is applied on $[0, \infty)$ with g the distribution function for a nonnegative random variable Y, and f the identity function. The two integrals in (iv) of Theorem 6 below then give the mathematical expectation of Y.

THEOREM 6. *Let f, g be functions on $L = [a, b)$ such that:*
(i) $df \geq 0$ and $f(a) = 0$,

(ii) $dg \geq 0, 0 \leq g \leq 1$, and $g(b-) = 1$,

(iii) f, g have no common left or right discontinuity.
Then
(iv) $0 \leq \int_L f dg = \int_L (1 - g) df \leq \infty$.

PROOF. Part of Theorem 5 applies to $f, 1-g$ to give $df dg = -df d(1 - g) = 0$. So

(22) $$d[(1 - g)f] = (1 - g)df - f dg.$$

But we cannot get integrability of (22) from Theorem 5 because f may be unbounded, $f(b-) = \infty$.

However we do have local integrability in (22). So we can define F, G on L by

(23) $$F(p) = \int_a^p (1 - g)df, \; G(p) = \int_a^p f dg$$

for all p in L. Then by (i) and (ii) (23) gives

(24) $dF = (1 - g)df \geq 0, \; dG = f dg \geq 0, \; F(a) = G(a) = 0.$

Since $f(a) = 0$ in (i) integration of (22) over $[a, p]$ gives by (i), (ii), (23)

(25) $$0 \leq [1 - g(p)]f(p) = F(p) - G(p)$$

for all p in L. By (23) and (24)

$$(26) \qquad F(b-) = \int_L (1-g)df, \; G(b-) = \int_L f dg.$$

By (25)

$$(27) \qquad 0 \le G(b-) \le F(b-) \le \infty.$$

If $G(b-) = \infty$ then (27) and (26) give (iv) with both integrals infinite.

If $G(b-) < \infty$ then since $g(b-) = 1$ in (ii), $df \ge 0$ in (i), and $dG = fdg$ in (24),

$$(28) \qquad \begin{aligned} [1 - g(p)]f(p) &= f(p) \int_{[p,b)} dg \\ &\le \int_{[p,b)} f dg = G(b-) - G(p). \end{aligned}$$

Now $G(b-) - G(p) \to 0$ as $p \to b-$. So $0 = [(1-g)f](b-) = F(b-) - G(b-)$ in (25). That is $0 \le G(b-) = F(b-) < \infty$ which by (26) is (iv) with both integrals finite. □

Fleissner [6] proved that the product of a derivative and a continuous function of bounded variation is a derivative. This is a corollary of our next theorem. The proof relies on integration by parts.

THEOREM 7. *Let f, g, G be functions on a cell K in $\mathbb{R}$ such that f is of bounded variation and $gdx = dG$. Then*

$$(29) \qquad fgdx = dF \text{ for some function } F \text{ on } K.$$

Moreover at every point p in K where f is continuous and $G'(p) = g(p)$,

$$(30) \qquad F'(p) = f(p)g(p).$$

PROOF. Since f is of bounded variation and G is continuous $f dG$ is integrable. So (29) holds.

Choose $c > |g(p)|$. Since $G'(p) = g(p)$ there exists $\delta > 0$ such that

(31) $|G(t) - G(p)| \le c|t - p|$ for all t in K such that $|t - p| < \delta$.

Consider any h in $\mathbb{R}$ such that $|h| < \delta$ and $p + h$ lies in K. By (29) and Theorem 1

$$F(p + h) - F(p) - f(p)[G(p + h) - G(p)] =$$
$$\int_p^{p+h} [f(t) - f(p)] dG(t) =$$

(32) $\qquad [f(p + h) - f(p)]G(p + h) - \int_p^{p+h} G(t) df(t) =$

$$[f(p + h) - f(p)][G(p + h) - G(p)] +$$
$$\int_p^{p+h} [G(p) - G(t)] df(t).$$

By (31)

(33) $|f(p+h) - f(p)||G(p+h) - G(p)| \le |f(p+h) - f(p)|c|h|.$

Since f is of bounded variation $|df| = dv$ for some function v on K. By (31)

(34) $\qquad \left| \int_p^{p+h} [G(p) - G(t)] df(t) \right| \le c|h| \, |v(p + h) - v(p)|.$

Since f is continuous at p so is v. Hence (32), (33), and (34) imply

(35) $\qquad F(p + h) - F(p) = f(p)[G(p + h) - G(p)] + o(h)$

as $h \to 0$ with $p + h$ in K. Since $G'(p) = g(p)$ (35) gives the conclusion (30). □

Exercises (§6.2).

1. Apply Theorem 5 with $f(t) = -\cos t$ and $g(t) = \frac{1}{t}$ to show that $\frac{\sin t}{t} dt$ is integrable on $[1, \infty)$. Why must this differential be integrable over $[0, 1]$? Why must it be integrable over $\mathbb{R}$?

2. Show that if g is of bounded variation on $[0, \infty)$ with limit $g(\infty-) = 0$, and $f dx$ is integrable on $[0, \infty)$ then so is $g f dx$. (Apply Theorem 5 to F, g with $dF = f dx$.)

3. Along with the hypothesis in Theorem 5 assume also that $dg \leq 0, f(a) = 0$, and $c \leq f \leq C$ dg-everywhere. Show that Theorem 5 then gives $cg(a) \leq \int_L g df \leq Cg(a)$.

4. (Dirichlet's Convergence Test.) Let $\langle a_n \rangle, \langle b_n \rangle$ be sequences in $\mathbb{R}$ such that $b_n \searrow 0$ and for some c, C in $\mathbb{R}$ and all n in $\mathbb{N}, c \leq a_1 + \cdots + a_n \leq C$. Prove:
(i) (Abel's Lemma.) $cb_1 \leq a_1 b_1 + \cdots + a_n b_n \leq Cb_1$ for all n,

(ii) $|a_n b_n + \cdots + a_{n+k} b_{n+k}| \leq (C - c) b_n$ for all n, k in $\mathbb{N}$,

(iii) The series $a_1 b_1 + a_2 b_2 + \cdots$ is convergent. (Apply Exercise 3 with $L = [0, \infty)$ and f, g defined by $f(0) = 0, f(t) = a_1 + \cdots + a_n$ for t in $(n-1, n]$, and $g(t) = b_n$ for t in $[n-1, n)$, for all n in $\mathbb{N}$.)

5. (The Second Mean Value Theorem.) Given f continuous and $dg \geq 0$ on $K = [a, b]$ prove that

$$\int_K g df = g(b)[f(b) - f(t)] + g(a)[f(t) - f(a)]$$

for some t in K. (Apply (4) and the First Mean Value Theorem, $\int_K f dg = f(t)[g(b) - g(a)]$ for some t in K.)

6. Given fdx and gdx integrable over $K = [a,b]$ in $\mathbb{R}$ with $F(t) = \int_a^t fdx, G(t) = \int_t^b gdx, c \le F \le C$, and $g \ge 0$ prove $cG(a) \le \int_K fGdx \le CG(a)$.

7. Prove that a function g on a cell K is regulated if and only if fdg is integrable for every function f of bounded variation on K. (For the direct implication apply (3) in Theorem 1 and Exercise 26 in §4.4.)

8. Show that $fgdh$ is integrable on a cell K if f is of bounded variation, h is regulated, and gdh is integrable. (Apply Exercise 7 and (ii) in Exercise 7 (§4.4).)

9. Apply (3) in Theorem 1 to show that (†) in Exercise 17 of §4.4 is equivalent to

$$\Phi(g) = F(a)g(a) + \int_K (g_- + Qdg_-)dF + \int_K (g - g_-)df.$$

10. Let f be a locally bounded function on an interval L in $\mathbb{R}$ such that $d(xf) = xdf + fdx$ on L. Show that f is continuous dx-everywhere on L. (Apply (3) to show $dfdx = 0$. Then apply Theorem 1 (§5.3).)

11. Let h be a function of bounded variation on a cell K such that h is continuous at the endpoints of K and $h(t) = \frac{1}{2}(h(t-) + h(t+))$ at every interior point of K. Let $dF = fdh$ and $dG = gdh$ on K. Prove:

    (i) $\int_K (fG + Fg)dh = \Delta(FG)(K)$. (See Exercise 10 in §3.1.)

    (ii) $\int_K 2FdF = \Delta(F^2)(K)$.

12. Let G be bounded on a cell K in $\mathbb{R}$ and continuous dx-everywhere on K. Let F be continuous on K and differentiable at all but countably many points in K. Prove:

    (i) $dGdx = 0$,

(ii) $dFdG = 0$ (Apply Theorem 4 (§6.1).),

(iii) FdG is integrable on K if and only if $GF'dx$ is integrable on K. (Apply (4).)

13. Given $p_k \geq 0$ in $\mathbb{R}$ for $k = 0, 1, \cdots$ with $p_0 + p_1 + \cdots = 1$ let $r_k = 1 - (p_0 + p_1 + \cdots + p_k)$. Prove

$$(36) \qquad 0 \leq \sum_{k=0}^{\infty} k p_k = \sum_{k=0}^{\infty} r_k \leq \infty.$$

(Apply Theorem 6 with $g(t) = 1 - r_k$ for t in $[k, k+1)$ and $f(t) = t$ on $[0, \infty)$. For an integer-valued random variable $Y \geq 0$ with p_k the probability that $Y = k$, r_k is the probability that $Y > k$, and (36) gives the expectation of Y.)

14. Let $F, g_1, g_2, \cdots$ be functions on $K = [a, b]$ such that F is continuous, $dg_n \geq 0$ with $g_n(a) = 0$ and $g_n(b) = 1$ for all n, and $g_n \to 0$ on $[a, b)$. Prove $\int_K g_n dF \to 0$. (*Outline of a proof:* $g_n dF = dG_n$ with $G_n(a) = 0$. Given $\varepsilon > 0$ choose p in $[a, b)$ such that $|F(b) - F(t)| < \varepsilon$ for all t in $[p, b]$. Use Exercise 5 to get s_n in $[a, p]$ and t_n in $[p, b]$ such that $G_n(p) = g_n(p)[F(p) - F(s_n)]$ and $G_n(b) - G_n(p) = F(b) - F(t_n)$. So $G_n(p) \to 0$ and $\overline{\lim}|G_n(b)| \leq \varepsilon$.)

15. Let fdx be integrable on $L = [0, \infty)$. Let $dg_n \geq 0$ on L with $g_n(0) = 0$ and $g_n(\infty-) = 1$, and $g_n \to 0$ on L. Show that $g_n fdx$ is integrable on L and $\int_L g_n fdx \to 0$. (Apply Exercise 14 with $K = [0, \infty]$ and $g_n(\infty) = 1, F(t) = \int_0^t fdx$.)

16. State and prove versions of Theorem 5 for the cases $L = (a, b]$ and $L = (a, b)$.

17. Prove:

(i) If $f_1, \cdots, f_n$ are locally bounded functions on $\mathbb{R}$ then

$$d(f_1 \cdots f_n) = \sum_{E \in \mathcal{N}_n} Q^{k(E)-1} \left(\Pi_{j \notin E} f_j \right) \left(\Pi_{i \in E} df_i \right)$$

where $\mathcal{N}_n$ is the set of all nonempty subsets of $\{1, \cdots, n\}$ and $k(E)$ is the number of elements in E.

(ii) If f is locally bounded on $\mathbb{R}$ and n is a positive integer then

$$d(f^n) = \sum_{k=1}^{n} Q^{k-1} \binom{n}{k} f^{n-k} (df)^k.$$

(iii) If f_i is continuous on $\mathbb{R}$ with df_i weakly archimedean for $i = 1, \cdots, n$ then

$$d(f_1 \cdots f_n) = \sum_{i=1}^{n} (\Pi_{j \neq i} f_j)\, df_i.$$

(iv) If f is continuous on $\mathbb{R}$ with df weakly archimedean then $d(f^n) = nf^{n-1}df$ for every positive integer n. ((iii) gives a proof that differs from the one in Exercise 7 of §6.1.)

§6.3 A Generalized Fundamental Theorem of Calculus.

Theorem 1 (§5.3) yields the following version of the Fundamental Theorem of Calculus.

THEOREM 1. *Let* $\sigma = [S]$ *and* $\tau = [T]$ *be differentials on a cell* K *such that* σ *is continuous wherever it is either left or right continuous, and* $|\sigma|$ *is dampable. Let* f *be a function on* K. *Then (i) is equivalent to (ii):*

(i) $\tau = f\sigma$

(ii) Every σ-*null set is* τ-*null, and for* σ-*all* t *in* K *the following condition holds:*

(1)   As $(I, t) \to t, S(I, t) \neq 0$ ultimately and $\dfrac{T(I, t)}{S(I, t)} \to f(t)$.

PROOF. Let (i) hold. Then every σ-null set is τ-null since $1_A\tau = 1_A f\sigma = f1_A\sigma = f0 = 0$ if $1_A\sigma = 0$.

Given S, T representing σ, τ define the summant R by

$$(2) \qquad R(I,t) = \begin{cases} \left|\frac{T(I,t)}{S(I,t)} - f(t)\right| \wedge 1 & \text{if } S(I,t) \neq 0 \\ 1 & \text{if } S(I,t) = 0. \end{cases}$$

Let $\rho = [R]$. Since $0 \leq R \leq 1, 0 \leq \rho \leq \omega$. So ρ is tag-finite. σ is tag-finite since $|\sigma|$ is dampable. So the product differential $\rho\sigma$ is defined. By (2) $|RS| \leq |T - fS|$. So $|\rho\sigma| \leq |\tau - f\sigma|$ which implies $\rho\sigma = 0$ since $\tau - f\sigma = 0$ by (i). Consequently ρ is continuous σ-everywhere by Theorem 1 (§5.3).

Consider any point p at which ρ is continuous, $1_p\rho = 0$. That is, $R(I,p) \to 0$ as $(I,p) \to p$. So (1) follows from (2) for $t = p$. Thus (i) implies (ii).

Conversely, let (ii) hold. So $1_E\sigma = \sigma$ for the set E of all t at which (1) holds. Since the complement of E is σ-null it is also τ-null by (ii). That is $1_E\tau = \tau$.

Given $\varepsilon > 0$ in $\mathbb{R}$ the definition of E implies through condition (1) that $1_E(t)|T(I,t) - f(t)S(I,t)| \leq \varepsilon|S(I,t)|$ ultimately as $(I,t) \to t$. Hence, $|\tau - f\sigma| = 1_E|\tau - f\sigma| \leq \varepsilon|\sigma|$ for all $\varepsilon > 0$. This implies(i) since $|\sigma|$, being dampable, is archimedean. So (ii) implies (i). $\square$

Note that (i) implies (ii) for *arbitrary* representatives S, T of σ, τ while the converse is valid if (ii) holds for *some* representatives S, T. Therefore, under the hypothesis in Theorem 1, if (ii) holds for *some* pair of representatives S, T of σ, τ it must hold for *all* pairs. Of course, the set E of all t at which (1) holds depends on the particular pair S, T. But $1_E\sigma = \sigma$ and $1_E\tau = \tau$ whatever the choice of S, T representing σ, τ.

The classical derivative is defined by $F'(t) = \lim\limits_{(I,t)\to t} \frac{\Delta F(I)}{\Delta x(I)}$ wherever the limit exists and is finite. This is the special case $G = x$ of the first two of the following derivatives.

For functions F, G on an interval L we define

$$\frac{dF}{dG}(t) = \lim_{(I,t)\to t}\frac{\Delta F(I)}{\Delta G(I)}, \quad \frac{dF}{|dG|}(t) = \lim_{(I,t)\to t}\frac{\Delta F(I)}{|\Delta G(I)|},$$

$$\frac{|dF|}{dG}(t) = \lim_{(I,t)\to t}\frac{|\Delta F(I)|}{\Delta G(I)}, \quad \frac{|dF|}{|dG|}(t) = \lim_{(I,t)\to t}\frac{|\Delta F(I)|}{|\Delta G(I)|}$$

wherever they exist. Their existence requires that $\Delta G(I) \neq 0$ ultimately as $(I, t) \to t$ in L. Under the restriction $(I, t) \to t-$ (or $(I, t) \to t+$) we get left derivatives (resp., right derivatives).

An immediate corollary of Theorem 1 for $\sigma = dG = [\Delta G]$ and $\tau = dF = [\Delta F]$ is the following more classical version of the Fundamental Theorem of Calculus.

THEOREM 2. *Let f, F, G be functions on a cell K such that G is continuous wherever it is either left or right continuous and $|dG|$ is dampable. Then (i) is equivalent to (ii):*
(i) $dF = fdG$,

(ii) Every dG-null set is dF-null, and the derivative $\frac{dF}{dG}(t)$ exists and equals $f(t)$ at dG-all t in K.

Using the remarks occuring after the proof of Theorem 1 we can draw a further conclusion. Given the hypothesis in Theorem 2 together with (i) (equivalently, (ii)) then for arbitrary summants S, T representing dG, dF

$$\frac{dF}{dG}(t) = \lim_{(I,t)\to t}\frac{T(I,t)}{S(I,t)} \text{ at } dG\text{-all } t \text{ in } K.$$

For $\sigma = |dG| = [|\Delta G|]$ and $\tau = dG = [\Delta G]$ Theorem 1 gives the following as an immediate corollary.

THEOREM 3. *Let G be a function on a cell K such that G is continuous wherever it is either left or right continuous, and $|dG|$ is dampable. Let f be a function on K. Then (i) is equivalent to (ii):*

(i) $dG = f|dG|$,

(ii) At dG-all t in $K, \Delta G(I) \neq 0$ ultimately as $(I,t) \to t$, and $\frac{dG}{|dG|}(t) = f(t)$.

Theorem 3 actually gives a Hahn decomposition for dG. (See Theorem 1 (§4.2).) In Theorem 3 (i) implies $|f| = 1$ dG-everywhere. So we can set $|f| = 1$ everywhere. Then for $A = f^{-1}(1)$ and $B = f^{-1}(-1)$ we have $1_A f = 1_A$ and $1_B f = -1_B$ with $f = 1_A - 1_B$. So (i) gives $1_A dG = 1_A|dG| \geq 0$ and $1_B dG = -1_B|dG| \leq 0$. The question of measurability of A, B will be resolved by Theorem 8.

If we apply Theorem 1 with $\tau = \sigma$ and $f = 1$ then (i) reduces to $\sigma = \sigma$ and we get the following result.

THEOREM 4. *Let σ be a differential on a cell K such that σ is continuous wherever it is either left or right continuous, and $|\sigma|$ is dampable. Let S and S' be arbitrary summants representing σ. Then for σ-all t in $K, S(I,t) \neq 0$ ultimately and $\frac{S'(I,t)}{S(I,t)} \to 1$ as $(I,t) \to t$.*

For a function F on an interval in $\mathbb{R}$, the derivative $F' = 0$ dx-everywhere if and only if $|dF| \wedge dx = 0$. This characterization of singular functions is the special case $\rho = dF = [\Delta F], \sigma = dx = [\Delta x]$ of the next theorem, another consequence of Theorem 1.

THEOREM 5. *Let $\rho = [R]$ and $\sigma = [S]$ be differentials on a cell K such that σ is continuous wherever it is left or right continuous, and $|\sigma|$ is dampable. Then (a) is equivalent to (b):*
(a) $|\rho| \wedge |\sigma| = 0$,

(b) For σ-all t in $K, S(I,t) \neq 0$ ultimately and $\frac{R(I,t)}{S(I,t)} \to 0$ as $(I,t) \to t$.

PROOF. Let $\tau = |\rho| \wedge |\sigma|$. Then $\tau = [T]$ for $T = |R| \wedge |S|$. Now $\left|\frac{T}{S}\right| = \frac{|R| \wedge |S|}{|S|} = \left|\frac{R}{S}\right| \wedge 1$ wherever $S \neq 0$. Hence, as

$(I,t) \to t$ with $S(I,t) \neq 0$ ultimately, $\frac{T(I,t)}{S(I,t)} \to 0$ if and only if $\frac{R(I,t)}{S(I,t)} \to 0$. Therefore, the equivalence of (i) and (ii) in Theorem 1 for $f = 0$ gives the equivalence of (a) and (b). □

If we strengthen the hypothesis of Theorem 2 we can combine it with the Radon-Nikodym Theorem (Theorem 1 (§4.5)) to get the following result.

THEOREM 6. *Let F, G be continuous functions on a cell K with both dF and dG dampable. Then the following three conditions are equivalent:*

(a) Every dG-null set is dF-null,

(b) $dF = fdG$ for some function f on K,

(c) $\frac{dF}{dG}(t)$ exists and is finite at dF-all t in K.

PROOF. Theorem 1 (§4.5) with $\sigma = dG$ and $\tau = dF$ gives the equivalence of (a) and (b). (b) implies (c) since (i) implies (ii) in Theorem 2. That (c) implies (b) follows from Theorem 5 (§6.1) applied with $\sigma = dG, S = \Delta G, \tau = dF$, and $T = \Delta F$. □

We remark that the demand in the hypothesis of Theorem 6 that F be continuous is redundant since it is implied by each of the conditions (a), (b), (c). In (c) "dF-all" cannot be replaced by "dG-all". (See Exercise 14.)

If F, G are functions of bounded variation and G is continuous then $\frac{dF}{dG}$ exists and is finite dG-everywhere. This is a consequence of the following version of the Lebesgue Decomposition Theorem .

THEOREM 7. *If F and G are functions of bounded variation on a cell K and G is continuous then there exist functions f, g, h on K such that: (a) g, h are of bounded variation and $F = g + h$, (b) $dg = fdG$, (c) $\frac{dh}{dG} = 0 \, dG$-everywhere, and (d) $\frac{dF}{dG} = \frac{dg}{dG} = f \, dG$-everywhere.*

PROOF. Apply Theorem 7 (§2.4) with

$$\sigma = dG \text{ and } \tau = dF$$

to get unique absolutely integrable ρ, θ such that:(i) $\tau = \rho + \theta$, (ii) $|\theta| \wedge |\sigma| = 0$, and (iii) $\phi \wedge |\rho| = 0$ for all ϕ such that $\phi \wedge |\sigma| = 0$. Since ρ, θ are absolutely integrable there exist functions g, h of bounded variation on K such that

$$\rho = dg \text{ and } \theta = dh.$$

So (i) is just $dF = dg + dh$. For $K = [a, b]$ we may set $g(a) = 0$ and $h(a) = F(a)$ to get $F = g + h$. So (a) holds.

For E σ-null $|1_E \rho| \wedge |\sigma| = |\rho| \wedge |1_E \sigma| = |\rho| \wedge 0 = 0$. So (iii) applied with $\phi = |1_E \rho|$ gives $|1_E \rho| = |1_E \rho| \wedge |\rho| = 0$. That is, every σ-null set E is ρ-null. In particular, ρ is continuous since σ is continuous. Theorem 1(§4.5) yields f such that $\rho = f\sigma$. That is, $dg = fdG$ which is just (b).

By (ii) $|dh| \wedge |dG| = |\theta| \wedge |\sigma| = 0$. So Theorem 5 with $S = \Delta G$, and $\theta = [\Delta h]$ in place of $\rho = [R]$, gives (c).

Finally Theorem 2 applied to (b) gives

$$\frac{dg}{dG} = f$$

dG-everywhere which together with (a) and (c) gives

$$\frac{dF}{dG} = f$$

dG-everywhere. That is, (d) holds.   □

In Theorem 2 each of the equivalent conditions (i), (ii) implies the existence of a measurable function $\overline{f}$ such that $\overline{f} = f \, dG$-everywhere on K. This follows from our next result.

THEOREM 8. *On a cell K let*

(a)
$$dF = fdG$$

where G is continuous wherever it is either left or right continuous, and $|dG|$ is dampable. Then

(b)
$dF = \overline{f}dG$ *for some measurable function $\overline{f}$ on K of the form $\overline{f} = 1_B f$ where B is measurable and $1_B dG = dG$.*

PROOF. By Theorem 2 the set E of all t in K where the condition $\frac{dF}{dG}(t) = f(t)$ fails is dG-null. Since $|dG|$ is dampable there exist $dv \geq 0$ and a damper $u > 0$ such that $u|dG| = dv$. dG and dv thereby have the same null sets. So Theorem 4 (§5.1) yields a measurable dG-null set A containing E. The complement B of A in K is measurable,

(3)
$$1_B dG = dG$$

and

(4)
$$\frac{dF}{dG}(t) = f(t) \text{ for all } t \text{ in } B.$$

Define

(5)
$$\overline{f} = 1_B f.$$

Then $dF = \overline{f}dG$ by (5), (a), and (3). So to prove (b) we need only show that $\overline{f}$ is measurable.

G has only countably many discontinuities since $|dG|$ is dampable. So G is measurable by Theorem 5 (§4.3). By (a) F is continuous wherever G is continuous, so F is measurable.

Using a suitable topological transformation we may assume that $K = [0, \infty]$. Define G_n on K by

$$(6) \qquad G_n(t) = G\left(t + \frac{1}{n}\right) \text{ for } n = 1, 2, \cdots$$

where $\infty + \frac{1}{n} = \infty$. A similar definition gives F_n. F_n and G_n are measurable since translates of measurable functions are measurable. So the sets $D_n = (G_n \neq G)$ are measurable. Clearly $D_n \subseteq [0, \infty)$ since $\infty \notin D_n$. Define the measurable functions f_n on K by

$$(7) \qquad f_n(t) = \begin{cases} \frac{F_n(t) - F(t)}{G_n(t) - G(t)} \text{ if } t \in B \cap D_n \\ f(\infty) \text{ if } t \in K - B \cap D_n. \end{cases}$$

The existence of $\frac{dF}{dG}(t)$ in (4) for all t in B implies $B \cap D_n \to B \cap [0, \infty)$ as $n \to \infty$. So (7) implies $f_n \to f$ on $B \cap [0, \infty)$ according to (4). Thus, since $f_n(\infty) = f(\infty)$ by (7), $1_B f_n \to 1_B f$ on K. So $\overline{f}$ in (5) is measurable since $1_B f_n$ is measurable, a consequence of Theorem 2 (§4.3). $\square$

Note that for F and G of bounded variation the continuity condition on G in Theorem 8 is not needed to get a measurable $\overline{f} = f$ dG-everywhere which in this case is given by Theorem 2 (§4.6).

If G is continuous then dampability of $|dG|$ is equivalent to dampability of dG. Indeed, $|dG|$ and dG have the same set of dampers. This follows from the next theorem.

THEOREM 9. *Let G be a continuous function on a cell K with $|dG|$ dampable. Then udG is absolutely integrable for every function $u \geq 0$ on K such that $u|dG|$ is integrable. So dG is dampable.*

PROOF. Consider any $u \geq 0$ such that $u|dG|$ is integrable. That is,

$$(8) \qquad dv = u|dG|$$

for some function v on K. By Theorem 2 $\frac{dv}{|dG|}(t) = u(t)$ at all t in a subset E of K such that $1_E dG = dG$. Existence of this derivative implies that at each t in E $\Delta G(I) \neq 0$ ultimately as $(I, t) \to t$. For such t the intermediate value property of the continuous function G implies that

(9)   either $\Delta G(I) > 0$ ultimately, or $\Delta G(I) < 0$ ultimately

as $(I, t) \to t-$, and (9) holds as $(I, t) \to t+$.

Now we contend that this condition implies that (9) holds as $(I, t) \to t$, for all but countably many t in E. Indeed, consider the set A of those t interior to K such that $\Delta G(I) < 0$ ultimately as $(I, t) \to t-$ and $\Delta G(I) > 0$ ultimately as $(I, t) \to t+$. Each t in A has a neighborhood L on which the restriction of G to L attains its minimum uniquely at t. Assign to each t in A such a neighborhood $L_t = (p, q)$ with rational endpoints p, q. So $G(t) < G(s)$ for all s in L such that $s \neq t$. This condition clearly implies that $L_s \neq L_t$ for $s \neq t$. So A is countable since there are only countably many intervals with rational endpoints. Since this argument applies also to $-G$ we can conclude that (9) holds as $(I, t) \to t$ for all but countably many t in E. Countable sets are dG-null since G is continuous. So (9) holds as $(I, t) \to t$ for dG- all t in K. That is, there is a function f on K with values ± 1 such that $\frac{dG}{|dG|}(t) = f(t)$ at dG-all t. By Theorem 3

(10) $dG = f|dG|$.

By Theorem 8 we may assume that f in (10) is measurable. By (10) and (8),

(11) $udG = fdv$.

Since $u \geq 0$ and $dv \geq 0$, (11) and (8) imply $|fdv| = dv$. So fdv is summable. Thus, since f is measurable, fdv is absolutely integrable by Theorem 3 (§4.3). In terms of (11) udG is absolutely integrable. □

Using Theorem 8 we get the following version of a Radon-Nikodym Theorem from Theorem 6. In particular it applies to the classical case $G = x$ on a bounded cell K. (For this case on $\mathbb{R}$ see Exercise 13.)

THEOREM 10. *Let G be a continuous function of bounded variation on a cell K. For any function F on K the following conditions are equivalent:*

(i) dF is dampable and every dG-null set is dF-null.

(ii) $dF = fdG$ for some function f on K.

PROOF. That (i) implies (ii) follows from Theorem 6 where (a) implies (b) for dampable dF. The converse follows similarly from Theorem 6 if we can show that (ii) implies that dF is dampable.

Given (ii) we may assume by Theorem 8 that f is measurable. Define the damper u on K by $u = \frac{1}{|f|}$ wherever $f \neq 0$ and $u = 1$ wherever $f = 0$. Then (ii) gives $udF = ufdG = (sgn f)dG$ which is absolutely integrable since $sgn f$ is measurable and bounded and dG is absolutely integrable. So dF is dampable. $\square$

Exercises (§6.3).

1. Given $dF = fdG$ on a cell $[a, b]$ with G regulated prove:

(i) If G is discontinuous from the left at p in $(a, b]$ then the left derivative

$$\left(\frac{dF}{dG}\right)_{-}(p) = \frac{F(p) - F(p-)}{G(p) - G(p-)} = f(p).$$

(ii) If G is discontinuous from the right at p in $[a, b)$ then the right derivative

$$\left(\frac{dF}{dG}\right)_{+}(p) = \frac{F(p+) - F(p)}{G(p+) - G(p)} = f(p).$$

(iii) If G is discontinuous from both the left and the right at p in (a, b) then

$$\frac{dF}{dG}(p) = f(p).$$

2. Let $dF_n = f_n dx$ on a cell K in $\mathbb{R}$ for $n = 1, 2, \cdots$ where f_n converges uniformly to f. Let B be the set of all t in K at which $F'_n(t) = f_n(t)$ for all n. Prove:

(i) $1_B dx = dx$ on K.

(ii) $dF_n \to dF$ on K where $dF = f dx$.

(iii) At each point t in B the functions $F_1, F_2, \cdots$ are equidifferentiable. (Given $\varepsilon > 0$ there exists a gauge δ on B such that $|\frac{\Delta F_n}{\Delta x}(I) - f_n(t)| < \varepsilon$ for all δ-fine (I, t) in K with t in B, and for all n.)

(iv) $F'(t) = f(t)$ for all t in B.

3. Apply the preceding exercise to prove the following well known theorem: If $F_1, F_2, \cdots$ is a sequence of differentiable functions on an interval L in $\mathbb{R}$ such that F_n converges to F and F'_n converges uniformly to f, then $F' = f$ on L.

4. Let $f dg$ and $1_D f dg$ be integrable on a cell K where g is of bounded variation and D is the set of all points at which g is discontinuous. Show that there is a measurable subset E of K such that $1_E dg = dg$, the function $\overline{f} = 1_E f$ is measurable, and $\overline{f} = f$ dg-everywhere. (*Hint:* Apply Theorem 8 to the continuous differential $dG = 1_C dg$ for $C = K - D$ to get B. Then let $E = B \cup D$.)

5. Given a continuous function G on a cell K with $|dG|$ dampable verify the following conclusions (citing the appropriate theorems):

(i) $dG = f|dG|$ for some measurable function f,

(ii) There exist complementary measurable subsets A, B of K such that $1_A dG = (dG)^+$ and $1_B dG = -(dG)^-$,

(iii) $\frac{dG}{|dG|}(t)$ exists at dG-all t in K,

(iv) For dG-all s in K there exist neighborhoods L of s such that either $G(r) < G(s) < G(t)$ for all r, t in L such that $r < s < t$, or $G(r) > G(s) > G(t)$ for all r, t in L such that $r < s < t$,

(v) The set of all points in K at which G has either a local minimum or a local maximum is dG-null.

6. Let F be a function on $\mathbb{R}$ such that F is of bounded variation on every bounded interval, and every dx-null subset of $\mathbb{R}$ is dF-null. Show that $dF = f dx$ for some function f on $\mathbb{R}$.

7. Let $g \leq f \leq h$ on a bounded cell K where f is measurable and both $g dx$ and $h dx$ are integrable. Show that $f dx$ is integrable. (Get a measurable function $\bar{g}$ from Theorem 8. Then apply Theorem 3 (§4.3) to the integrand $f - \bar{g}$ with $\sigma = dx$.)

8. Given $f dx$ integrable on $\mathbb{R}$ show that for dx-all p in $\mathbb{R}$,

$$\frac{1}{h} \int_p^{p+h} [f(t+h) - f(t)] dt \to 0 \text{ and}$$

$$\frac{1}{h} \int_0^h f(p+t) dt \to f(p) \text{ as } h \to 0.$$

9. Let f be measurable on a neighborhood of p in $\mathbb{R}$ and differentiable at p. Since f is continuous at p it is bounded on some neighborhood of p. So $f dx = dF$ for some Lipschitz function F on an interval K about p with $F' = f$ dx-everywhere on K. In particular $F'(p) = f(p)$ by continuity of f at p. Prove:

(i) $\frac{1}{h} \int_0^h \frac{f(p+t) - f(p)}{t} dt \to f'(p)$ as $h \to 0$.

(ii) $\frac{2}{h^2} \int_0^h [f(p+t) - f(p)]dt \to f'(p)$ as $h \to 0$.

(iii) $F(p+h) = F(p) + hf(p) + \frac{h^2}{2}f'(p) + o(h^2)$ as $h \to 0$.
(Use (ii) to prove this generalization of the quadratic Taylor formula.)

(iv) $f'(p)$ is the Schwarzian second derivative of F at p,

$$F(p+2h) - 2F(p+h) + F(p) = h^2 f'(p) + o(h^2)$$

as $h \to 0$. (Use (iii).)

(v) $f'(p)$ is the symmetric Schwarzian second derivative of F at p,

$$F(p+h) - 2F(p) + F(p-h) = h^2 f'(p) + o(h^2)$$

as $h \to 0$. (Use (iii).)

10. Given $dF \geq 0$ on an interval L in $\mathbb{R}$ show that F is differentiable dx-everywhere on L and $F'dx \leq dF$ on L. (Apply Theorem 7.)

11. Show that if $dG \leq dH$ on an interval L in $\mathbb{R}$ and one of the functions G, H is differentiable dx-everywhere on L then so is the other. (*Hint:* $d(H - G) \geq 0$.)

12. Show that if $gdx \leq dF \leq hdx$ on an interval L in $\mathbb{R}$ and either of the differentials gdx, hdx is integrable, then $dF = fdx$ for some function f on L. (Use Theorem 2.)

13. Given a function F on $\mathbb{R}$ apply Theorem 10 to prove the equivalence of the following conditions:

(i) dF is dampable and every dx-null set is dF-null.

(ii) $dF = fdx$ for some function f on $\mathbb{R}$.

14. In Theorem 6 let (d) be condition (c) with "dF-all" replaced by "dG-all". Prove:

(i) (a) implies (d), (ii) (d) does not imply (a). (Consider $F(t) = t$ and $G(t) = t^+$ on $[-1,1]$.)

15. Let f be a function on a cell K in $\mathbb{R}$ and g a function on $f(K)$ such that $df = (g \circ f)dx$ on K. Show that f is monotone (i.e. either $df \geq 0$ on K or $df \leq 0$ on K). (*Outline of Proof*: Let A consist of all t in K where at least one of the following holds: (i) $f'(t)$ does not exist, (ii) $f'(t)$ exists and equals 0, (iii) $f'(t)$ exists and does not equal $g \circ f(t)$. Then A is df-null, so $f(A)$ is Lebesgue-null. Suppose f *not* monotone. Then there would exist c in $f(K) - f(A)$ and p, q in $f^{-1}(c)$ such that $p < q$ and either $f > c$ on (p, q) or $f < c$ on (p, q). This yields the contradictory conclusion $f'(p)f'(q) \leq 0$ and $f'(p) = f'(q) \neq 0$.)

16. Prove that Theorem 10 remains valid of we adjoin a third condition as follows:

    (iii) If A is a measurable subset of K such that every measurable subset B of A with $1_B dF$ absolutely integrable is dG-null, then A is dF-null.

    (See Exercise 9 in §4.3.)

§6.4 L'Hôpital's Rule and the Limit Comparison Test Using Essential Limits.

The differentiation formulas of calculus were needed for evaluation of derivatives because derivatives are of the indeterminate form $\frac{0}{0}$. L'Hôpital's rule exploits differentiation to evaluate limits of the form $\frac{0}{0}$ or $\frac{\infty}{\infty}$. This rule, devised originally by J. Bernoulli, allows one to substitute the derivatives F', G' for the functions F, G on (a, b) to evaluate the limit of $\frac{F(t)}{G(t)}$ as $t \to a+$ (or as $t \to b-$) when this ratio has the limiting form $\frac{0}{0}$. To deal with $\frac{F'(t)}{G'(t)}$ as $t \to a+$ one must have $G'(t) \neq 0$ ultimately. So G must be strictly monotone ultimately since a derivative G' has the intermediate value property on intervals.

Now $\frac{F'(t)}{G'(t)}$ is the derivative $\frac{dF}{dG}(t)$. Given $F(a+) = G(a+) = 0$ we can extend F, G to $[a, b)$ with continuity at a by setting $F(a) = G(a) = 0$. Then

$$\frac{dF}{dG}(a) = \lim_{t \to a+} \frac{F(t) - F(a)}{G(t) - G(a)} = \lim_{t \to a+} \frac{F(t)}{G(t)}$$

if this limit exists. So L'Hôpital's rule says in effect that for G strictly monotone on $[a, b)$ with F, G continuous at a

(1) $$\frac{dF}{dG}(a) = \frac{dF}{dG}(a+)$$

whenever the right-hand side exists.

Now if $\frac{dF}{dG}(t)$ exists and is finite on (a, b) with G strictly monotone then

(2) $$dF = \frac{dF}{dG} dG \text{ on } (a, b).$$

Thus a more general formulation of (1) is $\frac{dF}{dG}(a) = f(a+)$ if $f(a+)$ exists for f a function such that

(3) $$dF = f dG \text{ on } (a, b).$$

Now the function f acting as a differential coefficient in (3) may be altered on a dG-null set. So the limit $f(a+)$ can be taken "dG-essentially". Let us explain this notion briefly.

Let σ be a differential on $L = (a, b)$. Let $\mathcal{F}$ be a filterbase of subsets J of L none of which is σ-null. (For $\sigma = dx$ and $t \to a+, \mathcal{F}$ consists of all intervals (a, r) with r in (a, b).) Let $\mathcal{F}_\sigma$ be the filterbase consisting of all $J \cap A$ such that J belongs to $\mathcal{F}$ and $1_A \sigma = \sigma$. For f a function on L the σ-**essential** lower and upper limits of $f(t)$ with respect to $\mathcal{F}$ are the lower and upper limits of $f(t)$ with respect to $\mathcal{F}_\sigma$. Clearly $\mathcal{F}_\sigma$ is finer that $\mathcal{F}$. That is, $\mathcal{F}_\sigma \supseteq \mathcal{F}$. So

$$-\infty \leq \underline{\lim}_{\mathcal{F}} f(t) \leq \underline{\lim}_{\mathcal{F}_\sigma} f(t) \leq \overline{\lim}_{\mathcal{F}_\sigma} f(t) \leq \overline{\lim}_{\mathcal{F}} f(t) \leq \infty.$$

(See §11.1 for filterbase lower and upper limits.)

THEOREM 1. *Let $dF = fdG$ on (a, b) with G strictly monotone. Let p, q respectively be the dG-essential lower and upper limits of $f(t)$ as $t \to a+$. Then (i) and (ii) hold:*

(i) As $s, t \to a+$ with $s < t$

$$(4) \qquad -\infty \leq p \leq \varliminf \frac{F(t) - F(s)}{G(t) - G(s)}$$

and

$$(5) \qquad \varlimsup \frac{F(t) - F(s)}{G(t) - G(s)} \leq q \leq \infty.$$

(ii) If $|F|(a+) = |G|(a+) = z$ where z is either 0 or ∞ then as $t \to a+$

$$(6) \qquad p \leq \varliminf \frac{F(t)}{G(t)}$$

and

$$(7) \qquad \varlimsup \frac{F(t)}{G(t)} \leq q.$$

(The theorem is also valid if $a+$ is replaced by $b-$.)

PROOF. We may assume $dG > 0$ since for the case $dG < 0$ we can substitute $-G, -F$ for G, F with $d(-F) = fd(-G)$ equivalent to $dF = fdG$. Moreover, we need only prove (4) and (6) since their application to $d(-F) = (-f)dG$ gives (5) and (7) respectively.

To prove (4) we may ignore the trivial case $p = -\infty$. (4) will then follow from

(iii) Given c in $(-\infty, p)$ there exists r in (a, b) such that $\Delta F(I) \geq c\Delta G(I)$ for every cell $I = [s, t]$ in (a, r).

So we must prove (iii).

Since p is the dG-essential lower limit of $f(t)$ as $t \to a+$ every c in $(-\infty, p)$ is a dG-essential ultimate lower bound of $f(t)$. That is, given c in $(-\infty, p)$ there exists r in (a, b) such that $f(t) \geq c$ for dG-all t in (a, r). This implies $dF \geq c\,dG$ on (a, r) since $dF = f\,dG \geq c\,dG$ for $dG > 0$. Integration of $dF \geq c\,dG$ over I gives the conclusion of (iii),

$$(8) \qquad F(t) - F(s) \geq c[G(t) - G(s)] \text{ for } a < s < t < r.$$

Let the hypothesis of (ii) hold with $z = 0$. That is, $F(a+) = G(a+) = 0$. Thus as $s \to a+$ (8) gives

$$(9) \qquad\qquad F(t) \geq cG(t) \text{ for } a < t < r.$$

Since $dG > 0$ and $G(a+) = 0, G(t) > 0$ for all t in (a, b). So we can conclude from (9) that given c in $(-\infty, p)$ these exists r in (a, b) such that $\frac{F(t)}{G(t)} \geq c$ for all t in (a, r). This proves (6) for the case $z = 0$.

For the case $z = \infty$ both $F(s)$ and $G(s)$ are ultimately nonzero as $s \to a+$ and their reciprocals $\frac{1}{F(s)}$ and $\frac{1}{G(s)}$ converge to 0. We invoke the identity

$$(10) \qquad \frac{F(s)}{G(s)} = \left[\frac{F(t) - F(s)}{G(t) - G(s)}\right] \left[\frac{\frac{G(t)}{G(s)} - 1}{\frac{F(t)}{F(s)} - 1}\right].$$

Under (8) the first factor on the right-hand side of (10) is bounded below by c while the second factor converges to 1 as $s \to a+$ with t fixed. So $\underline{\lim}\frac{F(s)}{G(s)} \geq c$ as $s \to a+$. Since this holds for all c in $(-\infty, p)$ we get (6). $\quad\square$

Our next result, a general version of L'Hôpital's rule, is the special case $p = q$ of Theorem 1. So it is just a corollary of Theorem 1.

THEOREM 2. *Let $dF = f dG$ on (a, b) with G strictly mono-tone. For some p in $[-\infty, \infty]$ let $f(t) \to p$ dG-essentially as $t \to a+$. Then (i) and (ii) hold:*
(i) As $s, t \to a+$ with $s < t$

(11)
$$\frac{F(t) - F(s)}{G(t) - G(s)} \to p.$$

(ii) If $|F|(a+) = |G|(a+) = z$ where z is either 0 or ∞ then

(12)
$$\frac{F(t)}{G(t)} \to p \ as \ t \to a+.$$

(The theorem is also valid if $a+$ is replaced by $b-$.)

Theorem 2 gives L'Hôpital's rule for series. This is our next result.

THEOREM 3. *Let $\langle a_n \rangle$ and $\langle b_n \rangle$ be real sequences with $b_n > 0$ for $n = 1, 2, \cdots$. Suppose that for some p in $[-\infty, \infty]$*

(i) $\frac{a_n}{b_n} \to p$ as $n \to \infty$.

Then
(ii) $\frac{a_m + a_{m+1} + \cdots + a_n}{b_m + b_{m+1} + \cdots + b_n} \to p$ as $m, n \to \infty$ with $m \leq n$.

Moreover, if both $\sum_1^\infty a_n$ and $\sum_1^\infty b_n$ are convergent series then

(iii) $\frac{a_m + a_{m+1} + \cdots}{b_m + b_{m+1} + \cdots} \to p$ as $m \to \infty$.

PROOF. Define the functions F, G, f on $[0, \infty)$ by

$$F(t) = \left(\sum_{k=1}^n a_k \right) + a_n(t - n),$$

$$G(t) = \left(\sum_{k=1}^n b_k \right) + b_n(t - n),$$

and

$$f(t) = \frac{a_n}{b_n} \text{ for } n - 1 \leq t < n \text{ where } n = 1, 2, \cdots .$$

F and G are continuous, piecewise linear functions with $F(0) = G(0) = 0$. Moreover, $dF = a_n dx$ and $dG = b_n dx > 0$ on $[n-1, n]$. So $dF = f dG$ on $[0, \infty)$. By (i) $f(t) \to p$ as $t \to \infty$. So Theorem 2 applies with $t \to b-$ for $b = \infty$. Since $F(n) = \sum_{k=1}^{n} a_k$ and $G(n) = \sum_{k=1}^{n} b_k$ (ii) is just $\frac{F(n)-F(m-1)}{G(n)-G(m-1)} \to p$ as $m, n \to \infty$ with $m \leq n$. This follows from (i) in Theorem 2.

If the two series are convergent we can apply (ii) of Theorem 2 to the functions $F - \sum_1^{\infty} a_n$, $G - \sum_1^{\infty} b_n$. Then (12) in Theorem 2 gives (iii) in Theorem 3. $\square$

The Limit Comparison Test for integrals (and for series as a special case) is based on the same ideas as those that give L'Hôpital's rule.

THEOREM 4. *Let* $dF = f dG$ *on* $[a, b)$ *with* $dG \geq 0$. *So* $-\infty < G(b-) \leq \infty$. *Let* p, q *be the* dG-*essential lower and upper limits of* $f(t)$ *as* $t \to b-$

(i) *If* $G(b-), p, q$ *are all finite then* $F(b-)$ *exists and is finite. So* dF *is integrable on* $[a, b)$ *and*

(13)
$$\int_{[a,b)} dF = F(b-) - F(a).$$

(ii) *If* $G(b-) = \infty$ *and either* $p > 0$ *or* $q < 0$ *then* $F(b-) = \infty$ *if* $p > 0$, *and* $F(b-) = -\infty$ *if* $q < 0$. *So* $\int_{[a,b)} dF = F(b-) = \pm\infty$.

PROOF. Under the hypothesis in (i) $G(b-) < \infty$ and there exists $c < \infty$ and r in (a, b) such that $|f(t)| < c$ for dG-all t in $[r, b)$. This implies $|dF| \leq c dG$ on $[r, b)$ since $dF = f dG$ and

$dG \geq 0$. So F is of bounded variation on $[r, b)$. Hence $F(b-)$ exists and is finite. So (13) holds.

To prove (ii) let $G(b-) = \infty$ and $p > 0$. The latter condition yields $\varepsilon > 0$ and r in (a, b) such that $f(t) > \varepsilon$ for all dG-all t in $[r, b)$. So

$$dF = f dG \geq \varepsilon dG \geq 0 \text{ on } [r, b).$$

So $dF \geq 0$ on $[r, b)$ and integration of $dF \geq \varepsilon dG$ over $[r, b)$ gives $F(b-) - F(r) \geq \varepsilon[G(b-) - G(r)] = \infty$ since $G(b-) = \infty$. So $F(b-) = \infty$.

For the case $q < 0$ apply the preceding argument to $d(-F) = (-f)dG$. The lower dG-essential limit of $-f(t)$ is $-q > 0$. So the previous case (for positive lower limit) gives $(-F)(b-) = \infty$. That is, $F(b-) = -\infty$.   □

(Some references for L'Hôpital's rule are [2], [10], [15].)
Exercises (§6.4).

Prove each of the following:

1.
$$\frac{\sin \frac{1}{n} + \sin \frac{1}{n+1} + \cdots + \sin \frac{1}{5n}}{\frac{1}{n} + \frac{1}{n+1} + \cdots + \frac{1}{5n}} \xrightarrow[n \to \infty]{} 1.$$

2. For all $c > 0$ in $\mathbb{R}$

$$\frac{\sum_{k=n}^{\infty} \left(\frac{c}{k}\right)^k}{\sum_{k=n}^{\infty} \left(\frac{c}{k+1}\right)^k} \xrightarrow[n \to \infty]{} e.$$

3. The function f defined on $\mathbb{R}$ by $f(t) = \frac{\sin t}{t}$ for $t \neq 0$ and $f(0) = 1$ is differentiable on $\mathbb{R}$ and $f'(0) = 0$.

4. (Limit Comparison Test.) Let f, g be functions on $L = [0, \infty)$ such that both $f dx$ and $g dx$ are integrable on $[0, c]$ for all c in L, $g > 0$, and the limit $\frac{f}{g}(\infty-) = r$ exists in $[-\infty, \infty]$. Prove:

(i) If $\int_L g\,dx < \infty$ and $|r| < \infty$ then $f\,dx$ is integrable on L,

(ii) If $\int_L g\,dx = \infty$ and $r \neq 0$ then $\int_L f\,dx = (sgn\, r)\infty$. (Apply Theorem 4 with $dG = g\,dx$, and f replaced by f/g.)

5. (Cesàro Means.) Given $a_n \to 0$ in $\mathbb{R}$ let $s_n = a_1 + \cdots + a_n$. Prove:

    (i) Given $\varepsilon > 0$ there exists a positive integer N such that $|s_n - s_N| < n\varepsilon$ for all $n > N$. (Apply Theorem 3 with $b_n = 1$.)

    (ii) $|s_n| < n\varepsilon + |s_N|$ for all $n > N$.

    (iii) $\frac{1}{n}s_n \to 0$.

    (iv) For any convergent sequence $c_n \to p$ in $\mathbb{R}$ the sequence $\frac{1}{n}(c_1 + \cdots + c_n) \to p$. (Let $a_n = c_n - p$.)

6. Let F be a continuous function on $\mathbb{R}$ which is differentiable at all but countably many points. Let c be a point in $\mathbb{R}$ where the dx-essential limits $F'(c-)$ and $F'(c+)$ of the derivative F' exist and are finite. Then these limits are equal if and only if F is differentiable at c.

7. Given a function F on $\mathbb{R}$ let S be the cell summant $\Delta F/\Delta x$.

    (i) If I, J are abutting cells in $\mathbb{R}$ then $S(I \cup J) = rS(I) + (1-r)S(J)$ for some r such that $0 < r < 1$.

    (ii) If F is differentiable at a point c then $S([s,t]) \longrightarrow F'(c)$ as $s \longrightarrow c-$ and $t \longrightarrow c+$.

§6.5 Differentiation Under the Integral Sign.

If $F' = f$ on an interval L in $\mathbb{R}$ then $dF = f\,dx$ on L. This version of the Fundamental Theorem of Calculus dispenses with the restrictive hypotheses such as continuity of f or absolute integrability of $f\,dx$ demanded in earlier versions. It thereby yields many results that formerly could only be proved by invoking the Theorem of the Mean. This diminishes the utility of the Theorem of the Mean. But this classical theorem is still of some use. We invoke it here to prove a version of the well known theorem on differentiation under the integral sign. In this version $\frac{\partial f}{\partial s}(s,t)dt$ need not be absolutely integrable, (2) with h replaced by $2h$ being weaker than the usual demand that $\left|\frac{\partial f}{\partial s}(s,t)\right| \leq h(t)$ for all s. Absolute integrability is not demanded in (i) either.

THEOREM 1. *Let f be a function on $J \times K$ where J, K are intervals in $\mathbb{R}$ such that (i), (ii), (iii) hold:*
(i) For each s in $J, f(s,t)dt$ is integrable on K,

(ii) At dx-all t in K

(1) $\qquad \dfrac{\partial f}{\partial s}(s,t)$ *exists and is finite for all s in J,*

(iii) There exists a function $h \geq 0$ on K such that $h(t)dt$ is integrable on K and for dx-all t in K

(2) $\qquad \left|\dfrac{\partial f}{\partial s}(s_1,t) - \dfrac{\partial f}{\partial s}(s_2,t)\right| \leq h(t)$ *for all s_1, s_2 in J.*

Then the function F defined on J by

(3) $$F(s) = \int_K f(s,t)dt$$

is differentiable on J and

(4) $$F'(s) = \int_K \frac{\partial f}{\partial s}(s,t)dt \text{ for all } s \text{ in } J.$$

PROOF. Given s in J consider any sequence $\lambda_n \to 0$ in $\mathbb{R}$ with $\lambda_n \neq 0$ and $s + \lambda_n$ in J for all n. To prove (4) we need only show that $\frac{\partial f}{\partial s}(s,t)dt$ is integrable on K and

$$(5) \qquad \frac{F(s + \lambda_n) - F(s)}{\lambda_n} \xrightarrow[n \to \infty]{} \int_K \frac{\partial f}{\partial s}(s, t)dt.$$

Define g_n on K by

$$(6) \qquad g_n(t) = \frac{f(s + \lambda_n, t) - f(s, t)}{\lambda_n}$$

for all t in K. By (6) and (ii)

$$(7) \qquad g_n(t) \xrightarrow[n \to \infty]{} \frac{\partial f}{\partial s}(s, t) \text{ for } dx\text{-all } t \text{ in } K.$$

By (3) and (6)

$$(8) \qquad \int_K g_n(t)dt = \frac{F(s + \lambda_n) - F(s)}{\lambda_n}.$$

For each t where (1) holds the Theorem of the Mean applied to (6) gives $g_n(t) = \frac{\partial f}{\partial s}(s_n, t)$ for some s_n between s and $s + \lambda_n$. So (2) gives

$$(9) \qquad |g_n(t) - g_1(t)| \leq h(t) \text{ at } dx\text{-all } t \text{ in } K.$$

The Dominated Convergence Theorem (Theorem 4(§2.8)) applied to (7), (8), and (9) gives (5) and (4). $\square$

Theorem 1 provides a useful tool for evaluating certain integrals. The exercises yield the evaluation $\int_{\mathbb{R}} \frac{\sin t}{t} dt = \pi$. The failed attempt in Exercise 1 shows the importance of (iii) in Theorem 1.

Exercises (§6.5).

1. For $J = K = (0, \infty)$ define f on $J \times K$ by $f(s, t) = \frac{\sin st}{t}$. Verify the following:

   (1°) $f(s, 0+) = s$ for all s in J.

   (2°) $F(s) = \int_K f(s, t)dt = \int_0^\infty \frac{\sin x}{x}dx$ with integrability given by (1°) and Exercise 5 (§3.4) or Exercise 1(§6.2).

   (3°) $\frac{dF}{ds} = 0$ on J.

   (4°) $\frac{\partial f}{\partial s}(s, t) = \cos st$ for all (s, t) in $J \times K$.

   (5°) $\int_0^x \frac{\partial f}{\partial s}(s, t)dt = f(x, s)$ whose values oscillate between $-\frac{1}{s}$ and $\frac{1}{s}$ as $x \to \infty$. $\int_0^\infty \frac{\partial f}{\partial s}(s, t)dt$ does not exist.

   (6°) Theorem 1 does not apply because (iii) fails: Given t in K there exist s_1, s_2 in J such that $|\cos s_1 t - \cos s_2 t| = 2$ and $\int_0^\infty h(t)dt = \infty$ for $h \geq 2$ a.e. on K.

2. Verify the validity of the following demonstration that $\int_0^\infty \frac{\sin t}{t}dt = \frac{\pi}{2}$: Let $F(s) = \int_0^\infty e^{-st}\frac{\sin t}{t}dt$ for $s > 0$. Then

$$F'(s) = -\int_0^\infty e^{-st}\sin t\, dt = \frac{e^{-st}(\cos t + s\sin t)}{1 + s^2}\Big|_{t=0}^\infty =$$

$$-\frac{1}{1 + s^2} = \frac{d}{ds}Arcctn\, s.$$

Thus, since $F(\infty-) = 0 = Arcctn\,\infty$, $F(s) = Arcctn\, s$ for all $s > 0$. So $F(0+) = Arcctn\, 0 = \frac{\pi}{2}$. By Exercise 15 (§6.2) $\int_0^\infty (1 - e^{-st})\frac{\sin t}{t}dt \to 0$ as $s \to 0+$. That is, $F(0) = F(0+) = \frac{\pi}{2}$.

3. Verify the correctness of the following demonstration that $\int_0^\infty \frac{\sin t}{t}dt = \frac{\pi}{2}$:

(1°) Let $F_n(s) = \int_0^{2\pi} e^{-s(t+a_n)}\frac{\sin t}{t+a_n}dt$ where $a_n = 2\pi n$ for $n = 0, 1, \cdots$ and $0 \le s < \infty$.

(2°) $F_n'(s) = -e^{-a_n s}\int_0^{2\pi} e^{-st}\sin t\, dt =$
$\frac{e^{-a_n s}[e^{-st}(\cos t + s\sin t)]}{1+s^2}\Big|_{t=0}^{2\pi} = \frac{e^{-a_{n+1}s} - e^{-a_n s}}{1+s^2}.$

(3°) $F_n(\infty-) = 0$ so $F_n(0) = \int_0^\infty \frac{e^{-a_n s} - e^{-a_{n+1}s}}{1+s^2}ds.$

(4°) $F_n(0) = \int_0^{2\pi}\frac{\sin t}{t+a_n}dt = \int_{a_n}^{a_{n+1}}\frac{\sin x}{x}dx.$

(5°) $\int_0^\infty \frac{\sin t}{t}dt = \sum_{n=0}^\infty F_n(0) = \int_0^\infty \frac{ds}{1+s^2} = Arctan\, s\Big|_0^\infty = \frac{\pi}{2}$. (See Exercises 4, 5 in §3.4 and Theorem 11 in §3.4.)

CHAPTER 7

ESSENTIAL PROPERTIES OF FUNCTIONS

Essential limits have been defined in §6.4 with respect to a differential σ. In this chapter $\sigma = dx$. So essential means dx-essential, measure is Lebesgue measure M, measurable is Lebesgue measurable, almost everywhere is dx-everywhere, etc. The essential properties of a function g on an interval L in $\mathbb{R}$ are those it shares with all functions equal a.e. to g on L. These are just the properties of g that are relevant to the differential $g\,dx$.

§7.1 Essentially Bounded Functions.

Let g be a function on a measurable subset E of $\mathbb{R}$ with $M(E) > 0$. An **essential upper bound** of g is a constant c in $[-\infty, \infty]$ such that $g \leq c$ a.e. on E. The least such c is the **essential supremum** of g on E. (See Exercise 1.) Dual definitions hold for **essential lower bounds** and the **essential infimum** of g. g is **essentially bounded** if $|g|$ has a finite essential upper bound.

If g is measurable and essentially bounded on an interval L in $\mathbb{R}$ then $g f\,dx$ is absolutely integrable for all functions f on L such that $f\,dx$ is absolutely integrable. The converse also holds. Indeed, Theorem 1 gives a seemingly stronger result. (See Exercise 2.) For its proof we need a measure-theoretic lemma.

LEMMA A. *Let g be measurable on a measurable subset L of $\mathbb{R}$ such that $g < \infty$ a.e. but g has no finite essential upper bound. Then there is a sequence of disjoint measurable subsets $E_1, \cdots, E_k, \cdots$ of L such that*

$$(1) \qquad\qquad 0 < M(E_k) < \infty$$

and $g > k$ on E_k.

PROOF. Since g is measurable the sets $A_n = (g > n)$ are measurable for $n = 1, 2, \cdots$. Also $A_{n+1} \subseteq A_n$. We contend that $M(A_n - A_{n+1}) > 0$ for infinitely many values of n. Suppose this were false. Then for n sufficiently large A_n would be the countable union $(g = \infty) + (A_n - A_{n+1}) + (A_{n+1} - A_{n+2}) + \cdots$ of sets of measure 0 making A_n of measure 0. But this implies by the definition of A_n that n is an essential upper bound of g, contradicting the hypothesis that no such bound exists.

So there is an increasing sequence of positive integers $n_1 < \cdots < n_k < \cdots$ such that $M(A_{n_k} - A_{n_k+1}) > 0$ for all k. For each k take a bounded interval I_k large enough so that (1) holds for $E_k = I_k \cap (A_{n_k} - A_{n_k+1})$. Clearly, the E_k's are disjoint and $g > n_k \geq k$ on E_k. $\quad\square$

THEOREM 1. *Let g be a function on an interval L in $\mathbb{R}$ such that $gf dx$ is integrable for every function f on L such that $f dx$ is absolutely integrable. Then g is measurable and essentially bounded.*

PROOF. The hypothesis applied to $f = \frac{1}{1+x^2}$ shows that $\frac{g}{1+x^2} dx$ is integrable. So $\frac{g}{1+x^2}$ is measurable by Theorem 8 (§6.3). Hence, its product with $1 + x^2$ is measurable. That is, g is measurable.

Suppose g were not essentially bounded. Say g has no finite essential upper bound. (Otherwise this case applies to $-g$.)

Apply Lemma A to get $E_1, E_2, \cdots$ and define f on L by

$$f = \sum_{k=1}^{\infty} a_k 1_{E_k} \quad \text{where } a_k = \frac{1}{k^2 M(E_k)}.$$

Then $\int_L f dx = \sum_{k=1}^{\infty} a_k \int_L 1_{E_k} dx = \sum_{k=1}^{\infty} \frac{1}{k^2} < \infty$ by Theorem 2 (§2.7). Thus, since $f \geq 0$, $f dx$ is absolutely integrable. Now $gf \geq \sum_{k=1}^{\infty} k a_k 1_{E_k}$ since $g > k$ on E_k by Lemma A. So $\int_L gf dx \geq \sum_{k=1}^{\infty} \frac{1}{k} = \infty$, contradicting integrability of $gf dx$. So g is essentially bounded. $\square$

Exercises (§7.1).

1. Given a nonzero differential σ on a cell K let $\mathcal{E}$ consist of all subsets E of K such that $1_E \sigma = \sigma$. Given a function g on K let $\mathcal{G}$ be the set of all images $g(E)$ of members E of $\mathcal{E}$. Prove:

   (i) $\mathcal{E}$ is a filter closed under countable intersection.

   (ii) $\mathcal{G}$ is a filterbase for which the intersection of countably many terminal sets is a terminal set.

   (iii) c in $[-\infty, \infty]$ is a σ-essential upper (lower) bound of g if and only if c is an ultimate upper (respectively, lower) bound of $\mathcal{G}$.

   (iv) σ-ess sup $g = \overline{\lim} \, \mathcal{G}$ and σ-ess inf $g = \underline{\lim} \, \mathcal{G}$.

   (v) $\overline{\lim} \, \mathcal{G}$ is the least ultimate upper bound of $\mathcal{G}$; $\underline{\lim} \, \mathcal{G}$ is the greatest ultimate lower bound of $\mathcal{G}$. (Apply (ii)).
   (See §11.1.)

2. Show directly that integrability of $gf dx$ on an interval L for all f on L such that $f dx$ is absolutely integrable implies absolute integrability of $gf dx$.

3. For g a function on an interval L in $\mathbb{R}$ prove the equivalence of the following conditions:

   (i) g is measurable and essentially bounded.

(ii) $g dx = dG$ for some Lipschitz function G on L.

(iii) $g = G'$ a.e. for some Lipschitz function G on L.

§7.2 Essentially Regulated Functions.

Let g be a function defined and finite almost everywhere on an open interval L in $\mathbb{R}$. g is **essentially regulated** on L if the essential limits

$$(1) \qquad \overline{g_-}(t) = \operatorname*{ess\,lim}_{s \to t-} g(s) \quad \text{and} \quad \overline{g_+}(t) = \operatorname*{ess\,lim}_{s \to t+} g(s)$$

exist and are finite at all t in L. g is **essentially continuous** at t in L if ess $\lim_{s \to t} g(s)$ exists and is finite. g is **essentially continuous** on L if it is essentially continuous at every point in L. The following conclusions are obvious:

(i) A continuous function is essentially continuous.

(ii) An essentially continuous function is essentially regulated.

(iii) A regulated function is essentially regulated.

(iv) If $g = h$ a.e. and h is regulated then g is essentially regulated.

The next theorem gives the converse of (iv).

THEOREM 1. *Let g be essentially regulated on an open interval L in $\mathbb{R}$. Then $g = h$ a.e. for some regulated function h on L. For all such h and all t in L*

$$(2) \qquad h(t-) = \overline{g_-}(t) \quad \text{and} \quad h(t+) = \overline{g_+}(t)$$

in terms of (1). There is a unique such h, namely $h = \overline{g_-}$, that is left continuous. There is a unique such h, namely $h = \overline{g_+}$, that is right continuous.

PROOF. Let

$$(3) \qquad\qquad\qquad h = \overline{g_-}.$$

Given t in L and $\varepsilon > 0$ choose $\delta(t) > 0$ such that $(t - \delta(t), t + \delta(t))$ lies in L,

(4) $|g(s) - \overline{g_+}(t)| < \varepsilon$ for almost all s in $J_+ = (t, t + \delta(t))$,

and by (3)

(5) $|g(s) - h(t)| < \varepsilon$ for almost all s in $J_- = (t - \delta(t), t)$.

Taking essential limits in (4) at r in J_+ as $s \to r-$ and as $s \to r+$ we get by (1), (3), (4)

(6) $$|h(r) - \overline{g_+}(t)| \leq \varepsilon$$

and

(7) $$|\overline{g_+}(r) - \overline{g_+}(t)| \leq \varepsilon$$

for all r in J_+. Similarly (5) gives

(8) $$|h(r) - h(t)| \leq \varepsilon$$

and

(9) $$|\overline{g_+}(r) - h(t)| \leq \varepsilon$$

for all r in J_-. (6) and (7) give

(10) $$h(t+) = \overline{g_+}(t) = \overline{g_+}(t+).$$

Similarly (8) and (9) give

(11) $$h(t-) = h(t) = \overline{g_+}(t-).$$

So h is regulated and left continuous. By (4) and (6),

$$(12) \qquad |g(s) - h(s)| \leq |g(s) - \overline{g_+}(t)| + |\overline{g_+}(t) - h(s)| < 2\varepsilon$$

for almost all s in J_+. Similarly (5) and (8) give

$$(13) \qquad |g(s) - h(s)| \leq |g(s) - h(t)| + |h(t) - h(s)| < 2\varepsilon$$

for almost all s in J_-. By (12) and (13),

$$(14) \qquad\qquad |g - h| \leq 2\varepsilon \quad \text{a.e.}$$

on every δ-fine tagged cell (I, t) in L. Since each cell in L has a δ-division, (14) holds on every cell K in L for all $\varepsilon > 0$. Thus $g = h$ a.e. on K, hence on L which is a countable union of cells.

So h is a regulated, left continuous function equal a.e. to g on L.

For any function h equal a.e. to g on L

$$(15) \qquad\qquad \overline{g_-} = \overline{h_-} \quad \text{and} \quad \overline{g_+} = \overline{h_+}.$$

If, moreover, h is regulated then

$$(16) \qquad\quad \overline{h_-}(t) = h(t-) \quad \text{and} \quad \overline{h_+}(t) = h(t+)$$

for all t in L. So (2) follows from (15) and (16) for every regulated h equal a.e. to g.

Let $h^\star$ be any regulated, left continuous function equal a.e. to g. Let $f = h - h^\star$ where h is given by (3). Then f is regulated, left continuous and equal a.e. to 0. Clearly $f = 0$. That is, $h^\star = h$. So h is the unique regulated, left continuous function equal a.e. to g. A similar proof holds for right continuity. $\square$

For g defined a.e. on an open interval L in $\mathbb{R}$ we have some immediate corollaries of Theorem 1:

(v) If g is essentially regulated then it is essentially continuous at all but countably many points, and Lebesgue-measurable; it is also essentially bounded on each cell in L.

(vi) If h is regulated on L and $g = h$ a.e. then g is essentially continuous at t in L if and only if $h(t-) = h(t+)$.

(vii) If h is continuous on L and $g = h$ a.e. then g is essentially continuous.

(viii) If g is essentially continuous then there is a unique continuous h on L such that $g = h$ a.e.

(ix) If g is essentially regulated then there exists a regulated $h = g$ a.e. such that h is continuous wherever g is essentially continuous.

Exercises (§7.2).

1. Verify $(i), \cdots , (ix)$.

2. On an open interval L in $\mathbb{R}$ let $dG = g\,dx$ with g essentially regulated. Prove that G has finite left and right derivatives at every point in L. (Hence, G is differentiable at all but countably many points in L. See Exercise 12 in §6.1.)

§7.3 Essential Variation.

Theorem 1 (§7.1) characterizes the multipliers g of absolutely integrable differentials of the form $f\,dx$. We seek a similar characterization for the multipliers g of integrable $f\,dx$. Such a characterization was given by Sargent [39] formulated in terms of the Denjoy integral. Our version is Theorem 2 below. It involves the concept of essential variation.

The **essential variation** of a function g on a bounded cell K is defined by

$$(1) \qquad \text{ess var } g = \inf_{h = g\,a.e.} \int_K |dh|.$$

This is finite if and only if g is equal a.e. to some function h of bounded variation. Since a function h of bounded variation is regulated it is essentially regulated. So a function g of bounded essential variation is essentially regulated.

The infimum in (1) is actually a minimum. This is obvious for ess var $g = \infty$ since we always have ess var $g \leq \int_K |dg|$. For the finite case we have the following theorem.

THEOREM 1. *Let g be of bounded essential variation on a bounded cell $K = [a, b]$. Then there exists a function h of bounded variation on K such that $h = g$ a.e. and $\int_K |dh| = ess$ var g. Moreover, h is unique under the additional restriction that h be continuous at a and left continuous on $(a, b]$.*

PROOF. Since g is of bounded essential variation it is essentially regulated, being equal a.e. to some function h of bounded variation on K. Given such h define $\overline{h}$ of bounded variation on K by

$$(2) \qquad \overline{h} = 1_a h_+ + 1_{(a,b]} h_-$$

in terms of (11) and (12) in §4.4. Since $h = g$ a.e. and $\overline{h} = h$ except at countably many points, $\overline{h} = g$ a.e. . Moreover, $\overline{h}$ is continuous at a and left continuous on $(a, b]$. So by Theorem 1 (§7.2) $\overline{h}$ is uniquely determined by (2) for the set of all functions h of bounded variation on K which equal g a.e. Now by (2)

$$(3) \qquad \int_K |d\overline{h}| = \int_{(a,b]} |d\overline{h}| = \int_{(a,b]} |dh_-| \leq \int_K |dh_-| \leq \int_K |dh|$$

with the last inequality being given by Theorem 4 (§4.4). Since (3) holds for all h of bounded variation on K which equal g a.e. the infimum in (1) is attained for $\overline{h}$. $\square$

For the proof of the next theorem we need the following simple lemma for divergent series.

LEMMA A. *Given $\sum_{n=1}^{\infty} \delta_n = \infty$ with $\delta_n \geq 0$ for all n there exists a sequence $\varepsilon_n \to 0$ such that $0 < \varepsilon_{n+1} \leq \varepsilon_n \leq 1$ and $\sum_{n=1}^{\infty} \varepsilon_n \delta_n = \infty$.*

PROOF. Set $n_1 = 1$. Having chosen n_i choose an integer $n_{i+1} > n_i$ large enough so that $\sum_{n \in E_i} \delta_n > i$ for E_i the set of all integers n such that $n_i \leq n < n_{i+1}$. Let $\varepsilon_n = \frac{1}{i}$ for all n in E_i. The conclusion of the lemma follows easily. $\square$

(The contrapositive of Lemma A is a special case of Exercise 7 (§3.4).)

THEOREM 2. *For any function g on a bounded cell $K = [a, b]$ the following conditions are equivalent:*

(i) g is of bounded essential variation.

(ii) $g dF$ is integrable on K for every function F on K such that every dx-null subset of K is dF-null.

(iii) $g f dx$ is integrable on K for every function f on K such that $f dx$ is integrable.

PROOF. $(i) \Rightarrow (ii)$. By (i) g is equal a.e. to some function h of bounded variation on K. By the weak absolute continuity condition in (ii) F is continuous. That is, $1_p dF = 0$ for all p in K since $1_p dx = 0$. So $h dF$ is integrable by Theorem 4 (§6.2). Since the set $(g \neq h)$ is dx-null it is dF-null. So $g dF = h dF$ which is integrable.

$(ii) \Rightarrow (iii)$. (à fortiori.) (ii) applies to $dF = f dx$.

$(iii) \Rightarrow (i)$. Given (iii) we contend first that g is essentially regulated.

By Theorem 1 (§7.1) g is measurable and essentially bounded. Suppose that g were to fail to have a left essential limit at some

p in $(a, b]$. Then there exist $\varepsilon > 0$ and c in $\mathbb{R}$ such that

(4)
$$\begin{cases} -\infty < \operatorname*{ess\,\underline{\lim}}_{s \to p-} g(s) < c - \varepsilon \\[2mm] \text{and} \\[2mm] c + \varepsilon < \operatorname*{ess\,\overline{\lim}}_{s \to p-} g(s) < \infty. \end{cases}$$

Let $p_0 = a$. Having chosen p_{n-1} in $[a, p)$ for some positive integer n choose p_n in (p_{n-1}, p) close enough to p so that $p_n > \frac{1}{2}(p_{n-1} + p)$ and the measurable sets $A_n = (p_{n-1}, p_n) \cap (g < c - \varepsilon)$ and $B_n = (p_{n-1}, p_n) \cap (g > c + \varepsilon)$ have positive measure. Such p_n must exist by (4). Induction yields a sequence $p_0 = a < p_1 < p_2 < \cdots < p_n < \cdots$ converging to p, and disjoint measurable subsets A_n and B_n of (p_{n-1}, p_n) such that $g < c - \varepsilon$ on A_n, $g > c + \varepsilon$ on B_n, $M(A_n) > 0$, and $M(B_n) > 0$. Define f on K by

(5)
$$f = \sum_{n=1}^{\infty} \frac{1}{n} \left[\frac{1}{M(B_n)} 1_{B_n} - \frac{1}{M(A_n)} 1_{A_n} \right]$$

where at each point in K the series has at most one nonzero term. Let $L_n = [p_{n-1}, p_n]$. Then $\int_{L_n} f \, dx = 0$ for $n = 1, 2, \cdots$ by (5). Also $|\int_{p_{n-1}}^{r} f \, dx| \leq \frac{1}{n}$ for all r in L_n. So for such r, $|\int_a^r f \, dx| = |\int_a^{p_{n-1}} f \, dx + \int_{p_{n-1}}^r f \, dx| = |0 + \int_{p_{n-1}}^r f \, dx| \leq \frac{1}{n}$. Hence, since $\frac{1}{n} \to 0$ and $1_p dx = 0$, $\int_a^p f \, dx = 0$ by Theorem 3 (§1.9). Thus, since $f = 0$ on $[p, b]$, $f \, dx$ is integrable on K and $\int_K f \, dx = 0$. So $gf \, dx$ is integrable on K by (iii).

Now $1_{B_n} gf \geq \frac{c+\varepsilon}{nM(B_n)} 1_{B_n}$, $1_{A_n} gf \geq -\frac{c-\varepsilon}{nM(A_k)} 1_{A_n}$, and $1_{L_n} f = (1_{B_n} + 1_{A_n})f$. Therefore $\int_{L_n} gf \, dx \geq \frac{c+\varepsilon}{n} - \frac{c-\varepsilon}{n} = \frac{2\varepsilon}{n}$. Hence $\int_a^{p_n} gf \, dx = \sum_{j=1}^n \int_{L_j} gf \, dx \geq 2\varepsilon \sum_{j=1}^n \frac{1}{j} \to \infty$ as $n \to \infty$. This implies $\int_a^p gf \, dx = \infty$ contradicting the integrability of $gf \, dx$.

So g must have finite left essential limits on $(a, b]$. By a similar proof g must have finite right essential limits on $[a, b)$. That is, g is essentially regulated. By Theorem 1 (§7.2) we may assume that g is regulated, continuous at a, and left continuous on $(a, b]$. For such g we shall show that g is of bounded variation, thereby proving (i).

Suppose g were of unbounded variation. Then the compactness of K yields a point p in K such that g is of unbounded variation on every neighborhood of p in K. So either $p > a$ and $\int_I |dg| = \infty$ on $I = [s, p]$ for all s in $[a, p)$, or $p < b$ and $\int_I |dg| = \infty$ on $I = [p, s]$ for all s in $(p, b]$. We treat only the first case since the second is similar.

For $n = 1, 2, \cdots$ choose cells $K_n = [a_n, b_n]$ in (a, b) such that $a_{n+1} > b_n$, $a_n \nearrow p$ (hence $b_n \nearrow p$), $\Delta g(K_n) > 0$, and $\sum_{n=1}^{\infty} \Delta g(K_n) = \infty$. Let $b_0 = a$. Choose a'_n and b'_n such that $b_{n-1} < a'_n < a_n < b'_n < b_n$ with the cells $A_n = [a'_n, a_n]$ and $B_n = [b'_n, b_n]$ small enough to ensure by left continuity of g that

(6)
$$|g - g(a_n)| < \frac{1}{n^2} \quad \text{on } A_n$$

and

(7)
$$|g - g(b_n)| < \frac{1}{n^2} \quad \text{on } B_n.$$

Apply Lemma A with $\delta_n = \Delta g(K_n)$ to get ε_n. Then define f on K by

(8)
$$f = \sum_{n=1}^{\infty} \varepsilon_n \left[\frac{1}{M(B_n)} 1_{B_n} - \frac{1}{M(A_n)} 1_{A_n} \right].$$

So

(9)
$$\int_{B_n} f \, dx = \varepsilon_n = - \int_{A_n} f \, dx$$

As with (5) definition (8) gives $\int_{b_{n-1}}^{b_n} f\,dx = 0$ and $|\int_{b_{n-1}}^{r} f\,dx| \le \varepsilon_n$ for all r in $[b_{n-1}, b_n]$. So as $r \to p-$, $\int_a^r f\,dx \to 0 = \int_K f\,dx$. Thus $f\,dx$ is integrable. By (iii) so is $gf\,dx$.

Since $f < 0$ on A_n (6) gives

(10) $$|gf - g(a_n)f| \le -\frac{1}{n^2}f \quad \text{on } A_n.$$

Since $f > 0$ on B_n (7) gives

(11) $$|gf - g(b_n)f| \le \frac{1}{n^2}f \quad \text{on } B_n.$$

By (9) the inequalities (10) and (11) respectively yield

$$\left| \int_{A_n} gf\,dx + \varepsilon_n g(a_n) \right| \le \frac{\varepsilon_n}{n^2} \le \frac{1}{n^2}$$

and

$$\left| \int_{B_n} gf\,dx - \varepsilon_n g(b_n) \right| \le \frac{1}{n^2}.$$

Thus, since $\int_{b_{n-1}}^{b_n} gf\,dx = \int_{A_n} gf\,dx + \int_{B_n} gf\,dx$ and also $g(b_n) - g(a_n) = \Delta g(K_n) = \delta_n$,

(12)
$$\left| \int_{b_{n-1}}^{b_n} gf\,dx - \varepsilon_n \delta_n \right| \le \left| \int_{A_n} gf\,dx + \varepsilon_n g(a_n) \right| +$$

$$\left| \int_{B_n} gf\,dx - \varepsilon_n g(b_n) \right| \le \frac{2}{n^2}.$$

Now $\int_a^{b_k} gf\,dx = \sum_{n=1}^{k} \int_{b_{n-1}}^{b_n} gf\,dx$ which with (12) gives

(13) $$\left| \int_a^{b_k} gf\,dx - \sum_{n=1}^{k} \varepsilon_n \delta_n \right| \le 2\sum_{n=1}^{\infty} \frac{1}{n^2} < \infty$$

for all k. Since $b_k \nearrow p$ and $f = 0$ on $[p, b]$ (13) implies $|\int_K gf\,dx - \sum_{n=1}^{\infty} \varepsilon_n \delta_n| < \infty$. But $\sum_{n=1}^{\infty} \varepsilon_n \delta_n = \infty$ in Lemma A contradicting the integrability of $gf\,dx$. So g must be of bounded variation. $\square$

THEOREM 3. *If g is continuous on a bounded cell $K = [a, b]$ then*

$$ess \ var \ g = \int_K |dg|.$$

PROOF. Since ess var $g \le \int_K |dg|$ we need only consider the case in which ess var $g < \infty$. For this case Theorem 1 yields h on K such that h is continuous at a, left continuous on $(a, b]$, equal to g a.e., and whose total variation $\int_K |dh|$ equals ess var g. For $a < t \le b$ we have $g(t) = g(t-) = ess \ lim_{s \to t-}$ $g(s) = ess \ lim_{s \to t-} h(s) = h(t-) = h(t)$. Similarly $g(a) = g(a+) = h(a+) = h(a)$. So $h = g$ on K. $\square$

Exercises §7.3.

1. For h a function of bounded variation on a bounded cell K prove the equivalence of the following conditions:

   (i) ess var $h = \int_K |dh|$.

   (ii) h is continuous at the endpoints of K and $|h_+ - h_-| = |h_+ - h| + |h - h_-|$ on the interior of K.

   (iii) h is continuous at the endpoints of K and $h(t)$ lies in $\overparen{h_-(t), h_+(t)}$ at every interior point t in K.

2. Show that a continuous function g on a cell K is of bounded variation if and only if Fdg is integrable for every continuous function F on K. (Use Theorem 2, Theorem 3, and integration by parts from Theorem 1 (§6.2).)

3. Let g be a continuous function on a cell K such that fdg is integrable for every function f on K. Prove $dg = 0$. (*Hints*: (i) By Exercise 2 g is of bounded variation. (ii) Use a Hahn decomposition (Theorem 1 (§4.2)) to reduce to the special case $dg \ge 0$. (iii) We may assume $g(a) = 0$ on $K = [a, b]$. Then if $g(b) > 0$ we have $g > 0$ dg-everywhere and $\int_K \frac{1}{g} dg = \infty$.)

4. Let f, g be functions of bounded essential variation on a bounded cell K. Show that ess var $(f - g) = 0$ on K if and only if $f = g + c$ a.e. for some constant c.

5. Show that a function g on a bounded cell K is both essentially continuous and of bounded essential variation if and only if $g = h$ a.e. for some continuous function h of bounded variation on K.

CHAPTER 8

ABSOLUTE CONTINUITY

§8.1 Various Concepts of Absolute Continuity for Differentials.

For arbitrary differentials σ, τ on a cell K there is a wide choice of conditions that may be used to define absolute continuity of τ with respect to σ. We shall list some of these conditions and determine how they are related. Later we shall introduce restrictions on σ, τ under which these conditions become equivalent. All of these conditions on σ, τ are conditions on $|\sigma|, |\tau|$. So the term "absolute continuity" is appropriate.

The first condition (AC_1) has been invoked in §4.5 (Theorem 1), §6.3 (Theorems 1,2,6), and §7.3 (Theorem 2).

$(\textbf{AC}_1)$ Every σ-null subset of K is τ-null.

If σ is not continuous this condition may not be hereditary. That is, it may hold on K but fail to hold on some cell J contained in K. (See Exercise 5.) Of course we can make it hereditary by demanding that it hold on every cell J in K. Alternatively we can use a somewhat stronger condition (AC_2) which *is* hereditary.

$(\textbf{AC}_2)$ $P\tau = 0$ for every indicator summant $P = P^2$ on K such that $P\sigma = 0$.

As we shall see in Theorem 1, (AC_2) has an equivalent lattice-theoretic formulation (AC_2') which was used previously. (See Theorem 7 in §2.4.)

249

(**AC$_2'$**) $\rho \wedge |\tau| = 0$ for every differential ρ on K such that $\rho \wedge |\sigma| = 0$.

Now (AC_1) can be strengthened in another direction to give the condition (AC_3).

(**AC$_3$**) Given $\varepsilon > 0$ in $\mathbb{R}$ there exists $\delta > 0$ in $\mathbb{R}$ such that $\nu(1_E \tau) < \varepsilon$ for every subset E of K for which $\nu(1_E \sigma) < \delta$.

An equivalent formulation is (AC_3') in terms of sequential convergence in differential norm ν.

(**AC$_3'$**) $1_{E_n} \tau \rightarrow 0$ for every sequence of subsets E_n of K such that $1_{E_n} \sigma \rightarrow 0$.

Like (AC_1) both (AC_3) and its equivalent (AC_3') may fail to be hereditary if σ is discontinuous. But just as (AC_1) was strengthened to the hereditary condition (AC_2) the nonhereditary (AC_3) can be strengthened to the hereditary condition (AC_4) which implies both (AC_2) and (AC_3).

(**AC$_4$**) Given $\varepsilon > 0$ in $\mathbb{R}$ there exists $\delta > 0$ in $\mathbb{R}$ such that $\nu(P\tau) < \varepsilon$ for every indicator summant $P = P^2$ on K for which $\nu(P\sigma) < \delta$.

Just as (AC_3) is equivalent to (AC_3') condition (AC_4) is equivalent to (AC_4').

(**AC$_4'$**) $P_n \tau \rightarrow 0$ for every sequence of indicator summants $P_n = P_n^2$ on K such that $P_n \sigma \rightarrow 0$.

A stronger condition than (AC_4) is (AC_5). (See Theorem 2 below.)

(**AC$_5$**) $|\tau| \wedge n|\sigma| \nearrow |\tau|$ as $n \nearrow \infty$.

Clearly (AC_5) is hereditary. It is commonly used in topological Riesz spaces.

Each figure A in K induces a cell summant $P = P^2$ defined by $P(I) = 1$ if $I \subseteq A$, $P(I) = 0$ otherwise. That is, P indicates containment in A. For every differential ρ on K and B the complementary figure $\overline{K - A}$ to A in K we have $\nu(P\rho) = \overline{\int_K} |P\rho| = \overline{\int_A} |P\rho| + \overline{\int_B} |P\rho| = \overline{\int_A} |\rho| + \overline{\int_B} 0 = \overline{\int_A} |\rho|$. So

(AC_4), equivalently (AC_4'), implies (AC_6) and its equivalent sequential formulation (AC_6').

$(\mathbf{AC_6})$ Given $\varepsilon > 0$ in $\mathbb{R}$ there exists $\delta > 0$ in $\mathbb{R}$ such that $\overline{\int_A}|\tau| < \varepsilon$ for every figure A in K for which $\overline{\int_A}|\sigma| < \delta$.

$(\mathbf{AC_6'})$ $\overline{\int_{A_n}}|\tau| \to 0$ for every sequence of figures A_n in K such that $\overline{\int_{A_n}}|\sigma| \to 0$.

These two equivalent conditions are clearly hereditary.

THEOREM 1. (AC_2) *is equivalent to* (AC_2').

PROOF. We may assume $\sigma, \tau \geq 0$.

Given (AC_2) and $\rho \wedge \sigma = 0$ we contend that $\rho \wedge \tau = 0$ proving (AC_2'). Take summants $R, S, T \geq 0$ representing ρ, σ, τ respectively. Let P indicate $S \leq R$. Then $0 \leq PS \leq R \wedge S$, so $0 \leq P\sigma \leq \rho \wedge \sigma = 0$. Hence $P\sigma = 0$. So $P\tau = 0$ by (AC_2). That is, $Q\tau = \tau$ for $Q = 1 - P$ which indicates $R < S$. Since $QR \leq S$, $Q\rho \leq \sigma$. Therefore $0 \leq Q\rho \leq \rho \wedge \sigma = 0$ since $Q\rho \leq \rho$. So $Q\rho = 0$. That is, $P\rho = \rho$ which together with $Q\tau = \tau$ implies $\rho \wedge \tau = P\rho \wedge Q\tau = 0$ since $PR \wedge QT = 0$.

Conversely given (AC_2') and $P\sigma = 0$ we contend $P\tau = 0$ proving (AC_2). To prove this contention we apply (AC_2') with $\rho = P\tau$. Indeed $\rho \wedge \sigma = (P\tau) \wedge \sigma = \tau \wedge (P\sigma) = \tau \wedge 0 = 0$. So $\rho \wedge \tau = 0$ by (AC_2'). That is, $0 = \rho \wedge \tau = (P\tau) \wedge \tau = P\tau$. $\square$

THEOREM 2. (AC_5) *implies* (AC_4).

PROOF. We may assume $\sigma, \tau \geq 0$. Given (AC_5) and $\varepsilon > 0$ apply (AC_5) to get n such that $\nu(\tau - n\sigma)^+ < \frac{\varepsilon}{2}$. We contend (AC_4) holds for $\delta = \varepsilon/2n$.

Let P be an indicator summant such that $\nu(P\sigma) < \delta$. Since $0 \leq \tau \leq (\tau - n\sigma)^+ + n\sigma, 0 \leq P\tau \leq (\tau - n\sigma)^+ + nP\sigma$. So $\nu(P\tau) \leq \nu(\tau - n\sigma)^+ + n\nu(P\sigma) < \frac{\varepsilon}{2} + n\delta = \varepsilon$. $\square$

In summary we have (AC_i) equivalent to (AC_i') for $i = 2, 3, 4, 6$ together with the implications $(AC_5) \Rightarrow (AC_4) \Rightarrow (AC_i)$ for $i = 2, 3, 6$ and $(AC_i) \Rightarrow (AC_1)$ for $i = 2, 3$.

Exercises (§8.1).

For $i = 1, \cdots, 6$ we use the abbreviation "$\tau(AC_i)\sigma$" for "τ is absolutely continuous with respect to σ in the sense of (AC_i)".

1. Verify that each of the absolute continuity relations (AC_i) is reflexive and transitive.

2. Show that $\tau(AC_1)\sigma$ if and only if $f\tau = 0$ for every function f on K such that $f\sigma = 0$.

3. Let $\tau_k \twoheadrightarrow \tau$ on K.

   (a) For $i = 3, 4, 5, 6$ show that if $\tau_k(AC_i)\sigma$ for all k then it holds uniformly for all k.

   (b) For $i = 1, \cdots, 6$ show that if $\tau_k(AC_i)\sigma$ for all k then $\tau(AC_i)\sigma$.

4. For the unit differential $\omega = [1]$ on K prove:

   (a) For $i = 1, 2, 6$, $\tau(AC_i)\omega$ for every differential τ on K.

   (b) For $i = 3, 4, 5$, $\tau(AC_i)\omega$ if and only if τ is tag-finite.

5. On $K = [-\infty, \infty]$ let $F = 1_{(0,\infty]}, G = 1_{[0,\infty]}$, and $g = F - G = -1_0$. Prove:

   (a) $dF = (dg)^+, dG = (dg)^-$, and so $dF \wedge dG = 0$.

   (b) $dF = 0$ on $[-\infty, 0]$ and $dG = 0$ on $[0, \infty]$.

   (c) For every subset E of K the following three conditions are equivalent:

$$1_E dF = 0, \quad 1_E dG = 0, \quad 0 \notin E.$$

   (d) $dF(AC_1)dG$ and $dG(AC_1)dF$ on K.

   (e) The first condition in (d) fails on $[0, \infty]$, the second on $[-\infty, 0]$.

(This exercise is related to Exercise 1 in §4.2.)

§8.2 Absolute Continuity for Restricted Classes of Differentials.

Under appropriate restrictions on σ, τ we gain some implicative relations on $(AC_1), \cdots, (AC_6)$ supplementing those already found in §8.1 for arbitrary σ, τ. We treat first the case of integrable $|\sigma|$ and $|\tau|$.

THEOREM 1. *Let σ, τ be differentials on a cell K such that $|\sigma|$ and $|\tau|$ are integrable. Then*

(i) (AC_2) is equivalent to (AC_5).

(ii) (AC_4) is equivalent to (AC_6).

So for such σ, τ the conditions (AC_i) for $i = 2, 4, 5, 6$ are all equivalent.

PROOF. (i). From (§8.1) we know that (AC_5) implies (AC_2). To prove the converse for integrable $|\sigma|, |\tau|$ we may assume $\sigma, \tau \geq 0$. By the Monotone Convergence Theorem (Theorem 4 and Exercise 4, §2.4) $(\tau - n\sigma)^+ \searrow \rho$ as $n \nearrow \infty$ for some integrable $\rho \geq 0$. Given (AC_2) we have its equivalent (AC_2'). To get (AC_5) we must prove $\rho = 0$. To do so we first prove by induction that

$$(1) \qquad\qquad n(\rho \wedge \sigma) \leq \tau$$

for $n = 0, 1, 2, \cdots$.

Now $\rho \wedge \sigma \leq \rho \leq (\tau - n\sigma)^+ = \tau - \tau \wedge n\sigma \leq \tau - \tau \wedge n(\rho \wedge \sigma) = [\tau - n(\rho \wedge \sigma)]^+$. That is,

$$(2) \qquad \rho \wedge \sigma \leq [\tau - n(\rho \wedge \sigma)]^+ \text{ for } n = 0, 1, 2, \cdots.$$

Since $\tau \geq 0$ (1) holds for $n = 0$. Now let (1) hold for some n. That is, $[\tau - n(\rho \wedge \sigma)]^+ = \tau - n(\rho \wedge \sigma)$ which together with (2) gives (1) with n replaced by $n + 1$. So (1) holds for $n = 0, 1, 2, \cdots$. This implies $\rho \wedge \sigma = 0$ since $\rho \wedge \sigma \geq 0$ and τ is summable, hence archimedean. So $\rho \wedge \tau = 0$ by (AC_2'). Therefore, $0 = \rho \wedge \tau = \rho$ since $\rho \leq \tau$. So (AC_5) holds.

(ii). (AC_4) implies (AC_6) according to §8.1. We need only prove the converse for integrable $\sigma, \tau \geq 0$.

Let (AC_6) hold for $\sigma = dG \geq 0$ and $\tau = dF \geq 0$. Given $\varepsilon > 0$ apply (AC_6) to get $\delta > 0$ such that

(3) $\displaystyle\int_A dF < \varepsilon$ for every figure A for which $\displaystyle\int_A dG < \delta.$

To verify (AC_4) consider any indicator summant $P = P^2$ such that

(4) $\displaystyle\overline{\int_K} PdG < \delta.$

We contend

(5) $\displaystyle\overline{\int_K} PdF \leq \varepsilon.$

By (4) there exists a gauge α on K such that

(6) $(P\Delta G)^{(\alpha)}(K) < \delta.$

Given an α-division $\mathcal{K}$ of K let A be the union of all cells I such that $(I, t) \in \mathcal{K}$ and $P(I, t) = 1$ for some t. Then $\int_A dG = (\Sigma P\Delta G)(\mathcal{K}) < \delta$ by (6). So $(\Sigma P\Delta F)(\mathcal{K}) = \int_A dF < \varepsilon$ by (3). That is, $(\Sigma P\Delta F)(\mathcal{K}) < \varepsilon$ for every α-division $\mathcal{K}$ of K. Hence (5) which proves (AC_4).   □

Our next theorem introduces conditions under which (AC_1) is equivalent to (AC_2).

THEOREM 2. *Let σ, τ be differentials on a cell K such that σ is continuous at every point in K at which it is either left or right continuous, $|\sigma|$ is integrable, and τ is weakly archimedean. For such σ and τ (AC_1) is equivalent to (AC_2).*

PROOF. Since (AC_2) implies (AC_1) we need only prove the converse under the given conditions. Given an indicator summant $P = P^2$ such that $P\sigma = 0$ we contend $P\tau = 0$. By Theorem 1 (§5.3) P is continuous σ-everywhere, hence τ-everywhere by (AC_1). Therefore, since τ is weakly archimedean, $P\tau = 0$. (See §3.3.) □

We can now impose conditions on σ, τ under which all six absolute continuity conditions are equivalent.

THEOREM 3. *Let σ, τ be differentials on a cell K such that σ is continuous at every point in K at which it is either left or right continuous, and both $|\sigma|$ and $|\tau|$ are integrable. For such σ and τ the conditions (AC_i) for $i = 1, \cdots, 6$ are all equivalent.*

PROOF. The theorem follows from the implications

$$(AC_1) \overset{Theorem2}{\Longrightarrow} (AC_2) \overset{Theorem1}{\Longrightarrow} (AC_5) \Rightarrow (AC_4)$$
$$\Rightarrow (AC_6) \overset{Theorem1}{\Longrightarrow} (AC_4) \Rightarrow (AC_3) \Rightarrow (AC_1)$$

where the implications with no labels come from §8.1.   □

Under suitable restrictions absolute continuity of τ with respect to a summable σ ensures summability of τ. We offer one such result for later (§8.4) use. It involves (AC_3).

THEOREM 4. *Let σ, τ be differentials on a cell K such that σ is summable, τ is tag-finite, and τ is absolutely continuous with respect to σ in the sense of (AC_3): $1_E\tau \to 0$ as $1_E\sigma \to 0$. Then τ is summable.*

PROOF. By Theorem 2 (§2.5) there is a monotone function v on K such that $dv \geq |\sigma|$. So τ is absolutely continuous (AC_3) with respect to dv. Let C be the set of points at which v is continuous, D the countable set where v is discontinuous. Since $\nu(\tau) \leq \nu(1_D\tau) + \nu(1_C\tau)$ we need only prove that both $1_D\tau$ and $1_C\tau$ are summable to conclude that τ is summable.

Now $\Sigma_{p \in D}\, \nu(1_p dv) = \Sigma_{p \in D} \int_K 1_p dv = \int_K 1_D dv \leq \nu(dv) < \infty$. Thus $\Sigma_{p \in D}\nu(1_p\tau) < \infty$ since $\nu(1_p\tau) < \infty$ for tag-finite τ

and for $D = \{p_1, p_2, \cdots\}$ the Cauchy criterion for series convergence

$$\sum_{i=m}^{n} \nu(1_{p_i}\tau) = \nu(1_{\{p_m,\ldots,p_n\}}\tau) \to 0 \text{ as } m, n \to \infty \text{ with } m \leq n$$

follows from the same criterion with τ replaced by dv since τ is absolutely continuous with respect to dv. So $\nu(1_D\tau) = \Sigma_{p \in D}\nu(1_p\tau) < \infty$. That is, $1_D\tau$ is summable.

To prove that $1_C\tau$ is summable we first note that $1_C\tau$ is absolutely continuous with respect to $1_C dv$. Now $1_C dv = dw$ for some function w on K since C is measurable and $dv \geq 0$ is integrable. Take $\delta > 0$ small enough to ensure that $\nu(1_{E \cap C}\tau) < 1$ for all E such that $\nu(1_E dw) < \delta$. Since w is monotone and continuous there exists a partition $\mathbb{K}$ of K such that $\nu(1_I dw) = \Delta w(I) < \delta$ for all members I of $\mathbb{K}$. So $\nu(1_{I \cap C}\tau) < 1$. That is, $\overline{\int_I}|1_C\tau| < 1$ since $\nu(1_{I \cap C}\tau) = \overline{\int_I}|1_C\tau|$ because absolute continuity of $1_C\tau$ with respect to the continuous dw ensures continuity of $1_C\tau$. Since $\mathbb{K}$ is finite, $\nu(1_C\tau) = \overline{\int_K}|1_C\tau| = \Sigma_{I \in \mathbb{K}}\overline{\int_I}|1_C\tau| < \infty$. That is, $1_C\tau$ is summable. $\quad\square$

Exercises (§8.2).

1. Show that for summable differentials σ, τ on a cell K the following two conditions are equivalent:

(i) Every measurable, σ-null subset of K is τ-null,

(ii) $1_{E_n}\tau \to 0$ for every sequence of measurable subsets E_n of K such that $1_{E_n}\sigma \to 0$.

(*Hint:* According to Exercise 3 in §4.3 $\nu(1_E\sigma)$ and $\nu(1_E\tau)$ are countably additive measures on the sigma-algebra of all measurable subsets E of K.)

2. Let τ be absolutely continuous with respect to σ on a cell K in the sense of either (AC_3) or (AC_6). Show that if σ is summable and continuous then so is τ.

§8.3 Absolutely Continuous Functions.

For the case of integrable $\tau = dF$ and absolutely integrable $\sigma = dG$ the absolute continuity condition (AC_6) is of particular interest, taking the equivalent forms (i) and (ii) below.

THEOREM 1. *Let F, G be a functions on a cell K such that G is of bounded variation on K and*

(i) $\int_{A_n} dF \to 0$ for every sequence $A_1, A_2, \cdots$ of figures in K such that

$$(1) \qquad \int_{A_n} |dG| \to 0.$$

Then F is of bounded variation on K and

(ii) $\int_{A_n} |dF| \to 0$ for every sequence $A_1, A_2, \cdots$ of figures in K for which (1) holds.

PROOF. Bounded variation of F is an immediate consequence of (i) by Theorem 2 (§2.6). The ε, δ version of (i) is the following.

(iii) Given $\varepsilon > 0$ there exists $\delta > 0$ such that $|\int_A dF| < \varepsilon$ for every figure A in K such that

$$(2) \qquad \int_A |dG| < \delta.$$

Application of Theorem 6 (§2.4) to each component of a figure A gives

$$(3) \qquad \int_A |dF| = \sup_{B, C \subseteq A} \left(\int_B dF - \int_C dF \right)$$

where the supremum is taken over all figures B, C in A. Now if (2) holds for A it holds for all figures in A. Hence (3) yields the conclusion that $\int_A |dF| \leq 2\varepsilon$ for every figure A satisfying (2). That is, (ii) holds. □

Let F be a function on an interval L in $\mathbb{R}$. F is **absolutely continuous** if $\int_A dF \to 0$ for figures A in L whose Lebesgue measure $\int_A dx \to 0$. Explicitly, given $\varepsilon > 0$ there exists $\delta > 0$ such that $|\int_A dF| < \varepsilon$ for every figure A in L for which $\int_A dx < \delta$. Under application of Theorem 1 to cells K in L, dF may be replaced by $|dF|$ in this definition. Since $L \subseteq \mathbb{R}$ every nonempty figure A in L is a bounded set whose convex closure $[\text{Min}A, \text{Max}A]$ is a cell in L. Thus, absolutely continuity of F on L is equivalent to uniform absolute continuity of F on I for all bounded intervals I in L.

Absolute continuity of F on L is equivalent to (AC_1) of dF with respect to dx, combined with a summability condition. This is given by our next result which also includes a special case of the Radon-Nikodym Theorem.

THEOREM 2. *For F a function on a bounded or unbounded open interval L in $\mathbb{R}$ the following three conditions are equivalent:*

(i) F is absolutely continuous,

(ii) Every dx-null set in L is dF-null, and F is of bounded variation on every open set in L of finite Lebesgue measure,

(iii) $dF = f dx$ for some function f on L, and $1_E dF$ is summable for every set E in L such that $1_E dx$ is summable.

PROOF. (i) $\Rightarrow$ (ii). Given a dx-null set E in L and $\varepsilon > 0$ apply (i) to get $\delta > 0$ such that $\int_A |dF| < \varepsilon$ for every figure A in L for which $\int_A dx < \delta$. Apply Theorem 3 (§5.1) locally on L to get U open in L such that $U \supseteq E$ and $\int_L 1_U dx < \delta$. Then $\int_A dx < \delta$ for any figure A that is the closure of a union of finitely many components of U. By absolute continuity $\int_A |dF| < \varepsilon$ for all such A. Hence $\int_L 1_U |dF| \leq \varepsilon$ by the Monotone Convergence Theorem. Since ε is arbitrary this implies $1_E dF = 0$. That is, every dx-null set E is dF-null.

To prove the bounded variation condition in (ii) consider

an open subset U of L such that $\int_L 1_U dx < \infty$. We contend $\int_L 1_U |dF| < \infty$. Suppose otherwise. Since each component U_i of U is bounded, $\int_L 1_{U_i} |dF| < \infty$ by absolute continuity (i) of F. Except for at most two components the closure K_i of U_i is a cell in L. For such K_i, $\Sigma_i \int_{K_i} dx \leq \int_L 1_U dx < \infty$ and $\Sigma_i \int_{K_i} |dF| = \infty$. For each term n in an increasing sequence of integers take a figure $A_n = K_n \cup K_{n+1} \cup \cdots \cup K_{m_n}$ such that $\int_{A_n} |dF| = \Sigma_{i=n}^{m_n} \int_{K_i} |dF| > 1$. Then $\int_{A_n} dx = \Sigma_{i=n}^{m_n} \int_{K_i} dx \leq \Sigma_{i \geq n} \int_{K_i} dx \searrow 0$ as $n \nearrow \infty$. By absolute continuity of F this implies $\int_{A_n} |dF| \to 0$ contradicting $\int_{A_n} |dF| > 1$. So the supposition that $\int_L 1_U |dF| = \infty$ is false.

(ii) $\Rightarrow$ (iii). That $dF = f dx$ for some f follows from the Radon-Nikodym Theorem (Theorem 1 (§4.5)) applied to each cell in a sequence of nonoverlapping cells whose union is L.

Given $1_E dx$ summable Theorem 3 (§5.1) applied locally on L yields an open set U such that $E \subseteq U \subseteq L$ and $\int_L 1_U dx < \infty$. Then $\overline{\int_L} 1_E |dF| \leq \int_L 1_U |dF| < \infty$ by (ii). So $1_E dF$ is summable.

(iii) $\Rightarrow$ (i). Let (iii) hold and suppose (i) false. We contend that this leads to a contradiction. If (i) is false then there exists $\varepsilon > 0$ and a sequence of figures $A_1, A_2, \cdots$ in L such that

(4)
$$\int_{A_n} dx < \frac{1}{2^n}$$

and

(5)
$$\int_{A_n} |dF| > \varepsilon$$

for all n. Since $dF = f dx$ continuity of x induces continuity of F. Thus $\int_A dx = \int_L 1_A dx$ and $\int_A |dF| = \int_L 1_A |dF|$ for every figure A in L. For the union $E = A_1 \cup A_2 \cup \cdots$ (4) implies

$\int_L 1_E dx \leq \Sigma_{n=1}^{\infty} \int_{A_n} dx < 1$. So $1_E dF$ is summable according to (iii). Therefore

$$\int_{A_1 \cup \cdots \cup A_n} |dF| = \int_L 1_{A_1 \cup \cdots \cup A_n} |dF| \nearrow \int_L 1_E |dF| < \infty$$

as $n \nearrow \infty$ by the Monotone Convergence Theorem. Now for $\tau = 1_E dF = 1_E f dx$ and $\sigma = 1_E dx$ we have $\tau(AC_1)\sigma$ which by Theorem 3 (§8.2) implies $\tau(AC_6)\sigma$. Hence, since $\int_{A_n} 1_E dx = \int_{A_n} dx \to 0$ by (4), $\int_{A_n} |dF| = \int_{A_n} 1_E |dF| \to 0$ contradicting (5). $\square$

Invoking Theorem 2 and three previous results we can easily prove a theorem of F. S. Cater [3].

THEOREM 3. *For any function f on a bounded cell K condition (a) is equivalent to (b):*

(a) $f dx$ is summable,

(b) $1_E f dx \to 0$ as $1_E dx \to 0$ for subsets E of K.

PROOF. Given (a) Theorem 1 (§4.6) yields $dF = \overline{f} dx$ such that $dF \geq |f| dx$. dF is summable since $dF \geq 0$. So (iii) holds in Theorem 2, hence (i) giving absolute continuity of F. So $1_E dF \to 0$ as $1_E dx \to 0$ by Theorem 3 (§8.2). This gives (b) since $|f dx| \leq dF$.

Since dx is summable on the bounded cell K, (b) implies (a) by Theorem 4 (§8.2). $\square$

Exercises (§8.3).

1. Let F be a continuous function of bounded variation on $\mathbb{R}$ with $I_1, I_2, \cdots$ a sequence of bounded intervals covering all but countably many points of $\mathbb{R}$ such that F is absolutely continuous on each I_i. Show that F is absolutely continuous on $\mathbb{R}$.

2. Show that a function F on $\mathbb{R}$ is absolutely continuous if and only if there exists a sequence of Lipschitz functions F_n on $\mathbb{R}$ ($|dF_n| \le c_n dx$) such that $1_E dF_n \to 1_E dF$ for every measurable subset E of $\mathbb{R}$ of finite Lebesgue measure ($\int_{\mathbb{R}} 1_E dx < \infty$).

3. Show that a function F of bounded variation on $\mathbb{R}$ is absolutely continuous if and only if $dF_n \to dF$ on $\mathbb{R}$ for some sequence of Lipschitz functions F_n of bounded variation on $\mathbb{R}$.

4. Let F be absolutely continuous on $\mathbb{R}$ and G be a Lipschitz function on $F(\mathbb{R})$. Show that $G \circ F$ is absolutely continuous on $\mathbb{R}$.

5. Let F, G be functions on $K = [0, 1]$ defined by $F(t) = (t \sin\frac{1}{t})^2$ for $t \ne 0$, 0 for $t = 0$ and $G(t) = t^{1/2}$. Prove:

(i) $F'(t) = 2t(\sin\frac{1}{t})^2 - \sin\frac{2}{t}$ for $t \ne 0$, and $F'(0) = 0$.

(ii) F satisfies the Lipschitz condition $|dF| \le 3dx$ on K.

(iii) G is absolutely continuous on K. (Apply Exercise 1.)

(iv) The function $G \circ F$ given by $G \circ F(t) = t|\sin\frac{1}{t}|$ for $t \ne 0$, 0 for $t = 0$ is not absolutely continuous on K.

6. Define F and f on $\mathbb{R}$ by $F(0) = 1$, $f(0) = 0$, and $F(t) = \frac{\sin t}{t}$, $f(t) = \frac{t \cos t - \sin t}{t^2}$ if $t \ne 0$. Prove:

(i) $F' = f$ and $dF = f dx$ on $\mathbb{R}$,

(ii) f is continuous and bounded on $\mathbb{R}$,

(iii) $\int_{-\infty}^{0} dF(t) = 1$,

(iv) $\int_{0}^{\infty} dF(t) = -1$,

(v) $\int_{-\infty}^{\infty} dF(t) = 0$,

(vi) $\int_{1}^{\infty} |dF(t)| = \infty$,

(vii) F is absolutely continuous on $\mathbb{R}$.

7. Define F on $[0, \infty)$ by $F(0) = 0$ and $F(t) = t \sin\frac{2\pi}{t}$ for $t > 0$. Prove:

(i) Given $a > 0$ F satisfies a Lipschitz condition $|dF| \le cdx$ on $[a, \infty)$,

(ii) $F(\frac{1}{n}) = 0$ and $F(\frac{4}{4n+1}) = \frac{4}{4n+1}$ for every positive integer n,

(iii) $\int_0^a |dF(t)| = \infty$ for all $a > 0$,

(iv) F is uniformly continuous but not absolutely continuous on $[0, \infty)$.

8. For g a function on a bounded cell K prove the equivalence of the following three conditions:

(i) gdF is integrable on K for every absolutely continuous function F on K,

(ii) g is Lebesgue measurable and essentially bounded on K,

(iii) gdF is absolutely integrable for every absolutely continuous function F on K.

9. Prove that every absolutely continuous function on a bounded, open interval L is of bounded variation on L.

10. Show that a function F on an interval L in $\mathbb{R}$ is absolutely continuous if and only if $|dF| = dH$ for some absolutely continuous function H on L.

11. Prove that the product of two bounded, absolutely continuous functions on an interval L in $\mathbb{R}$ is a bounded, absolutely continuous function on L.

12. Given a function F on $\mathbb{R}$ prove the equivalence of the following three conditions:

(i) F is Lipschitzian on $\mathbb{R}$,

(ii) $dF = fdx$ for some bounded function f on $\mathbb{R}$,

(iii) The upper and lower derivatives of F are bounded on $\mathbb{R}$.

§8.4 The Vitali Convergence Theorem.

For a sequence of differentials τ_n and any given $i = 3, 4, 5, 6$ we can define not only absolute continuity in the sense of (AC_i)

of τ_n with respect to a differential σ for all n, but also uniform absolute continuity. (See Exercise 3 in §8.1.) For $i = 3$ we have τ_n **uniformly absolutely continuous** with respect to σ on a cell K for all n if given $\varepsilon > 0$ there exists $\delta > 0$ such that $\nu(1_E\tau_n) < \varepsilon$ for all n and all subsets E of K for which $\nu(1_E\sigma) < \delta$. Uniformity here lies in the independence of δ from n.

This particular concept of uniform absolute continuity is relevant to the following version of Vitali's Convergence Theorem. In this version $\tau_n = f_n\sigma$ may fail to be integrable, but $\tau_m - \tau_n = (f_m - f_n)\sigma$ is absolutely integrable for all m, n. Also $\sigma = dg$ for some function g of bounded variation since σ is absolutely integrable.

THEOREM 1. *Let σ be an absolutely integrable differential on a cell K and $f_1, f_2, \cdots$ be a sequence of measurable functions on K such that $f_n \to f$ σ-everywhere and $(f_n - f_1)\sigma$ is uniformly absolutely continuous with respect to σ for all n. Then $f_n\sigma \to f\sigma$.*

PROOF. Given $\varepsilon > 0$ define for $k = 1, 2, \cdots$ the complementary measurable subsets of K,

$$(1) \quad D_k = \bigcap_{n>k}(|f_n - f| < \varepsilon) \quad \text{and} \quad E_k = \bigcup_{n>k}(|f_n - f| \geq \varepsilon)$$

Then $1_{D_k}|f_m - f_n| \leq 1_{D_k}|f_m - f| + 1_{D_k}|f_n - f| < 2\varepsilon$ for $m, n > k$. Thus since $f_m - f_n = 1_{D_k}(f_m - f_n) + 1_{E_k}(f_m - f_n)$,

$$|(f_m - f_n)\sigma| \leq 2\varepsilon|\sigma| + |1_{E_k}(f_m - f_n)\sigma| \quad \text{for } m, n > k.$$

So for all $m, n > k$

$$(2) \quad \nu((f_m - f_n)\sigma) \leq 2\varepsilon\nu(\sigma) + \nu(1_{E_k}(f_m - f_n)\sigma).$$

Since $f_n \to f$ σ-everywhere, $E_k \searrow \emptyset$ σ-everywhere by (1). Hence $1_{E_k}\sigma \to 0$ since E_k is measurable and σ is absolutely integrable.

Now $(f_m - f_n)\sigma$ is uniformly absolutely continuous with respect to σ for all m, n since $(f_m - f_n)\sigma = (f_m - f_1)\sigma - (f_n - f_1)\sigma$. Therefore for k sufficiently large

(3) $\nu(1_{E_k}(f_m - f_n)\sigma) < \varepsilon$ for all m, n

since $1_{E_k}\sigma \to 0$. For such k (2) and (3) give $\nu((f_m - f_n)\sigma) \leq (2\nu(\sigma) + 1)\varepsilon$ for all $m, n > k$. This is just the Cauchy criterion,

(4) $(f_m - f_n)\sigma \to 0$ as $m, n \to \infty$.

Now $g\sigma$ is tag-finite for any function g on K since $\nu(1_p g\sigma) = g(p)\int_K |1_p \sigma| < \infty$ for σ absolutely integrable. In particular, $(f_m - f_n)\sigma$ is tag-finite, hence summable according to Theorem 4 (§8.2). So $(f_m - f_n)\sigma$ is integrable by Theorem 3 (§4.3) since $f_m - f_n$ is measurable. Thus Theorem 2 (§2.8) applies to give the conclusion that $f_n\sigma \to f\sigma$.   □

Recall from §8.3 that a function F on an interval L in $\mathbb{R}$ is absolutely continuous if given $\varepsilon > 0$ there exists $\delta > 0$ such that

(5) $\int_A |dF| \leq \varepsilon$ for every figure A in L of measure $\int_A dx < \delta$. We contend that (5) implies

(6) $\nu(1_E dF) \leq \varepsilon$ for every subset E of L of outer measure $\nu(1_E dx) < \delta$.

To verify this let $\nu(1_E dx) < \delta$ and apply Theorem 3 (§5.1) to get an open subset B of L such that $B \supseteq E$ and $\int_L 1_B dx < \delta$. Since B is open there is an ascending sequence $A_1 \subseteq A_2 \subseteq \cdots$ of figures in B whose interiors cover B, $A_i^\circ \nearrow B$. Now $\int_{A_i} dx = \int_L 1_{A_i} dx \leq \int_L 1_B dx < \delta$. So $\int_{A_i} |dF| \leq \varepsilon$ by (5). Hence $\int_L 1_{A_i^\circ} |dF| \leq \varepsilon$. Thus, since the monotone convergence $1_{A_i^\circ} \nearrow 1_B$ implies $\int_L 1_{A_i^\circ} |dF| \nearrow \int_L 1_B |dF|$, $\int_L 1_B |dF| \leq \varepsilon$. Since $E \subseteq B$ this implies $\nu(1_E dF) \leq \varepsilon$. That is, condition (5) implies condition (6).

We go back to these ideas from §8.2 because the equivalence of (5) and (6) is relevant to uniform absolute continuity. Explicitly, a sequence $G_1, G_2, \cdots$ of functions on an interval L in $\mathbb{R}$ is **uniformly absolutely continuous** if given $\varepsilon > 0$ there exists $\delta > 0$ such that (5) (equivalently (6)) holds for $F = G_n$ where $n = 1, 2, \cdots$. This leads to the following formulation of the Vitali Convergence Theorem. It differs from the usual formulation by requiring uniform absolute continuity of $F_n - F_m$ for all m, n rather than of F_n. So although $dF_n = f_n dx$ on the cell K it may only be conditionally integrable.

THEOREM 2. *For* $n = 1, 2, \cdots$ *let* $dF_n = f_n dx$ *on a bounded cell* K *where* $f_n \to f dx$-*everywhere and the functions* $F_n - F_1$ *are uniformly absolutely continuous for all* n. *Then* $dF_n \to dF$ *where* $dF = f dx$.

PROOF. By Theorem 8 (§6.3) we may assume that f_n is measurable for all n. Apply Theorem 1 with $\sigma = dx$ to conclude that $dF_n \to f dx$. Since dF_n is integrable this implies $f dx$ is integrable. That is, $f dx = dF$ for some function F on K. $\square$

Exercises (§8.4).

1. Show that given the conditions in Theorem 1 there is a subset E of K such that $1_E \sigma = \sigma$ and for $dg = \sigma$ the summants $1_E f_n \Delta g$ are uniformly integrable on K for all n. (Apply Theorem 5 (§2.8).)

2. Show that Theorem 1 yields Theorem 4 (§2.8).

(*Caution:* Theorem 4 (§2.8) applies to all $\sigma \geq 0$ while Theorem 1 applies only to absolutely integrable σ.)

3. Let $dF_n = f_n dx$ on $\mathbb{R}$ for $n = 1, 2, \cdots$ where $f_n \to f$ dx-everywhere and on any given cell K in $\mathbb{R}$ the functions F_n are uniformly absolutely continuous.

(*i*) Show that $f dx = dF$ for some function F on $\mathbb{R}$.

(ii) Give an example of this situation in which each dF_n is absolutely integrable on $\mathbb{R}$, but dF fails to be integrable on $\mathbb{R}$.

4. Let $dF_n = f_n dx$ on $\mathbb{R}$ for $n = 1, 2, \cdots$. Show that if $dF_n \to 0$ on $\mathbb{R}$ then for some k the functions $F_k, F_{k+1}, \cdots$ are uniformly absolutely continuous on $\mathbb{R}$.

CHAPTER 9

CONVERSION OF LEBESGUE-STIELTJES INTEGRALS INTO LEBESGUE INTEGRALS

§9.1 Banach's Indicatrix Theorem.

The **Banach indicatrix** of a function h is the function N on $\mathbb{R}$ with $N(y)$ the number of points t in the domain of h such that $h(t) = y$. So $0 \leq N(y) \leq \infty$ with $N = 0$ outside the range of h. Let x be the identity $x(y) = y$ for all y in $\mathbb{R}$.

THEOREM 1 (BANACH [1], [38]). *Let h be a continuous function on a cell K with Banach indicatrix N. Then*

$$(1) \qquad \int_{\mathbb{R}} N dx = \int_K |dh| \leq \infty$$

(The integral over $\mathbb{R}$ is actually over $h(K) = (N > 0)$.)

PROOF. Since we can imbed K topologically in $\mathbb{R}$ we may assume that $K = [a, b]$ is bounded.

Take a sequence $\langle \mathbb{K}_j \rangle$ of partitions of K such that for all $j = 1, 2, \cdots$

$$(2) \qquad \mathbb{K}_{j+1} \text{ refines } \mathbb{K}_j,$$

$$(3) \qquad \text{each member } I \text{ of } \mathbb{K}_j \text{ is of length less than } 1/j,$$

and

$$(4) \qquad (\sum |\Delta h|)(\mathbb{K}_j) \to \int_K |dh| \quad \text{as } j \to \infty.$$

Let $N_j(y)$ be the number of cells I belonging to $\mathbb{K}_j$ such that y belongs to $h(I)$. That is,

$$(5) \qquad\qquad N_j(y) = \sum_{I \in \mathbb{K}_j} 1_{h(I)}(y).$$

Let E be the set of all endpoints of members of $\mathbb{K}_1 \cup \mathbb{K}_2 \cup \cdots$. Since E is countable so is $h(E)$.

Let $A = \mathbb{R} - h(E)$. Given y in A it is clear from (2) and (5) that for all j

$$(6) \quad 0 \le N_j(y) \le N_{j+1}(y) \le N(y) \le \infty \quad \text{and} \quad N_j(y) < \infty.$$

Since A contains all but countably many points y in $\mathbb{R}$ (6) holds for all j at dx-all y in $\mathbb{R}$.

Given y in A with $N(y) > 0$ consider any finite set $t_1 < \cdots < t_m$ of points t_i in $h^{-1}(y)$. For j large enough so that $1/j < t_{i+1} - t_i$ for $1 \le i < m$ (3) implies that each member I of $\mathbb{K}_j$ contains at most one t_i. So

$$(7) \qquad\qquad m \le N_j(y) \quad \text{for } j \text{ sufficiently large.}$$

Thus if $N(y) < \infty$ then for $m = N(y)$ (6) and (7) imply $N_j(y) = N(y)$ ultimately as $j \to \infty$. On the other hand if $N(y) = \infty$ then (7) holds for arbitrarily large m. That is, $N_j(y) \nearrow \infty$ as $j \nearrow \infty$. So in any case

$$(8) \qquad\qquad N_j(y) \nearrow N(y) \quad \text{as } j \nearrow \infty$$

for all y in A. Thus (8) holds for dx-all y in $\mathbb{R}$.

A cell I in K is compact and connected. Therefore so is $h(I)$ since h is continuous. That is, $h(I)$ is either a cell or a single point. In either case its length is its diameter given by $\int_{\mathbb{R}} 1_{h(I)}dx$. So (5) gives

$$(9) \qquad \int_{\mathbb{R}} N_j dx = \sum_{I \in \mathbb{K}_j} \int_{\mathbb{R}} 1_{h(I)} dx = \sum_{I \in \mathbb{K}_j} \text{diam } h(I).$$

Clearly for every cell I in K

$$(10) \qquad |\Delta h(I)| \leq \text{ diam } h(I) \leq \int_I |dh|.$$

Summing (10) over all members I of $\mathbb{K}_j$ and applying (9) we get

$$(11) \qquad (\sum |\Delta h|)(\mathbb{K}_j) \leq \int_{\mathbb{R}} N_j dx \leq \int_K |dh|$$

for all j. By the Monotone Convergence Theorem (Theorem 3 (§2.7)) and the remarks following the proof of Theorem 3 (§2.8), the conditions (4), (8), and (11) give (1).   □

In Banach's proof of Theorem 1 he had to show explicitly that N is measurable in order to conclude from Lebesgue theory that the integral $\int_{\mathbb{R}} N dx$ exists. Using the modern theory we do not have to do this in our proof. Moreover, we get an important generalization in §9.2.

(Note that (3) implies (4) by Exercise 4 (§2.6).)

Exercises (§9.1).

Let h be a continuous function on a cell K with Banach indicatrix N. Prove:

1. $N(y) = \frac{1}{2} \int_K 1_{h^{-1}(y)} \omega$ at dx-all y in $\mathbb{R}$. (See Exercise 7 in §3.1.)

2. If N is finite and constant on $h(K)$ then $N = 1$ on $h(K)$ and h is strictly monotone. (Consider $N(y)$ at the endpoints y of the cell $h(K)$. Use the Intermediate Value Theorem.)

3. If h is piecewise strictly monotone then N is a step function on $h(K)$.

4. That the converse of Exercise 3 is false is shown by the following counterexample. Let $f(t) = t^+ \wedge |t - 1| \wedge \frac{1}{2}$ for all t in $\mathbb{R}$. For $n = 0, 1, 2, \cdots$ let $f_{n+1}(t) = 2^{-n} f(2^n (t - c_n))$ where $c_n = 3(1 - 2^{-n})$. Let $h = \sum_{n=0}^{\infty} f_{n+1}$ on $K = [0, 3]$. Then

(i) f_{n+1} is a continuous, piecewise linear function on $\mathbb{R}$, $df_{n+1} = 0$ on the complement of $[c_n, c_{n+1}]$, and $\int_K |df_{n+1}| = 3(\frac{1}{2})^{n+1}$,

(ii) h is continuous, $dh = \sum_{n=0}^{\infty} df_{n+1}$, $|dh| = \sum_{n=0}^{\infty} |df_{n+1}|$ where both series converge in differential norm,

(iii) $\int_K |dh| = 3$ and $h(K) = [0, 1]$,

(iv) $N = 3$ on $(0, 1)$, $N(0) = 2$, and $N(1) = 1$,

(v) h is not piecewise monotone on K.

5. Let D be an open subset of K and $N_D(y)$ be the number of points t in D such that $h(t) = y$. Then $\int_K 1_D |dh| = \int_{\mathbb{R}} N_D dx$. (Apply Theorem 1 to the closure of each component of D. This exercise anticipates (26) in Theorem 2 of §9.2.)

§9.2 A Generalization of the Indicatrix Theorem with Applications.

Banach's Indicatrix Theorem (Theorem 1 (§9.1)) is the special case $f = 1$ of the following theorem [27].

THEOREM 1. *Let h be a continuous function of bounded variation on a cell K. Given a function f on K the function $\hat{f}$ defined by*

(1) $$\hat{f}(y) = \sum_{t \in h^{-1}(y)} f(t)$$

exists and is finite for dx-all y in $\mathbb{R}$. Moreover, if $f|dh|$ is absolutely integrable on K then $\hat{f} dx$ is absolutely integrable on

$\mathbb{R}$ *(effectively on $h(K)$ since $\hat{f} = 0$ elsewhere on $\mathbb{R}$) and*

$$(2) \qquad \int_{\mathbb{R}} \hat{f} dx = \int_{K} f|dh|.$$

PROOF. Finiteness of (1) follows from the Indicatrix Theorem. Let $dV = |dh|$ on K and $dF = f dV$ which is absolutely integrable.

Case 1. $0 \le f \le k$ for some positive integer k.

Define the cell summant S on K by

$$(3) \qquad S(I) = \begin{cases} 0 & \text{if } \Delta V(I) = 0 \\ \dfrac{\Delta F}{\Delta V}(I) & \text{if } \Delta V(I) > 0. \end{cases}$$

By Case 1 $0 \le dF \le k dV$ since $dF = f dV$. Integration of this inequality over any cell in K gives

$$(4) \qquad 0 \le \Delta F \le k \Delta V.$$

From (3) and (4) we get

$$(5) \qquad 0 \le S \le k$$

and

$$(6) \qquad S \Delta V = \Delta F.$$

Let T be the cell summant defined by

$$(7) \qquad T(I) = \text{diam } h(I) = \int_{\mathbb{R}} 1_{h(I)} dx.$$

Thus

$$(8) \qquad |\Delta h| \le T \le \Delta V.$$

By (5), (6), and (8) $|\Delta F - ST| = S(\Delta V - T) \leq S(\Delta V - |\Delta h|) \leq k(\Delta V - |\Delta h|)$. That is,

$$(9) \qquad |\Delta F - ST| \leq k(\Delta V - |\Delta h|).$$

We may assume that the cell K is bounded. For $\mathbb{K}_1, \mathbb{K}_2, \cdots$ a sequence of partitions of K satisfying (2), (3), (4) in §9.1, (9) gives

$$(10) \qquad (\sum ST)(\mathbb{K}_j) \to \Delta F(K) \quad \text{as } j \to \infty$$

since $\Delta F(K) = (\sum \Delta F)(\mathbb{K}_j)$.

For each positive integer j define the function f_j on $\mathbb{R}$ by

$$(11) \qquad f_j(y) = \sum_{I \in \mathbb{K}_j} S(I) 1_{h(I)}(y).$$

Since f_j is a linear combination of indicators of bounded intervals $h(I)$, $f_j dx$ is integrable on $\mathbb{R}$ and by (11) and (7),

$$(12) \qquad \int_{\mathbb{R}} f_j dx = (\sum ST)(\mathbb{K}_j).$$

By (10) and (12)

$$(13) \qquad \int_{\mathbb{R}} f_j dx \to \Delta F(K) \quad \text{as } j \to \infty.$$

Recall that the Banach indicatrix $N(y)$ is the number of points in $h^{-1}(y)$ and that $N_j(y)$ in (5) of §9.1 is the number of cells I belonging to $\mathbb{K}_j$ such that y belongs to $h(I)$. By (5) of §9.1 and (5), (11) here

$$0 \leq f_j \leq k N_j \quad \text{for all } j.$$

Thus, since $N_j \leq N$ dx-everywhere according to (6) of §9.1,

$$(14) \qquad 0 \leq f_j(y) \leq kN(y) \quad \text{for } dx\text{-all } y \text{ in } \mathbb{R}.$$

By the Indicatrix Theorem (Theorem 1 (§9.1))

$$(15) \qquad \int_{\mathbb{R}} N dx = \Delta V(K) < \infty.$$

Let D be the set of all y in $\mathbb{R}$ for which at least one of the following conditions holds:

(i) $N(y) = \infty$,

(ii) $y = h(t)$ for some t that is an endpoint of some member I of $\mathbb{K}_j$ for some j,

(iii) $y = h(t)$ for some t such that the condition

$$(16) \qquad \frac{dF}{dV}(t) = f(t)$$

fails to hold.

The set of all y satisfying (i) is dx-null by (15). The set of all y satisfying (ii) is countable, hence dx-null. By Theorem 2 (§6.3) (16) holds for dV-all (that is, dh-all) t in K. Moreover, the h-image of a dh-null set is dx-null by Theorem 2 (§5.2). So the set of all y satisfying (iii) is dx-null. In summary, D is dx-null.

Consider any y in $\mathbb{R} - D$. $h^{-1}(y)$ is a finite set covered for each j by the interiors of the cells belonging to $\mathbb{K}_j$. So if y lies in the range $h(K)$ of h then $h^{-1}(y)$ is a nonempty finite set $t_1 < \cdots < t_m$. So for j sufficiently large each member I of $\mathbb{K}_j$ contains at most one point t_i of $h^{-1}(y)$. For such j (11) takes the form

$$(17) \qquad f_j(y) = \sum_{i=1}^{m} S(I_{i,j})$$

where $I_{i,j}$ is the unique member I of $\mathbb{K}_j$ whose interior I° contains t_i. Since $\dfrac{dF}{dV}(t_i) = f(t_i)$ because y in $\mathbb{R} - D$ violates (iii), (3) of §9.1 and (3) here imply

(18) $$S(I_{i,j}) \to f(t_i) \quad \text{as } j \to \infty.$$

By the definition (1) of $\hat{f}$, (17), and (18)

(19) $$f_j(y) \to \hat{f}(y) \quad \text{as } j \to \infty$$

for dx-all y in $\mathbb{R}$.

Clearly, $f_j = \hat{f} = 0$ on $\mathbb{R} - h(K)$.

The Dominated Convergence Theorem (Theorem 4 (§2.8)) applied with (14), (15), and (19) gives

(20) $$\int_{\mathbb{R}} f_j dx \to \int_{\mathbb{R}} \hat{f} dx \quad \text{as } j \to \infty.$$

So for Case 1 (2) follows from (13) and (20) since $dF = f dV = f|dh|$.

Case 2. $0 \le f(t) < \infty$ for all t in K.

Apply Case 1 to $f_k = k \wedge f$ for each positive integer k. By the Monotone Convergence Theorem (Theorem 3(§2.7))

(21) $$\int_K f_k dV \nearrow \int_K f dV \quad \text{as } k \nearrow \infty$$

since $f_k \nearrow f$. For $\widehat{f_k}$ defined by (1) Case 1 gives

(22) $$\int_{\mathbb{R}} \widehat{f_k} dx = \int_K f_k dV.$$

Since $f_k \nearrow f$, $\widehat{f_k} \nearrow \hat{f}$ dx-everywhere as $k \nearrow \infty$ under definition (1). So

(23)
$$\int_{\mathbb{R}} \widehat{f_k} dx \nearrow \int_{\mathbb{R}} \hat{f} dx \quad \text{as } k \nearrow \infty$$

by the Monotone Convergence Theorem. Finally, equation (2) follows from (21), (22), and (23).

Case 3. $-\infty < f(t) < \infty$ for all t in K.

Apply Case 2 to f^+ and f^-, noting that (1) gives $\hat{f} = \widehat{f^+} - \widehat{f^-}$, to get $\int_{\mathbb{R}} \hat{f} dx = \int_{\mathbb{R}} \widehat{f^+} dx - \int_{\mathbb{R}} \widehat{f^-} dx = \int_K f^+ dV - \int_K f^- dV = \int_K f dV.$ $\square$

The special case $f = 1_E$ yields the following extension of Banach's Indicatrix Theorem.

THEOREM 2. *Let h be a continuous function of bounded variation on a cell K. For E any subset of K and y any real number let $N_E(y)$ be the number of points t in E such that $h(t) = y$. In terms of (1)*

(24)
$$N_E = \widehat{1_E}$$

and the Banach indicatrix of h is

(25)
$$N = N_K = \hat{1}.$$

Let E be a measurable subset of K. Then $N_E dx$ is integrable on $\mathbb{R}$ and

(26)
$$\int_{\mathbb{R}} N_E dx = \int_K 1_E |dh|.$$

Moreover, the set $h(E)$ is dx-measurable and

(27)
$$\int_{\mathbb{R}} 1_{h(E)} dx \leq \int_K 1_E |dh| \leq \int_{\mathbb{R}} 1_{h(E)} N dx.$$

PROOF. (24) and (25) are obvious. (26) is (2) with $f = 1_E$ according to (24). Since N_E is a dx-measurable function and $h(E) = (N_E > 0)$, the set $h(E)$ is dx-measurable. So (27) follows from (26) and the easily verified relations

$$1_{h(E)} \leq N_E = 1_{h(E)}N_E \leq 1_{h(E)}N.$$

$\square$

According to Theorem 2 (§5.2) $h(E)$ is dx-null for every dh-null subset E of K. For h a continuous function of bounded variation the converse holds according to our next result.

THEOREM 3. *Let h be a continuous function of bounded variation on a cell K. Let D be a dx-null subset of $\mathbb{R}$. Then $h^{-1}(D)$ is dh-null.*

PROOF. By Theorem 4 (§5.1) there is a measurable dx-null subset A of $h(K)$ such that $A \supseteq D \cap h(K)$. Let $E = h^{-1}(A)$. Then $E \supseteq h^{-1}(D)$. Also, E is measurable since A is measurable and h is a measurable function. Moreover, $h(E) = A$. So (27) in Theorem 2 yields

$$\int_K 1_E |dh| \leq \int_{\mathbb{R}} 1_A N dx = 0$$

since $1_A dx = 0$. Thus, E is dh-null which implies that its subset $h^{-1}(D)$ is dh-null. $\square$

Using the Hahn Decomposition (Theorem 1 (§4.2)) and the Fundamental Theorem of Calculus (Theorem 2 (§6.3)) we get the following result from Theorem 1.

THEOREM 4. *Let h be a continuous function of bounded variation on a cell K. Let A, B be complementary measurable subsets of K such that*

(28) $$(dh)^+ = 1_A dh \quad and \quad (dh)^- = -1_B dh.$$

For dx-all y in $\mathbb{R}$ the number $N(y)$ of intersections with the horizontal line $Y = y$ by the graph $(t, h(t))$ of h as t advances through K is finite and consists of $N_A(y)$ upward crossings alternating with $N_B(y)$ downward crossings of the line $Y = y$ by the graph of h.

Let f be a function on K such that $f|dh|$ is absolutely integrable. Then in terms of (1)

$$(29) \qquad \int_K f(dh)^+ = \int_{\mathbb{R}} \widehat{1_A f}\, dx, \quad \int_K f(dh)^- = \int_{\mathbb{R}} \widehat{1_B f}\, dx,$$

and

$$(30) \qquad \int_K f\, dh = \int_{\mathbb{R}} \widehat{wf}\, dx \quad \text{for } w = 1_A - 1_B.$$

Moreover, absolute integrability holds for all the integrals in (29) and (30).

PROOF. For $w = 1_A - 1_B$ (28) gives

$$(31) \quad (dh)^+ = 1_A|dh|, \ (dh)^- = 1_B|dh|, \quad \text{and } dh = w|dh|.$$

By Theorem 1 (§9.1) $N(y) < \infty$ for dx-all y since h is of bounded variation. By Theorem 2 (§6.3) (31) implies

$$(32) \qquad \frac{dh}{|dh|}(t) = w(t) = \pm 1$$

at all t in $K - E$ where E is dh-null. By Theorem 2 (§5.2) $h(E)$ is dx-null. Therefore for dx-all y (specifically for all y in $\mathbb{R} - h(E)$) (32) holds at all t in the finite set $h^{-1}(y)$. For such t the graph of h crosses the line $Y = y$ from below the line to above if (32) is positive, from above to below if (32) is negative. Since h is continuous its graph cannot pass from below the line $Y = y$ to above it, or from above the line to below it, without crossing the line. Hence, successive crossings must alternate between upward and downward.

Since $f|dh|$ is absolutely integrable and the sets A, B are measurable the differentials $f(dh)^+, f(dh)^-$, and fdh are absolutely integrable since $f(dh)^+ = 1_A f|dh|$ and $f(dh)^- = 1_B f|dh|$ by (31), and $fdh = f(dh)^+ - f(dh)^-$. So Theorem 1 applied to $1_A f$ and $1_B f$ gives (29) with absolute integrability by (2). The first equation in (29) minus the second gives (30) since the transform defined in (1) is linear. $\square$

In the special case $f = 1$ (29) reduces under (24) to

$$(33) \qquad \int_K (dh)^+ = \int_{\mathbb{R}} N_A dx, \quad \int_K (dh)^- = \int_{\mathbb{R}} N_B dx$$

and (30) reduces to

$$(34) \qquad \int_K dh = \int_{\mathbb{R}} \hat{w} dx = \int_{\mathbb{R}} (N_A - N_B) dx$$

since

$$(35) \qquad \hat{w} = \widehat{1_A} - \widehat{1_B} = N_A - N_B.$$

Of course, (33) agrees with (26) applied to (31).

For $K = [a, b]$ in Theorem 4 the following hold at dx-all y in $\mathbb{R}$: (i) If $h(a)$ and $h(b)$ are both less than or both greater than y then $N_A(y) = N_B(y)$ so $\hat{w}(y) = 0$ by (35); (ii) If $h(a) < y < h(b)$ then $\hat{w}(y) = 1$; (iii) If $h(a) > y > h(b)$ then $\hat{w}(y) = -1$. Therefore, for dx-all y in $\mathbb{R}$

$$(36) \qquad \hat{w}(y) = \text{sgn}\,(h(b) - h(a)) 1_J(y)$$

where $J = \widehat{h(a), h(b)}$. This is consistent with (34).

We can now prove two formulas on integration by substitution (change of variables).

THEOREM 5. *Let h be a continuous function of bounded variation on a cell $K = [a, b]$ with Banach indicatrix N. Let g be a function on $\mathbb{R}$ such that $(g \circ h)dh$ is absolutely integrable on K. Then gdx is absolutely integrable on $h(K)$ and*

$$(37) \qquad \int_a^b g(h(t))dh(t) = \int_{h(a)}^{h(b)} g(y)dy.$$

Moreover, $gNdx$ is absolutely integrable on $\mathbb{R}$ (effectively on $h(K)$) and

$$(38) \qquad \int_a^b g(h(t))|dh(t)| = \int_{-\infty}^{\infty} g(y)N(y)dy.$$

PROOF. For $f = g \circ h$ and E any subset of K (1) and (24) give $\widehat{1_E f} = \widehat{1_E g} = N_E g$. So by (35) $\widehat{wf} = \widehat{1_A f} - \widehat{1_B f} = (N_A - N_B)g = \hat{w}g$. Hence (37) follows from (30) and (36).

To prove (38) apply Theorem 1 with $f = g \circ h$. (1) gives $\hat{f} = Ng$ which reduces (2) to (38).   □

For $g = 1$ (38) is just Banach's indicatrix formula (1) in §9.1, and (37) is the algorithm $\int_a^b dh(t) = h(b) - h(a)$.

Theorem 5 yields the following extension of Theorem 3. (We dismiss the trivial case of constant h.)

THEOREM 6. *Let h be a nonconstant, continuous function of bounded variation on a cell $K = [a, b]$ with Banach indicatrix N. Let D be a subset of $\mathbb{R}$ such that $1_D dx$ is integrable on the cell $h(K)$. Then $1_{h^{-1}(D)}dh$ is absolutely integrable on K and*

$$(39) \qquad \int_K 1_{h^{-1}(D)}dh = \int_{h(a)}^{h(b)} 1_D(y)dy.$$

Moreover, $1_D N dx$ is integrable on $\mathbb{R}$ (effectively on $h(K)$) and

$$(40) \qquad \int_K 1_{h^{-1}(D)}|dh| = \int_{-\infty}^{\infty} 1_D(y)N(y)dy.$$

PROOF. By Theorem 4 (§5.1) there is a measurable set A containing D such that $A - D$ is dx-null on $h(K)$, since $1_D dx$ is integrable on $h(K)$. By Theorem 3 the set $h^{-1}(A) - h^{-1}(D) = h^{-1}(A - D)$ is dh-null on K. So $1_{h^{-1}(D)}dh = 1_{h^{-1}(A)}dh$ which is absolutely integrable on K since $h^{-1}(A)$ is measurable.

To complete the proof apply Theorem 5 with $g = 1_D$. This gives $g \circ h = 1_{h^{-1}(D)}$. So (39) and (40) follow respectively from (37) and (38). $\square$

In Theorem 5 absolute integrability of $(g \circ h)dh$ implies absolute integrability of $gN dx$. We can now prove the converse using Theorem 6.

THEOREM 7. *Let h be a nonconstant, continuous function of bounded variation on a cell K with Banach indicatrix N. Let g be a function on the cell $h(K)$ in $\mathbb{R}$. Then the following hold:*

(i) If g is dx-measurable on $h(K)$ then $g \circ h$ is dh-measurable on K,

(ii) If $gN dx$ is absolutely integrable on $h(K)$ then g is dx-measurable on $h(K)$ and $(g \circ h)dh$ is absolutely integrable on K.

PROOF. To prove (i) let g be dx-measurable on $h(K)$. Given c in $\mathbb{R}$ the set $(g < c)$ is a dx-measurable subset of $h(K)$. By Theorem 6 $h^{-1}(D)$ is dh-measurable for every dx-measurable subset D of $h(K)$. So $h^{-1}(g < c)$ is dh-measurable. Now $h^{-1}(g < c) = (g \circ h < c)$. So $(g \circ h < c)$ is dh-measurable for all c in $\mathbb{R}$. That is, $g \circ h$ is a dh-measurable function on K. This proves (i).

To prove (ii) let $gNdx$ be absolutely integrable on $h(K)$. Now Ndx is also absolutely integrable by Banach's Indicatrix Theorem (Theorem 1 (§9.1)). So both gN and N are dx-measurable functions according to Theorem 3 (§4.3). Since such functions form a linear space $cN - gN$ is dx-measurable for every constant c in $\mathbb{R}$. Hence, $(cN - gN > 0)$ is a dx-measurable subset of $h(K)$. Now, since $N > 0$ on $h(K)$, $(cN - gN > 0) = (g < c)$. So $(g < c)$ is a dx-measurable set for all c in $\mathbb{R}$. That is, g is dx-measurable. So $g \circ h$ is dh-measurable by (i).

. To prove that $(g \circ h)dh$ is absolutely integrable we may assume $g \geq 0$ since this applies to g^+ and g^- in the general case. Let $g_n = g \wedge n$ for $n = 1, 2, \cdots$. Since g is dx-measurable so is g_n in the Riesz space of dx-measurable functions on $h(K)$. By (i) $g_n \circ h$ is dh-measurable. Therefore, since $g_n \circ h$ is bounded and dh is absolutely integrable, $(g_n \circ h)dh$ is absolutely integrable. Thus Theorem 5 applies to g_n yielding (38) in the form

$$\int_K (g_n \circ h)|dh| = \int_{h(K)} g_n N dx \leq \int_{h(K)} gN dx < \infty.$$

By the Monotone Convergence Theorem (Theorem 3 (§2.7)) this gives (38) for g since $g_n \nearrow g$ and $g_n \circ h \nearrow g \circ h$ as $n \nearrow \infty$. □

Exercises (§9.2). In Exercises $1, \cdots, 6$ let h be a continuous function of bounded variation on a cell K with Banach indicatrix N and $dv = |dh|$.

1. Show that for all subsets E of K the image $h(E)$ is dx-null if and only if $v(E)$ is dx-null.

(Use Theorem 4 in §5.2 and Theorem 5 in §9.2.)

2. Show that (1) in Theorem 1 is equivalent to

$$\hat{f}(y) = \frac{1}{2} \int_K 1_{h^{-1}(y)} f\omega \quad \text{for } dx\text{-all } y \text{ in } \mathbb{R}.$$

3. Show that for every real number y the set $(h = y)$ is dh-null and
$$d(h \wedge y) = 1_{(h<y)}dh \quad \text{on } K.$$

4. Show that $dH = Ndx$ on $\mathbb{R}$ for H defined on $\mathbb{R}$ by $H(y) = \int_K |d(h \wedge y)|$.

5. Given that $dh \neq 0$ let G be a Lipschitz function on the bounded cell $h(K)$. Verify that G is differentiable dx-everywhere on $h(K)$, $G \circ h$ is of bounded variation on K, $d(G \circ h) = (G' \circ h)dh$ on K, and

$$\int_K (G' \circ h)|dh| = \int_{h(K)} N dG = \int_{h(K)} NG' dx.$$

6. Let E be the set of all t in K such that h attains either a local maximum or local minimum at t. Show that $h(E)$ is dx-null and E is dh-null.

(This is a weakened version of Exercise 5 (v) in §6.3 but its proof is easier using the results in §9.2.)

7. Let h be continuous on an interval L in $\mathbb{R}$ with h of bounded variation on every cell in L. Let D be a subset of $h(L)$. Show that D is dx-null if and only if $h^{-1}(D)$ is dh-null.

(Apply Theorem 3 (§9.2) and Theorem 2 (§5.2) locally.)

8. Let h be continuous on an interval L in $\mathbb{R}$. Show that h is absolutely continuous on L if and only if the following two conditions hold:

(i) $h(E)$ is dx-null for every dx-null subset E of L, and

(ii) h is of bounded variation on every open subset of L of finite Lebesgue measure. (See Theorem 2 (§8.3). In the special case where L is a cell Exercise 8 is the Banach-Zarecki Theorem [38]. In that case (ii) is just bounded variation of h on L.)

9. (i) Show that (1) in Theorem 1 gives $\widehat{1_D f} = 1_E \hat{f}$ for every subset E of $\mathbb{R}$ with $D = h^{-1}(E)$.

(ii) Show that if E is a measurable subset of $\mathbb{R}$ then the conclusions (2) in Theorem 1 and (30) in Theorem 4 can be replaced respectively by

(41)
$$\int_K 1_{h^{-1}(E)} f |dh| = \int_{\mathbb{R}} 1_E \hat{f} dx$$

and

(42)
$$\int_K 1_{h^{-1}(E)} f dh = \int_{\mathbb{R}} 1_E \widehat{wf} dx.$$

(iii) Verify that (41), (42) give (2), (30) if E contains $h(K)$.

10. Invoking the relevant results from §9.2 confirm the validity of the following application to physics. Let F be a fluctuating force field on the y-axis Y with $F(y, t)$ the downward force on a test particle P situated at y at time t. Let P travel along a continuous path $y = h(t)$ of finite length in Y (possibly with infinitely many reversals of direction) during a time interval $K = [a, b]$ so that the work done by P is $\int_a^b F(h(t), t) dh(t)$ with the total exchange of energy $\int_a^b |F(h(t), t) dh(t)| < \infty$.

Then there exists a force field G on Y fixed in time such that if a test particle Q makes a single upward passage through $h(K)$ then for any interval L in $h(K)$ the work done by Q against G as it passes through L equals the work done by P against F during all the time spent by P in L. (Use (42) with $E = L$.)

11. Let h be a continuous function of bounded variation on a cell K. Show that for every positive integer n:

(i) $(h^{-1}h)^n(E)$ is dh-null for every dh-null subset E of K,

(ii) $(hh^{-1})^n(D)$ is dx-null for every dx-null subset D of $\mathbb{R}$.

12. Let $f(t) = t^2$ and $h(t) = 1 - t^2$ for t in $K = [-1, 1]$. Verify the following:

(i) $dh(t) = -2t dt$. So $dh \geq 0$ on $A = [-1, 0]$ and $dh \leq 0$ on $B = [0, 1]$.

(ii) $\int_K f(dh)^- = \int_0^1 2t^3 dt = \frac{1}{2} = \int_K f(dh)^+.$

(iii) $\int_K f|dh| = 1$ and $\int_K f dh = 0.$

(iv) For each y in $h(K) = [0,1]$ the function f is constant on $h^{-1}(y) = \pm\sqrt{1-y}.$

(v) $\widehat{1_A f}(y) = \widehat{1_B f}(y) = 1 - y$ on $h(K).$

(vi) $\widehat{f}(y) = 2(1-y)$ on $h(K)$ and $\widehat{wf} = 0.$

(vii) $\int_0^1 \widehat{f}(y) dy = 1$ and $\int_0^1 \widehat{wf}(y) dy = 0.$

(viii) The results (iii), (vii) are consistent with (2) in Theorem 1 and (29), (30) in Theorem 5.

CHAPTER 10

SOME RESULTS ON HIGHER DIMENSIONS

§10.1 Integral and Differential on n-Cells.

Extension to n-dimensions of the 1-dimensional theory expounded in Chapters 1 and 2 is for the most part quite straightforward. In this chapter we shall assemble enough of the higher dimensional theory to prove a version of Green's Theorem in the plane. So our treatment is quite cursory and selective. All cells are assumed to be bounded.

In $\mathbb{R}^n$, which consists of all $t = (t_1, \cdots, t_n)$ with entries t_i in $\mathbb{R}$, the partial ordering $s \leq t$ is defined by $s_i \leq t_i$ for $i = 1, \cdots, n$. Similarly $s < t$ is defined by $s_i < t_i$ for $i = 1, \cdots, n$. Given $a < b$ in $\mathbb{R}^n$ the **n-cell** $[a, b]$ consists of all t in $\mathbb{R}^n$ such that $a \leq t \leq b$. The **interior** (a, b) of $[a, b]$ consists of all t such that $a < t < b$. A point t is a **vertex** of $[a, b]$ if each of its entries t_i equals either a_i or b_i. A **tagged n-cell** (I, t) consists of an n-cell I and a vertex t of I chosen from the 2^n vertices of I.

With "cell" replaced by "n-cell" the definitions and theorems in Chapter 1 remain valid under the appropriate interpretations for dimension n. For example, where the bounded 1-dimensional cell (I, t) is δ-fine if the length of I is less than $\delta(t)$ we replace "length" by "diameter". The results of Chapter 2 also remain valid although in §2.5 the generalization of Δg is given by an inclusion-exclusion formula (§11.7) as follows.

For g a function on an n-cell $K = [a, b]$ we define

$$(1) \qquad \Delta^{(n)}g(K) = \sum_{t \in \mathcal{V}_K} (-1)^{\mathcal{N}_K(t)} g(t),$$

a sum of 2^n terms where $\mathcal{V}_K$ is the set of all vertices $t = (t_1, \cdots, t_n)$ of K and $\mathcal{N}_K(t)$ is the number of coordinate entries in $(t_1, \cdots, t_n)$ such that $t_i = a_i$. $\Delta^{(n)}g$ is an additive summant on K (Exercise 1) and it defines the integrable differential

$$(2) \qquad d^{(n)}g = [\Delta^{(n)}g].$$

This is compatible with the definitions for the case $n = 1$.

To throw some light on (1) let $K = [a, b]$ be an n-cell. Let $h(j) = (h_1(j), \cdots, h_n(j))$ where $h_j(j) = b_j - a_j$ and $h_i(j) = 0$ for $i \neq j$. For each $j = 1, \cdots, n$ let Δ_j be the difference operator

$$(3) \qquad \Delta_j g(t) = g(t + h(j)) - g(t)$$

for all functions g on $\mathbb{R}^n$. The Δ_j's all commute. Moreover, for all functions g on K

$$(4) \qquad \Delta^{(n)}g(K) = \Delta_1 \cdots \Delta_n g(a).$$

To prove (4) we have $\Delta_j = E_j - I$ where I is the identity operator and $E_j g(t) = g(t + h(j))$. The E_j's commute and expansion of $(E_1 - I) \cdots (E_n - I)$ applied to g at a gives the sum in (1).

A more general formulation of (1) is useful especially for degenerate cases. It is expressed in terms of projections into the coordinate hyperplanes which pass through a given base point a.

Given a in $\mathbb{R}^n$ define the projection P_A for A a subset of $\{1, \cdots, n\}$ by $P_A x = y$ where $y_i = a_i$ if $i \in A$, and $y_i = x_i$ if

$i \notin A$. So $P_\emptyset$ is the identity operator I. Let $\#A$ be the number of points in A. Let K be a subset of $\mathbb{R}^n$ closed under P_A for all A belonging to $\mathcal{A}$, the set of all subsets of $\{1, \cdots, n\}$. Let the operator $\mathbb{D}_a$ acting on all functions g on K be defined by

(5) $$\mathbb{D}_a g(x) = \sum_{A \in \mathcal{A}} (-1)^{\#A} g(P_A x)$$

By (5) and (1)

(6) $\qquad \Delta^{(n)} g(K) = \mathbb{D}_a g(b) \quad$ for K an n-cell $[a, b]$.

Adjointly we can treat P_A as an operator on functions g by setting

(7) $$(P_A g)(x) = g(P_A x).$$

Then (5) can be expressed concisely as

(8) $$\mathbb{D}_a = \sum_{A \in \mathcal{A}} (-1)^{\#A} P_A$$

in terms of linear operators acting on all functions g on K.

For each $i = 1, \cdots, n$ define the operator $D_i = I - P_i$ where I is the identity operator and $P_i = P_{\{i\}}$. Then

(9) $$\mathbb{D}_a = D_1 \cdots D_n.$$

To verify (9) expand $(I - P_1) \cdots (I - P_n)$ to get the sum in (8), noting that

$$P_A = P_{j_1} \cdots P_{j_k} \quad \text{for } A = \{j_1, \cdots, j_k\} \text{ with } \#A = k.$$

For degenerate cases (9) gives (10).

(10) $\qquad$ If $x_i = a_i$ for some i then $\mathbb{D}_a g(x) = 0$

since $D_i g(x) = 0$ and $D_1, \cdots, D_n$ all commute.

In the 1-dimensional case $dg = 0$ if and only if g is constant. To generalize this we need the following result.

THEOREM 1. *Let g be a function on an n-cell $K = [a, b]$. Let $\mathcal{A}_i$ consist of all nonempty subsets A of $\{1, \cdots, n\}$ whose first member is i. For $i = 1, \cdots, n$ define for base point a*

(11) $$g_i = - \sum_{A \in \mathcal{A}_i} (-1)^{\#A} P_A g.$$

Then $g_i(x)$ is independent of x_i for all x in K and

(12) $$g = g_1 + \cdots + g_n + \mathbb{D}_a g.$$

PROOF. The set $\mathcal{A}$ in (5) is partitioned into $\{\emptyset\}, \mathcal{A}_1, \cdots, \mathcal{A}_n$. The term in (5) with $A = \emptyset$ is $g(x)$. So (11) gives by summation $g - (g_1 + \cdots + g_n) = \mathbb{D}_a g$ which gives (12). In (5) the i-th coordinate of $P_A x$ is a_i for all A in $\mathcal{A}_i$. So $g_i(x)$ in (11) is independent of x_i according to (7).

$\square$

THEOREM 2. *Let g be a function on an n-cell $K = [a, b]$. Then $d^{(n)}g = 0$ on K if and only if*

(13) $$g = g_1 + \cdots + g_n$$

where $g_1, \cdots, g_n$ are functions on K such that $g_i(x)$ is independent of x_i for all x in K and $i = 1, \cdots, n$.

PROOF. Let $d^{(n)}g = 0$. Then $\mathbb{D}_a g(x) = \Delta^{(n)} g[a, x] = \int_{[a,x]} d^{(n)}g = 0$ for all $x > a$ in K by (6) and the additivity of $\Delta^{(n)}g$. Together with (10) this gives $\mathbb{D}_a g = 0$ on K. So in Theorem 1 (12) gives (13).

Conversely let (13) hold with the given independence condition. Consider any n-cell $J = [c, d]$ in K. Apply our constructions to the base point c in place of a. Since for any base point $D_i g_i = 0$ by the independence condition, and the D_j's in (9) commute, $\mathbb{D}_c g = 0$ in (13). Apply (6) to J to get $\Delta^{(n)} g(J) = \mathbb{D}_c g(d) = 0$. Thus, since J is an arbitrary n-cell in K, $d^{(n)}g = 0$ by (2). $\square$

THEOREM 3. *Given an integrable differential σ on an n-cell $K = [a, b]$ define g on K by*

(14) $$g(x) = \int_{[a,x]} \sigma \quad \text{for all } x \text{ in } K.$$

(So $g(x) = 0$ for $x \not> a$.) Then:
  (i) $d^{(n)}g = \sigma$.
  (ii) If $\int_{[a,x]} \sigma = 0$ for all $x > a$ in K, then $\sigma = 0$.

PROOF. Define the additive function S on the boolean algebra of n-figures B in K by

(15) $$S(B) = \int_B \sigma.$$

Consider any n-cell $J = [c, d]$ in K. With c as base point, $P_j x = y$ with $y_j = c_j$ and $y_i = x_i$ for $i \neq j$. For $J_j = [a, P_j d]^\star$ let B be the n-figure $J_1 \cup \cdots \cup J_n$. (See Exercise 11.) Since S is additive the inclusion-exclusion formula gives

(16) $$S(B) = \sum_{m=1}^{n} (-1)^{m-1} s_m$$

where

(17) $$s_m = \sum_{1 \leq j_1 < \cdots < j_m \leq n} S(J_{j_1} \wedge \cdots \wedge J_{j_m}).$$

Now $J_{j_1} \wedge \cdots \wedge J_{j_m} = [a, P_A d]^\star$ for base point c where $A = \{j_1, \cdots, j_m\}$. So (17) is just $s_m = \sum_{A \in \mathcal{A}, \#A = m} P_A g(d)$ which in (16) yields $S(B) = -\sum_{\emptyset \neq A \in \mathcal{A}} (-1)^{\#A} P_A g(d)$. That is,

(18) $$S(B) = g(d) - \mathbb{D}_c g(d)$$

by (5) for base point c.

Now $[a, d] = B \cup J$ with $B \wedge J = \emptyset$. So

$$(19) \qquad\qquad S(J) = g(d) - S(B)$$

by (14), (15), and additivity of S. Therefore

$$(20) \qquad\qquad S(J) = \mathbb{D}_c g(d) = \Delta^{(n)} g(J)$$

by (18), (19), and (6). Since $\sigma = [S]$ and J is arbitrary, (20) proves (i).

(ii) follows from (14) and (i).   □

Given an n-cell $K = [a, b]$ let K^+ consist of all x such that $a < x \leq b$. From Theorem 3 we can conclude that the lower corners of K, the n-cells $[a, x]$ with x in K^+, form a set of independent generators of the boolean algebra $\mathcal{F}$ of all n-figures in K in the following sense. Given any function g on K^+ there is a unique additive function S on $\mathcal{F}$ such that

$$(21) \qquad\qquad S[a, x] = g(x) \quad \text{for all } x \text{ in } K^+.$$

To get S from g extend g to K by setting $g = 0$ on the lower boundary $K - K^+$ of K and defining $S(B) = \int_B d^{(n)} g$ for all B in $\mathcal{F}$. For $B = [a, x]$ this gives (21) since $\int_B d^{(n)} g = \Delta^{(n)} g(B) = g(x)$.

Most of the results in Chapters 3 and 4 extend to higher dimensions although there are different ways in which the concepts of continuous differential, regulated function, and step function can be defined in n-dimensions and be consistent with their definitions for $n = 1$. In Chapters 5 and 6 results that depend on the Vitali Covering Theorem may fail to hold in dimensions higher than 1. Moreover integrability of $F' dx$ on a 1-cell extends to higher dimensions only under imposition of conditions on F beyond differentiability. Attempts (e.g. [32], [35], [36], [16], [17]) to remove this impediment by altering the integration process succeed only by losing other parts of the present theory.

Exercises (§10.1).

1. ($\Delta^{(n)}g$ is additive.) Let $K = [a, b]$ be an n-cell. The j-th **lower face** K_j^- of K consists of all x in K such that $x_j = a_j$; the j-th **upper face** K_j^+ consists of all x in K such that $x_j = b_j$. Suppression of the j-th coordinate makes $(n-1)$-cells of these two faces. For g a function on K prove:

(i) $\Delta^{(n-1)}g(K_j^-) = (\prod_{i \neq j} \Delta_i)g(a)$. (Apply (4).)

(ii) $\Delta^{(n-1)}g(K_j^+) = (\prod_{i \neq j} \Delta_i)g(a + h(j))$.

(iii) $\Delta^{(n)}g(K) = \Delta^{(n-1)}g(K_j^+) - \Delta^{(n-1)}g(K_j^-)$. (Apply Δ_j to (i).)

(iv)   If $K = I \cup J$ where I, J are abutting n-cells with $I_j^+ = J_j^-$ then $\Delta^{(n)}g(K) = \Delta^{(n)}g(I) + \Delta^{(n)}g(J)$.

(v) $\int_I d^{(n)}g = \Delta^{(n)}g(I)$ for every n-cell I in K.

2. (Opposing vertices.) For s, t vertices of an n-cell $K = [a, b]$ prove the equivalence of the following conditions:

(i) $s \wedge t = a$ and $s \vee t = b$.

(ii) $s \wedge t < s \vee t$.

(iii) $|s - t| > 0$.

(iv) $s + t = a + b$.

(v) $\mathcal{N}_K(s \wedge t) = n$ and $\mathcal{N}_K(s \vee t) = 0$.

3. Show that for all vertices s, t of an n-cell K,
$\mathcal{N}_K(s \vee t) + \mathcal{N}_K(s \wedge t) = \mathcal{N}_K(s) + \mathcal{N}_K(t)$.

4. Let g be a function on a 2-cell K. Let $(x + h, y + k)$ and (x, y) be any pair of opposing vertices of K. For each of the cases $(h, k) > (0, 0)$, $(h, k) < (0, 0)$, $h < 0 < k$, and $k < 0 < h$ prove $(\mathrm{sgn}\, hk)[g(x+h, y+k) - g(x+h, y) - g(x, y+k) + g(x, y)] = \Delta^{(2)}g(K)$.

5. Let $g(x, y) = x \wedge y$ for all (x, y) in $\mathbb{R}^2$. Let $K = K_1 \times K_2$ where K_i is a 1-cell $[a_i, b_i]$ for $i = 1, 2$. Prove $\Delta^{(2)}g(K) =$

$(b_1 \wedge b_2 - a_1 \vee a_2)^+ = \text{diam}(K_1 \cap K_2) = 2^{-\frac{1}{2}} \text{diam}(K \cap (y = x))$.
(Draw sketches.)

6. Let f be a function on $\mathbb{R}^2$ whose mixed partial derivative f_{xy} exists and is finite on $\mathbb{R}^2$. Prove:

(i) For every 2-cell $K = [a, b]$ with $a = (a_1, a_2)$ and $b = (b_1, b_2)$,

$$\Delta^{(2)} f(K) = \int_{a_1}^{a_2} \int_{b_1}^{b_2} f_{xy}(x, y) dy dx.$$

(ii) $d^{(2)} f = 0$ on $\mathbb{R}^2$ if and only if $f_{xy} = 0$ on $\mathbb{R}^2$.

7. Let f be a function on $\mathbb{R}^2$ whose partial derivative f_x exists and is finite on $\mathbb{R}^2$.

(i) Prove $d^{(2)} f \geq 0$ on $\mathbb{R}^2$ if and only if $f_x(x, y) \leq f_x(x, z)$ for all x, y, z in $\mathbb{R}$ with $y \leq z$.

(ii) Prove (ii) of Exercise 6.

8. Let $K = [(0, \cdots, 0), (1, \cdots, 1)]$ in $\mathbb{R}^n$ where $n > 1$. Define g on K by $g(t_1, \cdots, t_n) = (t_1 \wedge \cdots \wedge t_{n-1}) \vee t_n$. Show that $g(s) \leq g(t)$ for all $s \leq t$ in $\mathbb{R}^n$, but $\Delta^{(n)} g(K) < 0$. (On $\mathcal{V}_K$ $g = 1$ at $(1, \cdots, 1, 0)$ and at all t with last coordinate $t_n = 1$; $g = 0$ elsewhere on $\mathcal{V}_K$. So $\Delta^{(n)} g(K) = -1 + \sum_{j=0}^{n-1} \binom{n-1}{j} (-1)^j = -1 + 0 = -1$.)

9. Show that for f, g functions on an n-cell $K = [a, b]$, $d^{(n)} f = d^{(n)} g$ if and only if $\mathbb{D}_a f = \mathbb{D}_a g$.

10. State and prove n-dimensional versions of Theorems 2 and 3 in §1.2.

11. (For this exercise the reader may find §11.2 - §11.5 helpful.) Let $K = [a, b]$ where $a \leq b$ in $\mathbb{R}^n$. Prove:

(i) The distance of a point y in $\mathbb{R}^n$ from the halfspace $(x_i < a_i)$ equals $(y_i - a_i)^+$.

(ii)   The distance of y from the halfspace $(x_i > b_i)$ equals $(b_i - y_i)^+$.

(iii)   The distance of y from the complement of K is the minimum value of $(y_i - a_i)^+ \wedge (b_i - y_i)^+$ for $i = 1, \cdots, n$.

(iv)   The interior of K is (a, b). It is nonempty if and only if $a < b$.

(v)   The regular closure K^* of K in $\mathbb{R}^n$ is K if $a < b, \emptyset$ if $a \not< b$.

§10.2 Direct Products of Summants.

For the Fubini Theorem we need the concept of direct product of two summants. Given an m-cell J and n-cell K the Cartesian product $J \times K$ is an $(m + n)$-cell L. The tagged $(m + n)$-cells in L are of the form

$$(1) \qquad (H \times I, (r, s)) = (H, r) \times (I, s)$$

where (H, r) is a tagged m-cell in J and (I, s) is a tagged n-cell in K. Given summants R on J and S on K we define the summant $T = RS$ on $J \times K$ by

$$(2) \qquad T(H \times I, (r, s)) = R(H, r)S(I, s).$$

T is well defined by (2) since for a given tagged cell in L (H, r) and (I, s) are uniquely determined by the Cartesian projections of $J \times K$ onto J and K. Thus (2) induces a differential $\tau = [T]$ on $J \times K$.

For $\rho = [R]$ on J and $\sigma = [S]$ on K we are tempted by (2) to write $\tau = \rho\sigma$, conforming to the traditional notation for multiple integrals. But we have no assurance that τ is determined by ρ and σ without regard for the particular representatives R, S of ρ, σ used in (2). Indeed, $R' \sim R$ on J and $S' \sim S$ on K do not in general imply $R'S' \sim RS$ on $J \times K$. However, under suitable restrictions on S we can conclude that $R'S \sim RS$ for all $R' \sim R$. This is the case $P = R' - R$ in the following theorem.

THEOREM 1. *Let P be a summant on an m-cell J such that $P \sim 0$. Let S be a tame summant on an n-cell K. Then $PS \sim 0$ on the $(m+n)$-cell $J \times K$.*

PROOF. S tame (§3.3) means $u|S| \leq W$ for some damper u and additive W on K. So we need only prove $PW \sim 0$ for $W \geq 0$ additive on K. Given $\varepsilon > 0$ we must find a gauge δ on $J \times K$ such that

$$(3) \qquad |PW|^{(\delta)}(J \times K) \leq \varepsilon.$$

Since $P \sim 0$ there is a gauge α on J making $|P|^{(\alpha)}(J)$ small enough to ensure that

$$(4) \qquad |P|^{(\alpha)}(J)W(K) < \varepsilon.$$

Define the gauge δ on $J \times K$ by

$$(5) \qquad \delta(r,s) = \alpha(r)$$

for all r in J and s in K. To prove (3) take any δ-division $\mathcal{L} = \{(H_i \times I_i, (r_i, s_i)) : i = 1, \cdots, p\}$ of $L = J \times K$. Since $I_1 \cup \cdots \cup I_p = K$ there is a partition $\mathbb{K} = \{K_1, \cdots, K_q\}$ of K such that each I_i is partitioned by a subset $\mathbb{K}_i$ of $\mathbb{K}$. For $i = 1, \cdots, p$ and $j = 1, \cdots, q$ let $\psi(i,j)$ indicate $I_i \supseteq K_j$, that is, $K_j \in \mathbb{K}_i$. By additivity of W

$$(6) \qquad W(I_i) = \left(\sum W\right)(\mathbb{K}_i) = \sum_{j=1}^{q} \psi(i,j)W(K_j)$$

for $i = 1, \cdots, p$. If $\psi(i,j) = \psi(k,j) = 1$ for some $i \neq k$ then both I_i and I_k contain K_j, hence must overlap. Therefore H_i and H_k cannot overlap since $H_i \times I_i$ and $H_k \times I_k$ do not overlap for $i \neq k$. Hence

$$(7) \qquad \sum_{i=1}^{p} \psi(i,j)|P(H_i, r_i)| \leq |P|^{(\alpha)}(J)$$

for $j = 1, \cdots, q$ since the H_i's with $\psi(i,j) = 1$ do not overlap, and (H_i, r_i) is α-fine by (5) because diam$H$ in $\mathbb{R}^m$ equals
diam$(H \times s)$ in $\mathbb{R}^{m+n}$ for all s in $\mathbb{R}^n$. By (6), (7), (4) and change
of summation order, $(\sum |PW|)(\mathcal{L}) = \sum_{i=1}^{p} |P(H_i, r_i)| W(I_i) =$
$\sum_{i=1}^{p} \sum_{j=1}^{q} |P(H_i, r_i)| \psi(i,j) W(K_j) \leq |P|^{(\alpha)}(J) \sum_{j=1}^{q} W(K_j)$
$= |P|^{(\alpha)}(J) W(K) < \varepsilon$. That is, $(\sum |PW|)(\mathcal{L}) < \varepsilon$ for every
δ-division $\mathcal{L}$ of L. This gives (3). $\square$

Note that Theorem 1 applies if $|S|$ is subadditive and inte-
grable since such S are tame. In particular, for g a function
on K $\Delta^{(n)}g$ is tame if $d^{(n)}g$ is summable (i.e., if $|d^{(n)}g|$ is inte-
grable). Then $f\Delta^{(n)}g$ is tame for all f on K.

The next result is a corollary of Theorem 1. It will be used
in our proof of the Fubini Theorem.

THEOREM 2. *Let $\rho = [R]$ on an m-cell J. Let S be a tame
summant on an n-cell K. Define $\tau = [RS]$ on $J \times K$ in terms
of (2). Then $E \times K$ is τ-null for every ρ-null subset E of J.*

PROOF. $1_{E \times K}(r,s) = 1_E(r)$ for all (r,s) in $J \times K$. So
$1_{E \times K} RS = (1_E R)S$. Apply Theorem 1 with $P = 1_E R$ to get
$1_{E \times K} RS \sim 0$. That is, $1_{E \times K} \tau = 0$. $\square$

THEOREM 3. *Let f be a function on an m-cell J, g a function on an n-cell K. Define h on $J \times K$ by the direct product*

$$(8) \qquad h(r,s) = f(r)g(s) \quad \text{for } r \text{ in } J, s \text{ in } K.$$

Then

$$(9) \qquad \Delta^{(m+n)}h(J \times K) = \Delta^{(m)}f(J)\Delta^{(n)}g(K).$$

PROOF. $J \times K = [(a_1, \cdots, a_{m+n}), (b_1, \cdots, b_{m+n})]$ with

$$J = [(a_1, \cdots, a_m), (b_1, \cdots, b_m)] \text{ and}$$
$$K = [(a_{m+1}, \cdots, a_{m+n}), (b_{m+1}, \cdots, b_{m+n})].$$

In the notation of (1) in §10.1 where $\Delta^{(n)}$ is defined

(10) $$\mathcal{V}_{J \times K} = \mathcal{V}_J \times \mathcal{V}_K$$

and

(11) $$\mathcal{N}_{J \times K}(t) = \mathcal{N}_J(r) + \mathcal{N}_K(s)$$

for $t = (r, s) = (r_1, \cdots, r_m, s_1, \cdots, s_n)$ with $r = (r_1, \cdots, r_m)$ and $s = (s_1, \cdots, s_n)$. The definition of $\Delta^{(n)}$ together with (10), (11), and (8) give

$$\Delta^{(m+n)} h(J \times K) = \sum_{t \in \mathcal{V}_{J \times K}} (-1)^{\mathcal{N}_{J \times K}(t)} h(t) =$$

$$\sum_{(r,s) \in \mathcal{V}_J \times \mathcal{V}_K} (-1)^{\mathcal{N}_J(r) + \mathcal{N}_K(s)} f(r) g(s) =$$

$$\left(\sum_{r \in \mathcal{V}_J} (-1)^{\mathcal{N}_J(r)} f(r) \right) \left(\sum_{s \in \mathcal{V}_K} (-1)^{\mathcal{N}_K(s)} g(s) \right) =$$

$$\Delta^{(m)} f(J) \Delta^{(n)} g(K)$$

which proves (9).   □

In general we cannot translate (9) into

(12) $$d^{(m+n)} h = d^{(m)} f d^{(n)} g$$

since the direct product of differentials may not be well defined. We cannot assume that RS represents $d^{(m+n)}h$ for *all* summants R representing $d^{(m)}f$ and S representing $d^{(n)}g$. But (12) can be made valid if we exclude all summants that are not tame. Under this restriction a differential is an equivalence class of tame summants. All such differentials are damper-summable. They form a Riesz space that is isomorphic to the space of all damper-summable differentials. Explicitly each damper-summable differential $[S]$ on K corresponds to the restricted differential $[S] \cap \mathbb{T}$ in the modified theory, where $\mathbb{T}$ is the set of all tame summants on K. (See Exercise 15 in §3.3.)

Given damper-summable differentials ρ on J and σ on K we can define the differential

$$(13) \qquad \rho\sigma = [RS] \quad \text{on } J \times K$$

where R, S are tame summants representing ρ, σ. But for $R' \sim R$ and $S' \sim S$ we can only substitute R' for R and S' for S in (13) using Theorem 1 if both R' and S' are tame. With this restriction $\rho\sigma$ is well defined.

For $i = 1, \cdots, n$ let X_i be the identity function on a 1-cell K_i in $\mathbb{R}$. Define the function V on the n-cell $K = K_1 \times \cdots \times K_n$ by the direct product

$$(14) \qquad V = X_1 X_2 \cdots X_n.$$

in terms of (8). That is, for x_i in K_i

$$(15) \qquad V(x_1, x_2, \cdots, x_n) = x_1 x_2 \cdots x_n.$$

By Theorem 3 applied inductively

$$(16) \qquad \Delta^{(n)} V(K) = \Delta X_1(K_1) \cdots \Delta X_n(K_n),$$

the n-dimensional volume of the n-cell K. The integral of $d^{(n)}V$ induces Lebesgue measure in $\mathbb{R}^n$. Under restriction to tame summants inductive application of (13) to (16) gives

$$(17) \qquad d^{(n)}V = dX_1 \cdots dX_n.$$

The terms "of measure zero" and "almost everywhere" with respect to Lebesgue measure in $\mathbb{R}^n$ correspond to "$d^{(n)}V$-null" and "$d^{(n)}V$-everywhere" respectively. Clearly each point in $\mathbb{R}^n$ is of measure zero.

The traditional typography uses "x_i" in place of "X_i" with the familiar notation "dx_i" in place of "dX_i" in formulas such as (14), (16), and (17).

Exercises (§10.2).

1. Let $L = J \times K$ be an $(n+1)$-cell formed from a 1-cell $J = [a,b]$ and an n-cell K. Given a function f on L and r in J define the function f_r on K by

$$f_r(s) = f(r,s) \quad \text{for all } s \text{ in } K.$$

Prove:

(i) $\mathcal{N}_L(a,s) = 1 + \mathcal{N}_K(s)$ and $\mathcal{N}_L(b,s) = \mathcal{N}_K(s)$ for all s in $\mathcal{V}_K$.

(ii) $\Delta^{(n+1)}f(L) = \Delta^{(n)}f_b(K) - \Delta^{(n)}f_a(K)$.

(iii) If $f_r = f_t$ for all r,t in J then $d^{(n+1)}f = 0$ on L.

(*Hint*: Use (10) and (11).)

2. Let J be an m-cell and K an n-cell. Prove:

(i) If $J \times K$ is cut by a coordinate hyperplane $X_j = c$ in $\mathbb{R}^{m+n}$ into two abutting $(m+n)$-cells $J_1 \times K_1$ and $J_2 \times K_2$ then either J is partitioned into J_1, J_2 and $K = K_1 = K_2$, or K is partitioned into K_1, K_2 and $J = J_1 = J_2$.

(ii) If R is an additive summant on J and S is an additive summant on K then the summant $T = RS$ is additive on $J \times K$.

3. Given c in $\mathbb{R}^n$, and n-cells $K = [a,b]$ and $K_c = [a+c, b+c]$ show that $\Delta^{(n)}V(K) = \Delta^{(n)}V(K_c)$.

(*Hint*: Use (16).)

4. Let f be a function on an m-cell J, g a function on an n-cell K, and h a function on $J \times K$ such that

$$\int_{H \times I} d^{(m+n)}h = \int_H d^{(m)}f \int_I d^{(n)}g$$

for every m-cell H in J and n-cell I in K. Apply Theorem 3 to prove $d^{(m+n)}h = d^{(m+n)}(fg)$ on $J \times K$.

5. For $T = RS$ in (2) show that if R and S are tame then so is T.

6. Let $d^{(m)}F = fd^{(m)}u$ on an m-cell J and $d^{(n)}G = gd^{(n)}w$ on an n-cell K where $d^{(m)}u$ and $d^{(n)}w$ are summable.

Prove:

(i) $d^{(m+n)}(FG) = fgd^{(m+n)}(uw)$ on $J \times K$.

(ii) Under the restriction to tame summants, $d^{(m+n)}(FG) = d^{(m)}Fd^{(n)}G = fgd^{(m)}ud^{(n)}w$.

§10.3 A Fubini Theorem.

THEOREM 1. *Let $\rho = [R]$ be a damper-summable differential on an m-cell J. Let $\sigma = [S]$ be a damper-summable differential on an n-cell K where S is tame. Define the differential $\tau = [T]$ on $L = J \times K$ by $T = RS$. Let f be a function on L such that $f\tau$ is integrable. Let $f_r(s) = f(r,s)$ for r in J and s in K. Then*

(1) $\qquad f_r\sigma$ *is integrable on K for ρ-all r in J.*

For the function g defined at ρ-all r in J by

(2) $$g(r) = \int_K f_r\sigma$$

the differential $g\rho$ is integrable on J and

(3) $$\int_J g\rho = \int_L f\tau.$$

PROOF. Given $c > 0$ let A be the set of all r in J such that given a gauge β on K there exist β-divisions $\mathcal{K}_{r,1}$ and $\mathcal{K}_{r,2}$ of K for which

(4) $\qquad c < (\sum f_r S)(\mathcal{K}_{r,1}) - (\sum f_r S)(\mathcal{K}_{r,2}).$

Since c may be arbitrarily small and a countable union of ρ-null sets is ρ-null we need only show that A is ρ-null to get the Cauchy criterion for integrability of $f_r S$, that is, the integrability of $f_r \sigma$, at ρ-all r. This will give (1).

To conclude that A is ρ-null it suffices to prove that given $\varepsilon > 0$ there is a gauge α on J such that

$$(5) \qquad c|1_A R|^{(\alpha)}(J) \leq 4\varepsilon.$$

For F the additive summant on L defined by

$$(6) \qquad F(M) = \int_M f\tau \quad \text{for every } (m+n)\text{-cell } M \text{ in } L$$

choose a gauge δ on L such that

$$(7) \qquad |fT - F|^{(\delta)}(L) < \varepsilon.$$

For each r in J the gauge δ induces a gauge δ_r on K given by

$$(8) \qquad \delta_r(s) = \delta(r, s) \quad \text{for all } s \text{ in } K.$$

For each r in A choose δ_r-divisions $\mathcal{K}_{r,1}$ and $\mathcal{K}_{r,2}$ of K satisfying (4). Choose a gauge α on J such that $\alpha(r)$ is the minimum value of $\delta_r(s)$ on the tags s of $\mathcal{K}_{r,1}$ and $\mathcal{K}_{r,2}$ if r belongs to A. We contend that (5) holds.

Consider any α-division $\mathcal{J}$ of J. For $i = 1, 2$ let $\mathcal{L}_i$ be the δ-division of L consisting of all $(H \times I, (r, s))$ such that (H, r) belongs to $\mathcal{J}$ and (I, s) to $\mathcal{K}_{r,i}$. Let $\mathcal{J}^+$ consist of all members of $\mathcal{J}$ at which $R \geq 0$, $\mathcal{J}^-$ all those at which $R < 0$. Let $\mathcal{L}_i^+ = \mathcal{J}^+ \otimes \mathcal{K}_{r,i}$ and $\mathcal{L}_i^- = \mathcal{J}^- \otimes \mathcal{K}_{r,i}$ for $i = 1, 2$ where $\mathcal{J} \otimes \mathcal{K}$ denotes the set of all $(H \times I, (r, s))$ with (H, r) in $\mathcal{J}$ and (I, s) in $\mathcal{K}$.

For each member (H, r) of $\mathcal{J}^+$ multiply (4) through by $1_A(r)R(H, r)$. Then sum over $\mathcal{J}^+$ to get for $T = RS$

(9)

$$
c\left(\sum 1_A R\right)(\mathcal{J}^+) \le \left(\sum fT\right)(\mathcal{L}_1^+) - \left(\sum fT\right)(\mathcal{L}_2^+) =
$$
$$
\left(\sum fT - F\right)(\mathcal{L}_1^+) + \left(\sum F - fT\right)(\mathcal{L}_2^+) < 2\varepsilon
$$

by (7) since $\mathcal{L}_1^+$ and $\mathcal{L}_2^+$ are δ-divisions of a figure $J^+ \times K$ where J^+ is the figure supporting $\mathcal{J}^+$. Similarly multiply (4) through by $-1_A(r)R(H, r)$ and sum over $\mathcal{J}^-$ to get

(10)

$$
-c\left(\sum 1_A R\right)(\mathcal{J}^-) \le \left(\sum fT\right)(\mathcal{L}_2^-) - \left(\sum fT\right)(\mathcal{L}_1^-)
$$
$$
= \left(\sum fT - F\right)(\mathcal{L}_2^-) + \left(\sum F - fT\right)(\mathcal{L}_1^-) < 2\varepsilon.
$$

Add (9) and (10) to get $c\left(\sum |1_A R|\right)(\mathcal{J}) < 4\varepsilon$. Since $\mathcal{J}$ is an arbitrary α-division of J this gives (5).

So (1) holds validating the definition (2) of g on $J - E$ where E is ρ-null. By Theorem 2 (§10.2) $E \times K$ is τ-null. So we can annihilate f on $E \times K$ without affecting the differential $f\tau$. This makes $f_r = 0$ on K for all r in E, thereby making $g = 0$ on E by (2). Since E is ρ-null this does not affect the differential $g\rho$.

To prove (3) let $\varepsilon > 0$ be given. Since ρ is damper-summable there is a damper u on J such that

(11) $$ 0 < u < \varepsilon \quad \text{and} \quad \overline{\int_J} u|\rho| < \varepsilon. $$

Choose a gauge δ on L such that (7) holds for F given by (6). Since $g(r) = f(r, s) = 0$ for all r in E and all s in K, (2) holds for all r in J. So for each r in J there exists a gauge β_r on K such that

$$(12) \qquad |g(r) - (\textstyle\sum f_r S)(\mathcal{K}_r)| < u(r)$$

for every β_r-division $\mathcal{K}_r$ of K. We may also assume β_r is small enough so that for δ_r given by (8)

$$(13) \qquad \beta_r < \delta_r \quad \text{for all } r \text{ in } J.$$

For each r in J choose a β_r-division $\mathcal{K}_r$ of K. Then choose a gauge α on J such that

$$(14) \qquad \alpha(r) < \beta_r(s) \quad \text{for every tag } s \text{ in } \mathcal{K}_r$$

and by (11),

$$(15) \qquad |uR|^{(\alpha)}(J) < \varepsilon.$$

Given an α-division $\mathcal{J}$ of J let $\mathcal{L}$ be the δ-division of L consisting of all $(H \times I, (r, s))$ with (H, r) in $\mathcal{J}$ and (I, s) in $\mathcal{K}_r$. Since $T = RS$ (12) and (15) yield

$$(16) \qquad |(\textstyle\sum gR)(\mathcal{J}) - (\textstyle\sum fT)(\mathcal{L})| \le (\textstyle\sum |uR|)(\mathcal{J}) < \varepsilon.$$

Thus

$$|(\textstyle\sum gR)(\mathcal{J}) - F(L)| \le$$
$$|(\textstyle\sum gR)(\mathcal{J}) - (\textstyle\sum fT)(\mathcal{L})| + |(\textstyle\sum fT)(\mathcal{L}) - F(L)| < 2\varepsilon$$

by (16) and (7). That is, $|(\sum gR)(\mathcal{J}) - F(L)| < 2\varepsilon$ for every α-division $\mathcal{J}$ of J. This gives (3) since $F(L) = \int_L f\tau$ by (6).
$\square$

Exercises (§10.3).

1. Let F be a function on a region U in the (x, y)-plane such that the mixed partial derivative $F_{xy} = f$ exists and is finite on U.

(i) Apply Theorem 1 to prove that if $fd^{(2)}(xy)$ is integrable on a 2-cell L in U then

(17)
$$\iint_L fd^{(2)}(xy) = \Delta^{(2)}F(L).$$

(See Exercise 6 in §10.1.)

(ii) Show that if for each point $z = (x, y)$ in U the operator $\mathbb{D}_z$ in (5) of §10.1 satisfies

(18)
$$\frac{1}{hk}\mathbb{D}_zF(z + p) \to f(z)$$

as $p = (h, k) \to (0, 0)$ with $hk \neq 0$, then $d^{(2)}F = fd^{(2)}(xy)$ on every 2-cell in U.

(See Exercise 4 in §10.1.) So (18) gives (17) without invoking Theorem 1.

(iii)   Show that (18) holds if $F(x, y) = \sum_{i=1}^k G_i(x)H_i(y)$ for all (x, y) in U where G_i and H_i are differentiable functions on $\mathbb{R}$ for $i = 1, \cdots, k$. (In particular, (18) holds if F is a polynomial in two variables.)

§10.4 Integration on Paths in $\mathbb{R}^n$.

A **parametrized path** in $\mathbb{R}^n$ is a pair (ϕ, K) where K is a 1-cell $[a, b]$ and ϕ is a continuous mapping of K into $\mathbb{R}^n$ such that ϕ is not constant on any 1-cell in K. So $0 < \int_I \|d\phi\| \leq \infty$ for every 1-cell I in K. (Throughout this section the norm is the Euclidean norm.)

Define the equivalence relation $(\phi_\alpha, K_\alpha) \sim (\phi_\beta, K_\beta)$ on the set of all parametrized paths in $\mathbb{R}^n$ to be the existence of a strictly increasing function $\psi_{\beta,\alpha}$ on K_α such that

(1) $$\psi_{\beta,\alpha}(K_\alpha) = K_\beta \quad \text{and} \quad \phi_\alpha = \phi_\beta \circ \psi_{\beta,\alpha}.$$

($\psi_{\beta,\alpha}$ is an order-preserving homeomorphism from K_α to K_β.) A **path** P in $\mathbb{R}^n$ is an equivalence class of parametrized paths in $\mathbb{R}^n$. The **pathway** $\phi(K)$ of P is invariant for all (ϕ, K) belonging to P. Also invariant are the **length** $\int_K \|d\phi\|$, the **initial point** $\phi(a)$, and the **terminal point** $\phi(b)$ of the path.

A path P is **rectifiable** if its length is finite. We may think of a rectifiable path P as being generated by a constantly moving particle which travels a finite distance in $\mathbb{R}^n$ starting at some initial point and ending at some terminal point. During this trip the position of the particle is tracked by observers α whose clocks (ϕ_α, K_α) can run at arbitrarily varying rates with arbitrary initial settings. The properties of P are just those which are the same for all observers.

For a rectifiable path there is a canonical parametrization (ϕ, K) which uses the distance travelled by the particle as its clock. Specifically $K = [0, c]$ with c the length of P, and $\phi = \phi_\alpha \circ \psi_\alpha^{-1}$ where the function ψ_α on $K_\alpha = [a_\alpha, b_\alpha]$ is defined by

(2) $$\psi_\alpha(t) = \int_{a_\alpha}^{t} \|d\phi_\alpha\| \quad \text{for all } t \text{ in } K_\alpha$$

and (ϕ_α, K_α) is any parametrization of P. Under (2), $d\psi_\alpha = \|d\phi_\alpha\|$ on K_α and $\psi_\alpha(a_\alpha) = 0$. For the canonical parametrization the particle at $\phi(s)$ at time s in K is moving with constant speed $\frac{\|d\phi\|}{ds} = 1$. However, the velocity $\frac{d\phi}{ds}$ need not exist since the particle moving at unit speed can make abrupt changes of direction. Existence and finiteness of the velocity requires ϕ to be differentiable.

A path is **closed** if its initial point is also its terminal point. Such a point is the **base point** of the closed path P. It is the ϕ-image of the endpoints of K for all parametrizations (ϕ, K) of P. (The role of base point can be suppressed but we shall suppress it only for the special case of directed simple closed curves.)

A **simple arc** A in $\mathbb{R}^n$ is any subset of $\mathbb{R}^n$ that is homeomorphic to a 1-cell K. A homeomorphism in this case is any continuous ϕ which maps K one-one onto A. These conditions imply continuity of ϕ^{-1} on A since K is a compact Hausdorff space. (ϕ, K) is a **parametrized simple arc**, a special type of parametrized path. The path it defines under the equivalence relation (1) is a **directed simple arc**. A simple arc A in $\mathbb{R}^n$ is the pathway for exactly two directed simple arcs distinguished by the direction of the linear ordering imposed on A. The ordering is determined by the order of the two endpoints of A which distinguishes between the initial and terminal points.

For integration on a 1-cell K the only relevant properties of K are those it has as a directed simple arc. So the entire integration process on K is exactly duplicated on any directed simple arc A. In the more general case of a path P our claims are more modest.

Given functions F, G on the pathway of a path P the differential $F dG$ on P is defined to be the set of all differentials $(F \circ \phi_\alpha) d(G \circ \phi_\alpha)$ on K_α for all parametrizations (ϕ_α, K_α) of P. The properties of $F dG$ are those which are common to all its members. For mappings $F = (F_1, \cdots, F_n)$ and $G = (G_1, \cdots, G_n)$ on the pathway into $\mathbb{R}^n$ we then have the differential $F \cdot dG = \sum_{i=1}^{n} F_i dG_i$ on P with addition of differentials on P defined by addition in each parametrization. The special case of this with G the identity X on $\mathbb{R}^n$ is of particular interest in §10.5. Here $X = (X_1, \cdots, X_i, \cdots, X_n)$ where $X_i(x_1, \cdots, x_i, \cdots, x_n) = x_i$.

There is a restricted type of closed path, the directed simple closed curve, on which the entire integration process can be applied using its application on directed simple arcs.

A **simple closed curve** C is any homeomorph of a circle. C is **directed** if every simple arc A in C is a directed simple arc where the ordering in A is the restriction to A of the ordering in any simple arc in C which contains A. The direction of C is determined by the direction of any simple arc A in C since the terminal point of A must be the initial point of $A' = \overline{C - A}$ (and the terminal point of A' must be the initial point of A).

For a simple closed curve C in the (X, Y)-plane $\mathbb{R}^2$ one of the two possible directions in "positive", the other is "negative". The positive direction in the circle C_0 with equation $x^2 + y^2 = 1$ is imposed by the parametrization $(x, y) = (\cos \theta, \sin \theta)$ which periodically traces C_0 in the positive direction as θ advances through $\mathbb{R}$. To characterize the positive direction of C_0 succinctly, the simple arc of C_0 in the quadrant $(x, y) \geq (0, 0)$ has initial point $(1, 0)$ and terminal point $(0, 1)$.

The positive direction in C_0 is transferred to C by homotopically deforming C_0 into C through a continuum of homeomorphisms of the plane. Such homotopies exist and all of them impose the same direction in C.

For integration around a directed simple closed curve C the 1-cells in C are just the simple arcs in C directed in accord with the direction in C. A tagged cell in C is a simple arc A in C paired with an endpoint of A. Although C itself is not a cell it can be partitioned into cells. A partition of C is any finite set of nonoverlapping simple arcs whose union is C. (A partition of C must have at least two members since C itself is not a simple arc.) The concepts of division, gauge, summant, integral, differential, etc. apply in C just as they do in a 1-cell K. So the integration process carries over to C.

To parametrize integration on C we take a continuous mapping ϕ on a 1-cell $K = [a, b]$ with $\phi(K) = C$ such that:

(i) $s < t$ in K and $\phi(s) = \phi(t)$ if and only if $s = a$ and $t = b$,

(ii) If $a < c < b$ then the initial point of the simple arc $\phi([a, c])$ directed by C is $\phi(a)$.

In the (X, Y)-plane integration on C in the positive direction is denoted by $\oint_c$, in the negative direction by $- \oint_c$.

For our proof of Green's Theorem in §10.5 we shall need the case $n = 2$ of the following result on Lebesgue measure M in the space $\mathbb{R}^n$.

THEOREM 1. *For $n \geq 2$ the pathway of any rectifiable path P in $\mathbb{R}^n$ is of n-dimensional Lebesgue measure zero.*

PROOF. Let (ϕ, K) be the canonical parametrization of P with $K = [0, c]$ and s the length of the path $(\phi, [0, s])$ for all $0 < s \leq c$. Given a positive integer k partition K into 1-cells $K_1, \cdots, K_k$ where $K_i = [c_{i-1}, c_i]$ with $c_i = i\frac{c}{k}$. Each K_i is of length $\frac{c}{k}$. So diam $\phi(K_i) \leq \frac{c}{k}$. Let D_i be the equilateral n-cell centered at $\phi(c_i)$ with each edge of length $\frac{2c}{k}$. Since $\phi(c_i) \in \phi(K_i)$ and diam $\phi(K_i) \leq \frac{c}{k}$, $\phi(K_i) \subseteq D_i$ for $i = 1, \cdots, k$. So the pathway $\phi(K)$ of P is contained in the n-figure $D = D_1 \cup \cdots \cup D_k$. Hence $M(\phi(K)) \leq M(D) \leq \sum_{i=1}^{k} M(D_i) = k(\frac{2c}{k})^n$. Thus, since $n \geq 2$, $M(\phi(K)) \leq \frac{1}{k}(2c)^n$ for every positive integer k. So $M(\phi(K)) = 0$. □

Rectifiability in Theorem 1 rules out pathological cases such as space filling curves.

For calculus on paths conditions weaker than rectifiability sometimes suffice. A path P in $\mathbb{R}^n$ is **weakly archimedean** if $\|d\phi\|$ is weakly archimedean, as defined in §3.3, for some (hence, all) member(s) (ϕ, K) of P. Equivalently, each component $d\phi_i$ of $d\phi = (d\phi_1, \cdots, d\phi_n)$ is weakly archimedean on K. A **region** in $\mathbb{R}^n$ is a connected, open subset of $\mathbb{R}^n$.

We can reformulate Theorem 1 (§6.1) for paths as follows.

THEOREM 2. *Let f be a continuous function on a region U in $\mathbb{R}^n$ such that f is differentiable at all but countably many points of U. Then*

$$df = \nabla f \cdot dX$$

on every weakly archimedean path P in U. Hence,

$$\oint_P \nabla f \cdot dX = 0$$

on every closed, weakly archimedean path P in U.

Our next theorem characterizes mappings F which are gradients, $F = \nabla f$. A **polygonal path** is a path P with some member (ϕ, K) such that ϕ is piecewise linear. That is, K has a partition $\{K_1, \cdots, K_m\}$ such that on $K_i = [a_i, b_i]$,

$$\phi(t) = \phi(a_i) + (\frac{t - a_i}{b_i - a_i})[\phi(b_i) - \phi(a_i)] \quad \text{for all } t \text{ in } K_i.$$

Each pair of points in a region U of $\mathbb{R}^n$ is joined by some polygonal path in U. (See Exercise 4.) Every polygonal path is rectifiable. (Exercise 5.)

THEOREM 3. *Let F map a region U of $\mathbb{R}^n$ into $\mathbb{R}^n$. The following conditions (A) and (B) are equivalent:*

(A)   (i) $F \cdot dX$ is integrable on every polygonal path P in the region U.

(ii) $\oint_P F \cdot dX = 0$ on every closed polygonal path P in U.

(iii)   At every point x in U

$$\int_0^1 F(x + th) \cdot h \, dt = F(x) \cdot h + o(\|h\|)$$

as $h \to 0$ in $\mathbb{R}^n$.

(B)   There exists a continuous function f on U such that $\nabla f = F$ on U.

(Note that (iii) is satisfied if F is continuous. See Exercise 6 to confirm this.)

PROOF. Given (A) choose an arbitrary point a in U and define f on U by

$$(3) \qquad f(x) = \int_a^x F \cdot dX$$

where the integral is taken over any polygonal path in U from a to x. The definition of f is effective by (i) and (ii). Indeed, given polygonal paths P, Q from a to x in U the composition (concatenation) of P with the reversal $-Q$ of Q gives a closed polygonal path $P - Q$. So $0 = \oint_{P-Q} F \cdot dX = \int_P F \cdot dX - \int_Q F \cdot dX$. That is, $\int_Q F \cdot dX = \int_P F \cdot dX$.

Given x in U take $\delta > 0$ such that $x + h$ lies in U for all h with $\|h\| < \delta$. (3) then gives

$$(4) \qquad f(x + h) - f(x) = \int_x^{x+h} F \cdot dX.$$

For P the directed line segment from x to $x + h$ parametrized by $\phi(t) = x + th$ with t in $[0, 1]$ we have

$$(5) \qquad \int_x^{x+h} F \cdot dX = \int_0^1 F(x + th) \cdot h \, dt.$$

By (4),(5), and (iii)

$$(6) \qquad f(x + h) - f(x) = F(x) \cdot h + o(\|h\|)$$

which gives (B). So (A) implies (B).

The converse follows from Theorem 2 and the characterization (6) of the gradient F of f.   □

Let E be a subset of $\mathbb{R}^n$. For G a function on a region U in $\mathbb{R}^n$ we call E dG-null on U if $1_E dG = 0$ on every rectifiable path in U. For F, G mapping U into $\mathbb{R}^m$ we call E $F \cdot dG$-null on U if $1_E F \cdot dG = 0$ on every rectifiable path in U.

THEOREM 4. *Let $E = E_1 \times \cdots \times E_n$ where E_i is a Lebesgue-null subset of $\mathbb{R}$ for $i = 1, \cdots, n$. Then E is dX-null on $\mathbb{R}^n$.*

PROOF. For any parametrization (ϕ, K) of a rectifiable path P in $\mathbb{R}^n$, $\phi = (\phi_1, \cdots, \phi_n)$ where ϕ_i is a continuous function of bounded variation on K for $i = 1, \cdots, n$. By Theorem 3 (§9.2) $\phi_i^{-1}(E_i)$ is $d\phi_i$-null in K since E_i is Lebesgue-null in $\mathbb{R}$. Therefore, since $\phi^{-1}(E) \subseteq \phi_i^{-1}(E_i)$, $\phi^{-1}(E)$ is $d\phi_i$-null for $i = 1, \cdots, n$. That is, $\phi^{-1}(E)$ is $d\phi$-null in K. So $\int_P 1_E |dX| = \int_K (1_E \circ \phi)|d\phi| = \int_K 1_{\phi^{-1}(E)} |d\phi| = 0$.   □

Exercises (§10.4).

1. Given a continuous function h on a 1-cell K define H on K into $\mathbb{R}^2$ by $H(t) = (t, h(t))$ for all t in K. Prove:

(i) H is a homeomorphism from K to the graph $H(K)$ of h.

(ii)   The simple arc (H, K) is rectifiable if and only if h is of bounded variation on K.

(*Hint*: Prove $|dh| \le \|dH\| \le dx + |dh|$ on K.)

2. (i) Verify that $(\phi_\alpha, K_\alpha) \sim (\phi_\beta, K_\beta)$ defined by (1) is an equivalence, $\phi_\alpha(K_\alpha) = \phi_\beta(K_\beta)$, $\int_{K_\alpha} \|d\phi_\alpha\| = \int_{K_\beta} \|d\phi_\beta\|$, and $\phi_\alpha(a_\alpha) = \phi_\beta(a_\beta)$ and $\phi_\alpha(b_\alpha) = \phi_\beta(b_\beta)$ for $K_\alpha = [a_\alpha, b_\alpha]$ and $K_\beta = [a_\beta, b_\beta]$. (ii) Show that for a rectifiable path (1) holds for $\psi_{\beta,\alpha} = \psi_\beta^{-1} \circ \psi_\alpha$ where ψ_α, ψ_β are given by (2).

3.   Let the path P in $\mathbb{R}^3$ be the cyclindrical helix with parametrization (ϕ, K) given by $\phi(t) = (3 \cos t, 3 \sin t, 4t)$ on $K = [0, 2\pi]$.

Prove: (i) $\|\frac{d\phi}{dt}\| = 5$; (ii) The canonical parametrization (ϕ_0, K_0) of P is given by

$$\phi_0(s) = \phi(\frac{s}{5}) \quad \text{on } K_0 = [0, 10\pi].$$

4. Given distinct points x, y in a region U of $\mathbb{R}^n$ prove that there is a polygonal simple arc in U from x to y.

5. Prove:

(a) If a path P in $\mathbb{R}^n$ is parametrized by (ϕ, K) with K bounded and ϕ linear then the length of P is $\|\Delta\phi(K)\|$.

(b) Every polygonal path P in $\mathbb{R}^n$ is rectifiable.

6. Prove:

(a) Condition (iii) in Theorem 3 is equivalent to the condition that given x in U the convergence

$$\frac{1}{\delta} \int_0^\delta (F(x + sh) - F(x)) \cdot h \, ds \to 0 \quad \text{as } \delta \to 0+$$

holds uniformly for all h in some neighborhood of 0 in $\mathbb{R}^n$.

(b) Every continuous F satisfies (iii) in Theorem 3.

7. Let (ϕ, K) be a rectifiable, parametrized path in $\mathbb{R}^n$ with K bounded. Apply Theorem 7 (§6.3) coordinatewise to get a Lebesgue decomposition $\psi + \theta = \phi$ with $d\psi = \phi' dx$ on K and $\theta' = 0$ a.e. on K. Show that

$$\|d\phi\| = \|\phi'\| dx + \|d\theta\|.$$

(*Outline of a Proof:* For $\theta = (\theta_1, \cdots, \theta_n)$ we have $|d\theta_i| \wedge dx = 0$ for each i by Theorem 5 (§6.3). Take $[S] = dx$ and $[T_i] = d\theta_i$ so that $ST = 0$ for $T = (T_1, \cdots, T_m)$. (See Exercise 4 in §2.2.) Using inner products show that $\|S\phi' + T\|^2 = (S\phi')^2 + T^2 = (|S|\|\phi'\| + \|T\|)^2$ which yields $\|d\phi\| = [\|S\phi' + T\|] = [|S|\|\phi'\| + \|T\|] = \|\phi'\| dx + \|d\theta\|$.)

§10.5 Green's Theorem.

A **topological disk** D is any homeomorph of the circular disk $(X^2 + Y^2 \leq 1)$, that is, the set of all (x, y) in the (X, Y)-plane such that $x^2 + y^2 \leq 1$. For a topological disk D in the (X, Y)-plane with its boundary C (a simple closed curve) positively directed the following hold:

(i)   If for some $\delta > 0$ the open interval $\{(x, 0) : -\delta < x < 0\}$ in the negative X-axis lies in the interior of D and the open interval $\{(x, 0) : 0 < x < \delta\}$ in the positive X-axis lies in the exterior (i.e., complement) of D, then C crosses the X-axis at $(0, 0)$ from the lower halfplane $(Y < 0)$ to the upper halfplane $(Y > 0)$.

(ii)   Moreover (i) holds for any coordinate frame (X', Y') obtained from (X, Y) by translation and rotation.

Let L be a horizontal line $(Y = y)$ in the (X, Y)-plane such that L intersects C at only finitely many points and C crosses L at each of these points. Let $L \cap C$ consist of the points $P_1, \cdots, P_k$ where $P_j = (x_j, y)$ for $j = 1, \cdots, k$ and $x_1 < \cdots < x_k$. The components of $L \cap D$ are the segments $S_1, \cdots, S_n$ in L where $S_i = [P_{2i-1}, P_{2i}]$ for $i = 1, \cdots, n$ with $2n = k$. Translation of the coordinate frame to P_{2i} yields the conclusion from (i) and (ii) that C crosses L at P_{2i} from $(Y < y)$ to $(Y > y)$. Translation of the coordinate frame to P_{2i-1} and rotation by a half revolution yields the conclusion that C crosses L at P_{2i-1} from $(Y > y)$ to $(Y < y)$.

Let X, Y be the functions on $\mathbb{R}^2$ defined by $X(x, y) = x$, $Y(x, y) = y$. So the identity $Z(x, y) = (x, y)$ on $\mathbb{R}^2$ is $Z = (X, Y)$ and $dZ = (dX, dY)$. For $F = (p, q)$ any mapping of $\mathbb{R}^2$ into $\mathbb{R}^2$, $F \cdot dZ = p \, dX + q \, dY$. In conformity with traditional notation we shall frequently use "x", "y" in place of "X", "Y".

THEOREM 1.   *Let D be a closed topological disk in the (X, Y)-plane with rectifiable boundary C. Let p and q be continuous* *functions on D which with the exception of at most countably many points have finite partial derivatives $\frac{\partial q}{\partial x}$ and $\frac{\partial p}{\partial y}$ on the interior D° of D such that $\frac{\partial q}{\partial x}d^{(2)}(xy)$ and $\frac{\partial p}{\partial y}d^{(2)}(xy)$ are integrable on D. Then*

(1) $$\iint_D (\frac{\partial q}{\partial x} - \frac{\partial p}{\partial y})d^{(2)}(xy) = \oint_C p\,dx + q\,dy.$$

PROOF. Parametrize the positively directed, simple closed curve C according to §10.4 by $\phi(t) = (g(t), h(t))$ for $0 \le t \le 1$ where the continuous functions g and h are of bounded variation on $I = [0,1]$ since C is rectifiable. By Theorem 1 (§4.2) and Theorem 2 (§6.3) there is a function w on I such that

(2) $dh = w|dh|$, and $\dfrac{dh}{|dh|}(t) = w(t) = \pm 1$ at dh-all t in I.

The Cartesian product $J \times K$ of the 1-cells $J = g(I)$ and $K = h(I)$ is a 2-cell containing D.

Since $h(0) = h(1)$ Theorem 4 (§9.2) yields a Lebesgue-null subset E of K such that given y in $K - E$ the following three conditions hold:

(a) C intersects the line $(Y = y)$ in a finite, nonempty set consisting of $(x_1, y), \cdots, (x_j, y), \cdots, (x_{2n}, y)$ where $x_1 < \cdots < x_j < \cdots < x_{2n}$ and $(x_j, y) = (g(t_j), h(t_j))$ for $j = 1, \cdots, 2n$.

(b) The intersection of the line $(Y = y)$ with D has components $S_1, \cdots, S_i, \cdots, S_n$ where S_i is the 1-cell in $(Y = y)$ with initial endpoint (x_{2i-1}, y) and terminal endpoint (x_{2i}, y).

(c) $w(t_j) = (-1)^j$ for $j = 1, \cdots, 2n$ where w is the chosen function on I satisfying (2).

Since C is rectifiable it is Lebesgue-null in the plane by Theorem 1 (§10.4). So $\frac{\partial q}{\partial x}$ and $\frac{\partial p}{\partial y}$ exist almost everywhere on D since the points at which either of these partial derivatives fails to exist lie in a Lebesgue-null subset $C \cup A$ of D where A is countable. By hypothesis we have the existence and finiteness of the integral

$$(3) \qquad \iint_D \frac{\partial q}{\partial x} d^{(2)}(xy) = \iint_{J \times K} 1_D(x,y) \frac{\partial q}{\partial x}(x,y) d^{(2)}(xy)$$

where $x \in J$ and $y \in K$. The Fubini Theorem (Theorem 1 (§10.3)) converts (3) into an iterated integral

$$(4) \qquad \iint_D \frac{\partial q}{\partial x} d^{(2)}(xy) = \int_K (\int_J 1_D(x,y) \frac{\partial q}{\partial x}(x,y) dx) dy$$

where the integral over J exists and is finite for almost all y in the 1-cell K.

Since q is continuous on S_i and $\frac{\partial q}{\partial x}$ exists and is finite at all but countably many points (x,y) in S_i, Theorem 2 (§6.3) gives

$$(5) \qquad \int_{S_i} \frac{\partial q}{\partial x}(x,y) dx = q(x_{2i}, y) - q(x_{2i-1}, y)$$

for $i = 1, \cdots, n$. By (b) summation of (5) gives

$$(6) \qquad \int_J 1_D(x,y) \frac{\partial q}{\partial x}(x,y) dx = \sum_{j=1}^{2n} (-1)^j q(x_j, y).$$

Define f on I by the composition

$$(7) \qquad f(t) = q(g(t), h(t)) \quad \text{for } 0 \le t \le 1.$$

This function f is continuous since g, h and q are continuous. Thus, since h is of bounded variation, $f\, dh$ and $f|dh|$ are absolutely integrable on I, Theorem 4 (§9.2) thus gives

$$(8) \qquad \int_I f\, dh = \int_K \widehat{wf}(y) dy$$

where by (1) in (§9.2)

(9) $$\widehat{wf}(y) = \sum_{t \in h^{-1}(y)} w(t)f(t)$$

for almost all y. For y in $K - E$ (a), (b), (c) and (7) enable us to express (9) as

(10) $$\widehat{wf}(y) = \sum_{j=1}^{2n} w(t_j)f(t_j) = \sum_{j=1}^{2n}(-1)^j q(x_j, y).$$

Under the parametrization of C

(11) $$\oint_C qdy = \int_I f \, dh$$

by (7). Finally (4), (6), (10), (8), and (11) give

(12) $$\iint_D \frac{\partial q}{\partial x} d^{(2)}(xy) = \oint_C q \, dy.$$

Apply (12) with p, q interchanged and x, y interchanged. The latter interchange reflects the plane about the diagonal $(Y = X)$. This reverses the direction in C giving

(13) $$\iint_D \frac{\partial p}{\partial y} d^{(2)}(xy) = -\oint_C pdx.$$

Subtract (13) from (12) to get (1).   □

Note that (12) and (13) are just the special cases $p = 0$ and $q = 0$ respectively of (1).

One advantage of Theorem 1 is that the only demand it makes on the boundary C of D is that it be rectifiable. The usual formulations of Green's Theorem impose more restrictive conditions on C.

Exercises (§10.5).

1. Let D be a closed topological disk in the (x, y)-plane with rectifiable boundary C. Let u, v be functions on the (x, y)-plane such that on D u is continuously differentiable and v is twice continuously differentiable. Prove

$$\iint_D \begin{vmatrix} \frac{\partial u}{\partial x} & \frac{\partial v}{\partial x} \\ \frac{\partial u}{\partial y} & \frac{\partial v}{\partial y} \end{vmatrix} d^{(2)}(xy) = \oint_C u\, dv.$$

(*Hint*: Apply Theorem 1 with $p = u\frac{\partial v}{\partial x}, q = u\frac{\partial v}{\partial y}$.)

2. Let $D = J \times K = [(a_1, a_2), (b_1, b_2)]$ be a 2-cell formed by 1-cells $J = [a_1, b_1], K = [a_2, b_2]$. Let C be the boundary of D. Let q be a continuous function on D such that $\frac{\partial q}{\partial x}$ exists and is finite at every point in $D^\circ - E$ where E is countable. Without assuming that $\frac{\partial q}{\partial x} d^{(2)}(xy)$ is integrable on D prove:

(i) $\oint_C q\,dy = \int_K [q(b_1, y) - q(a_1, y)]dy = \int_K (\int_J \frac{\partial q}{\partial x} dx)dy$.

(ii) If moreover $E = \emptyset$ then $\oint_C q\,dy = \frac{\partial q}{\partial x}(\bar{x}, \bar{y})\Delta x(J)\Delta y(K)$ for some point $(\bar{x}, \bar{y})$ in D°.

3. Let D be a topological disk in the (x, y)-plane with rectifiable boundary C. Prove:

(a) The area $A = \iint_D d^{(2)}(xy)$ of D is given by $A = \oint_C x\,dy = -\oint_C y\,dx = \oint_C \frac{1}{2}(x\,dy - y\,dx)$.

(b) If D lies in the right halfplane ($x \geq 0$) then the volume $V = 2\pi \iint_D x\,d^{(2)}(xy)$ of revolution of D about the Y-axis is given by $V = \pi \oint_C x^2 dy = -2\pi \oint_C xy\,dx$.

(c) If C is polygonal with vertices $P_i = (x_i, y_i)$ in the positive direction for $i = 1, \cdots, \kappa$ then the area of the polygon D is given by $A = \frac{1}{2}\sum_{i=1}^{\kappa}(x_i\Delta y_i - y_i\Delta x_i) = \frac{1}{2}\sum_{i=1}^{\kappa}(x_i y_{i+1} - y_i x_{i+1})$ where $\Delta x_i = x_{i+1} - x_i$ and $\Delta y_i = y_{i+1} - y_i$ with $P_{\kappa+1} = P_1$. (Apply (a).)

4. For each integer $n > 1$ let D_n be the polygon with vertices $P_i = (\cos i\frac{\pi}{n}, \sin i\frac{\pi}{n})$ for $i = 1, \cdots, 2n$. Then in Exercise 3(c).

(a) $x_i y_{i+1} - y_i x_{i+1} = \sin \frac{\pi}{n}$.

(b) The area A_n of D_n equals $n \sin \frac{\pi}{n}$.

(c) $A_n \to \pi$ as $n \to \infty$.
(Explain this result geometrically.)

CHAPTER 11.

MATHEMATICAL BACKGROUND

§11.1 Filterbases, Lower and Upper Limits.

A **filterbase** in a set Y is a nonempty set $\mathcal{F}$ of nonempty subsets of Y such that given members A, B of $\mathcal{F}$ there is some member C which is contained in both A and B. By induction the intersection of finitely many members of $\mathcal{F}$ contains a member of $\mathcal{F}$. A subset E of Y is **cofinal** for $\mathcal{F}$ if E meets every member of $\mathcal{F}$. E is **terminal** for $\mathcal{F}$ if E contains a member of $\mathcal{F}$. E is cofinal if and only if its complement $Y - E$ is not terminal. A property of points in Y holds **frequently** if it holds at every point in some cofinal set; it holds **ultimately** if it holds at every point in some terminal set. These two concepts behave like existential and universal quantifiers. They reduce to these for the trivial filterbase having Y as its only member.

A **filter** is a filterbase to which every terminal set belongs. The terminal sets for a filterbase $\mathcal{F}$ form a filter containing $\mathcal{F}$. An **ultrafilter** is a maximal filterbase in Y. (The only filterbase in Y containing an ultrafilter $\mathcal{F}$ in Y is $\mathcal{F}$ itself.) Equivalently, an ultrafilter is a filterbase to which every cofinal set belongs. By the axiom of choice every filterbase in Y extends to some ultrafilter in Y.

Filterbases $\mathcal{A}, \mathcal{B}$ in Y are contained in some filter $\mathcal{C}$ in Y if and only if every member of $\mathcal{A}$ meets every member of $\mathcal{B}$. Under the latter condition the set of all $A \cap B$ where $A \in \mathcal{A}$

319

and $B \in \mathcal{B}$ is a filterbase whose terminal sets form a filter containing both $\mathcal{A}$ and $\mathcal{B}$.

Given a filterbase $\mathcal{E}$ in a set X and a mapping $\phi: X \rightarrow Y$ the images $\phi(E)$ of the members E of $\mathcal{E}$ form a filterbase $\mathcal{F}$ in Y which we denote by $\mathcal{F} = \phi(\mathcal{E})$. Here are some examples:

(i) Let $\mathcal{E}$ consist of all sets $\{n, n+1, \cdots\}$ where $n \in \mathbb{N}$, the set of all natural numbers. The nest $\mathcal{E}$ is a filterbase in $\mathbb{N}$ whose cofinal sets are the infinite subsets of $\mathbb{N}$, and terminal sets are the complements of finite subsets of $\mathbb{N}$. A sequence $\phi(1), \phi(2), \cdots$ in $\mathbb{R}$ is just a function $\phi : \mathbb{N} \rightarrow \mathbb{R}$. The filterbase $\mathcal{F} = \phi(\mathcal{E})$ consists of all sets $\{\phi(n), \phi(n+1), \cdots\}$ with n in $\mathbb{N}$. $\mathcal{F}$ determines the limiting behavior of $\phi(n)$ as $n \rightarrow \infty$.

(ii) The set of all open intervals in $\mathbb{R}$ which contain the point p is a filterbase $\mathcal{E}$ whose terminal sets are the neighborhoods of p. A cofinal set is a set whose closure contains p. For a function ϕ on $\mathbb{R}$ the limiting behavior of $\phi(x)$ as $x \rightarrow p$ is determined by the filterbase $\mathcal{F} = \phi(\mathcal{E})$.

A filterbase $\mathcal{F}$ in $[-\infty, \infty]$ has a **lower limit** p and **upper limit** q defined by

$$(1) \qquad p = \underline{\lim}\mathcal{F} = \sup_{A \in \mathcal{F}}(\inf A), q = \overline{\lim}\mathcal{F} = \inf_{B \in \mathcal{F}}(\sup B).$$

Thus

$$(2) \qquad\qquad -\infty \leq p \leq q \leq \infty.$$

Indeed, given A, B in $\mathcal{F}$ take C in $\mathcal{F}$ such that $C \subseteq A$ and $C \subseteq B$. Then, since $C \neq \emptyset, \inf A \leq \inf C \leq \sup C \leq \sup B$. So by (1) $p \leq \sup B$ for all B in $\mathcal{F}$, hence $p \leq q$. If $p = q$ we define $\lim \mathcal{F} = p = q$ to be the **limit** of $\mathcal{F}$.

A point r in $[-\infty, \infty]$ is an **ultimate lower bound** of $\mathcal{F}$ if r is a lower bound of some member A of $\mathcal{F}$, that is, if $[r, \infty]$ is a terminal set. Dually r is an **ultimate upper bound** of $\mathcal{F}$ if r is an upper bound of some member B of $\mathcal{F}$, that is, if $[-\infty, r]$

is a terminal set. The set L of ultimate lower bounds of $\mathcal{F}$ is an interval containing $-\infty$ such that $[-\infty, p) \subseteq L \subseteq [-\infty, p]$ for some p in $[-\infty, \infty]$. Moreover, $p = \underline{\lim}\mathcal{F}$. The set U of ultimate upper bounds of $\mathcal{F}$ is an interval containing ∞ such that $(q, \infty] \subseteq U \subseteq [q, \infty]$ for some q in $[-\infty, \infty]$. Moreover, $q = \overline{\lim}\mathcal{F}$.

Given a filterbase $\mathcal{E}$ in a set X and real-valued functions ϕ, ψ on X we have for the filterbases $\phi(\mathcal{E}), \psi(\mathcal{E})$, and $(\phi + \psi)(\mathcal{E})$:

$$\text{(3)} \qquad \underline{\lim}\phi + \underline{\lim}\psi \le \underline{\lim}(\phi + \psi)$$

and

$$\text{(4)} \qquad \overline{\lim}(\phi + \psi) \le \overline{\lim}\phi + \overline{\lim}\psi$$

ignoring cases involving the indeterminate form $\infty - \infty$. Thus, since $\underline{\lim}(-\psi) = -\overline{\lim}\psi$, we also have $\underline{\lim}(\phi + \psi) \le \underline{\lim}\phi + \overline{\lim}\psi \le \overline{\lim}(\phi + \psi)$ again ignoring $\infty - \infty$.

A **directed set** is a nonempty set X with a transitive binary relation $\succ$ such that given members j, k of X there is some member i of X such that both $i \succ j$ and $i \succ k$. Such a relation induces a filterbase $\mathcal{E} = \{E_j : j \in X\}$ in X where $E_j = \{i : i \succ j\}$. A mapping $\phi : X \to Y$ on a directed set X is called a **net** in Y. It carries $\mathcal{E}$ onto a filterbase $\mathcal{F}$ in Y.

Exercises (§11.1).

1. Verify the following for a filterbase $\mathcal{F}$ in a set Y and C, D subsets of Y:

   (i) If C contains a cofinal set then C is cofinal.

   (ii) If C contains a terminal set then C is terminal.

   (iii) If C and D are terminal then so is $C \cap D$.

   (iv) If $C \cup D$ is terminal then either C is terminal or D is cofinal.

2. Prove that a set $\mathcal{E}$ of subsets of a nonempty set Y can be extended to a filterbase if and only if $\mathcal{E}$ has the finite intersection property: Every finite subset of $\mathcal{E}$ has nonempty intersection.

3. For $\mathcal{F}$ a filterbase in $[-\infty, \infty]$ with lower limit p and upper limit q prove:

   (i) $[p, q]$ is the intersection of all terminal, closed intervals in $[-\infty, \infty]$.

   (ii) Ignoring the indeterminate case $p = q = \pm\infty$, $q - p = \inf_{E \in \mathcal{F}}(\operatorname{diam} E)$ where $\operatorname{diam} E = \sup E - \inf E$.

   (iii) p and q are both finite if and only if some bounded subset of $\mathbb{R}$ belongs to $\mathcal{F}$.

   (iv) (The Cauchy Criterion.) $\lim \mathcal{F}$ exists and is finite if and only if $\mathbb{R}$ is terminal and given $\varepsilon > 0$ there is some member E of $\mathcal{F}$ such that $\operatorname{diam} E < \varepsilon$.

4. Using (3), (4), (5) show that if $\lim \phi = c$ then (ignoring indeterminate forms):

   (i) $\underline{\lim}(\phi + \psi) = c + \underline{\lim}\psi$
   and
   (ii) $\overline{\lim}(\phi + \psi) = c + \overline{\lim}\psi$.

§11.2 Metric Spaces.

A **metric** on a set X is a function δ on X^2 such that for all x, y, z in X:

(1) $\delta(x, y) \geq 0$ with equality if and only if $x = y$,

(2) $$\delta(x, y) = \delta(y, x),$$

(3) $$\delta(x, z) \leq \delta(x, y) + \delta(y, z).$$

A metric δ defines a **diameter** $\delta[A]$ for each subset A of X given by

(4)
$$\begin{cases} \delta[A] = \sup_{x,y \in A} \delta(x,y) \text{ if } A \neq \emptyset \\ \text{and} \\ \delta[\emptyset] = 0. \end{cases}$$

So $0 \leq \delta[A] \leq \infty$. A is **bounded** if $\delta[A] < \infty$. Clearly

(5)
$$\delta[A] \leq \delta[B] \text{ if } A \subseteq B.$$

Also

(6) $\delta[A] = 0$ if and only if A has at most one point.

Since $\delta[A] = \delta(x,y)$ for $A = \{x,y\}$ diameter is an extension of the metric. δ can also be extended to give the distance $\delta(A,B)$ between subsets A and B of X by defining

(7)
$$\begin{cases} \delta(A,B) = \inf_{a \in A, b \in B} \delta(a,b) \\ \text{with} \\ \delta(A,B) = \infty \text{ if at least one of the sets } A,B \text{ is empty.} \end{cases}$$

This extends δ since $\delta(A,B) = \delta(a,b)$ for $A = \{a\}, B = \{b\}$. Clearly for all subsets A, B, C of X

(8)
$$\delta(A \cup B, C) = \delta(A,C) \wedge \delta(B,C).$$

The triangle inequality (3) has two extensions,

(9)
$$\delta(A,C) \leq \delta(A,B) + \delta[B] + \delta(B,C)$$

and

(10)
$$\delta[A \cup B] \leq \delta[A] + \delta(A,B) + \delta[B]$$

for all subsets A, B, C of X. For $B = \{y\}$ (9) reduces to

$$(11) \qquad \delta(A, C) \leq \delta(A, y) + \delta(y, C).$$

For $A = \{x\}$ and $C = \{z\}$ (11) reduces to (3).

A metric δ on X defines an equivalence relation on sequences in X,

$$(12) \qquad \langle x_n \rangle \sim \langle y_n \rangle \text{ whenever } \delta(x_n, y_n) \to 0.$$

For the special case $y_n \equiv y$ we have sequential convergence defined by

$$(13) \qquad x_n \to y \text{ whenever } \delta(x_n, y) \to 0.$$

The **uniform invariants** of (X, δ) are those properties which can be defined in terms of sequential equivalence (12). These include the **topological invariants**, properties definable in terms of sequential convergence (13). The Cauchy criterion is a uniform invariant. A sequence $\langle x_n \rangle$ in X is **Cauchy** if its tail diameters $\delta[x_n, x_{n+1}, \cdots] \to 0$ as $n \to \infty$. In terms of (12) $\langle x_n \rangle$ is Cauchy if it is equivalent to all its subsequences. X is **complete** if every Cauchy sequence in X converges to a limit in X.

For metric spaces (X, δ) and (X', δ') a mapping $f : X \to X'$ is **uniformly continuous** if it preserves sequential equivalence: $\delta(x_n, y_n) \to 0$ in X implies $\delta'(f x_n, f y_n) \to 0$ in X'. f is **continuous at** x in X if $x_n \to x$ in X implies $f x_n \to f x$ in X'. f is **continuous** if it is continuous at every x in X. For $X' = X$ the metric δ' is uniformly continuous with respect to δ if the identity mapping from X to X' is uniformly continuous. That is, $\delta(x_n, y_n) \to 0$ implies $\delta'(x_n, y_n) \to 0$. δ and δ' are **uniformly equivalent** if each of these metrics is uniformly continuous with respect to the other. δ' is **continuous** with respect to δ if the identity mapping from X to X' is continuous.

δ and δ' are **topologically equivalent** if each is continuous with respect to the other.

Alternatively the uniform invariants of (X, δ) are those properties whose dependence on δ rests only on the proximity relation defined in terms of (7) by

$$(14) \qquad\qquad \delta(A, B) = 0.$$

For example $f : X \rightarrow X'$ is uniformly continuous if and only if for all subsets A, B of X

$$(15) \qquad \delta(A, B) = 0 \text{ implies } \delta'(fA, fB) = 0.$$

δ and δ' on X are uniformly equivalent if and only if for all subsets A, B of X

$$(16) \qquad \delta(A, B) = 0 \text{ if and only if } \delta'(A, B) = 0.$$

The topological invariants of (X, δ) depend only on the restriction of (14) to the case in which A or B is a single point. For example $f : X \rightarrow X'$ is continuous at x in X if and only if $\delta'(fx, fA) = 0$ for all subsets A of X such that $\delta(x, A) = 0$.

The uniform concept (12) of sequential equivalence can be expressed in terms of (14) since $\delta(x_n, y_n) \rightarrow 0$ if and only if $\delta(A, B) = 0$ for all $A = \{x_n : n \in \mathbb{M}\}, B = \{y_n : n \in \mathbb{M}\}$ with $\mathbb{M}$ an infinite subset of $\mathbb{N}$.

A subset D of X is **uniformly discrete** if there is some $\varepsilon > 0$ such that $\delta(x, y) > \varepsilon$ for all $x \neq y$ in D. Uniform discreteness is a uniform invariant. In terms of (12) it says that $\langle x_n \rangle \sim \langle y_n \rangle$ in D implies $x_n = y_n$ ultimately as $n \rightarrow \infty$. In terms of (14) it says that for all subsets A, B of $D, \delta(A, B) = 0$ implies A meets B. X is **totally bounded** if every uniformly discrete subset D of X is finite. Equivalently, X is totally bounded if every sequence in X has a Cauchy subsequence. This in turn is equivalent to the condition that every uniformly continuous function on X is bounded, where the metric on $\mathbb{R}$ is $|s - t|$ for all s, t in $\mathbb{R}$.

Some topological invariants of X are the following: The **closure** $\overline{A}$ of a subset A of X is the set of all x in X such that $\delta(x, A) = 0$, the **exterior** $A^{\perp}$ is the set of all x such that $\delta(x, A) > 0$, the **interior** A° is the set of all x such that $\delta(x, A^{\sim}) > 0$ where $A^{\sim}$ is the complement of A in X, and the **boundary** $A^{\bullet}$ is the set of all x such that $\delta(x, A) = \delta(x, A^{\sim}) = 0$. A is **closed** if $A = \overline{A}$, **open** if $A = A^{\circ}$.

Exercises (§11.2).

Let (X, δ) be a metric space and A, B be subsets of X.

1. Prove that $\delta(x, B) \leq \delta(x, A)$ for all x in X if and only if $\delta(y, B) = 0$ for all y in A.

2. Verify that the relation $\delta(x, A) = 0$ satisfies:

   (i) If $\delta(x, A) = 0$ then $A \neq \emptyset$,

   (ii) If $x \in A$ then $\delta(x, A) = 0$,

   (iii) $\delta(x, A \cup B) = 0$ if and only if either $\delta(x, A) = 0$ or $\delta(x, B) = 0$,

   (iv) If $\delta(x, A) = 0$ and $\delta(y, B) = 0$ for all y in A then $\delta(x, B) = 0$.

3. Show that $\delta[\overline{A}] = \delta[A]$ and $\delta(\overline{A}, \overline{B}) = \delta(\overline{A}, B) = \delta(A, B)$.

4. Verify that (12) is an equivalence relation (reflexive, symmetric, transitive).

5. Given nonempty, complete, and totally bounded sets A, B prove:

   (i) $\delta[A] = \delta(x, y)$ for some x, y in A,

   (ii) $\delta(A, B) = \delta(x, y)$ for some x in A and y in B.

6. Given a nonempty metric space (X, δ) let B be the set of all bounded, uniformly continuous functions on X. Choose

x_0 in X and for each x in X define the function f_x on X by $f_x(z) = \delta(x, z) - \delta(x_0, z)$ for all z in X. Prove:

(i) B is a complete metric space under the metric $\delta'(f, g) = \sup_{x \in X} |f(x) - g(x)|$.

(ii) $f_x \in B$ for all x in X. (Show $|f_x(z)| \leq \delta(x_0, x)$ and $|f_x(y) - f_x(z)| \leq 2\delta(y, z)$ for all x, y, z in X.)

(iii) The mapping $\psi : X \to B$ defined by $\psi(x) = f_x$ is an isometry from X to $\psi(X)$. That is,

$$\delta'(f_x, f_y) = \delta(x, y) \text{ for all } x, y \text{ in } X.$$

(iv) The closure $\overline{\psi(X)}$ in B is complete.

7. Let $f : X \to X'$ where (X, δ) and (X', δ') are metric spaces. Define the function δ_f on $X \times X$ by

$$\delta_f(x, y) = \delta'(fx, fy) \text{ for all } x, y \text{ in } X.$$

The mapping f is **nonexpansive** if $\delta_f \leq \delta$. In some situations we can gain this condition by an admissible change of metric on X. Let $\delta^* = \delta + \delta_f$. Verify the following:
(i) δ^* is a metric on X such that $\delta_f \leq \delta^*$,

(ii) If f is continuous then δ^* is topologically equivalent to δ on X,

(iii) If f is uniformly continuous then δ^* is uniformly equivalent to δ on X.

§11.3 Norms and Inner Products.

A **norm** on a linear space X is a function $\| \cdot \|$ on X such that for all x, y in X and all real c:

(1) $\|x\| \geq 0$ with equality only for $x = 0$,

(2) $\|cx\| = |c|\|x\|,$

and

(3) $\|x + y\| \le \|x\| + \|y\|.$

A norm induces a metric δ on X defined by

(4) $\delta(x, y) = \|x - y\|.$

Such a metric has the characteristic properties

(5) $\delta(x + z, y + z) = \delta(x, y)$

and

(6) $\delta(cx, cy) = |c|\delta(x, y)$

for all x, y, z in X and real c. Given any metric δ on a linear space X satisfying (5) and (6) the definition

(7) $\|x\| = \delta(x, 0)$

yields a unique norm on X satisfying (4). X is a **Banach space** if it is complete under the norm-induced metric (4).

   Completeness of a normed linear space X can be formulated in terms of series convergence. X is complete if given $x_1, x_2, \cdots$ in X such that $\sum_{i=1}^{\infty} \|x_i\| < \infty$ there exists y in X such that $y = x_1 + x_2 + \cdots$, that is $\|(x_1 + \cdots + x_n) - y\| \to 0$ as $n \to \infty$.

   A **Hilbert space** is a Banach space in which the norm satisfies the parallelogram law

(8) $\|x + y\|^2 + \|x - y\|^2 = 2\|x\|^2 + 2\|y\|^2$

for all x, y in X. Such a norm induces a (real-valued) inner product

(9) $x \cdot y = \frac{1}{4}(\|x + y\|^2 - \|x - y\|^2)$

for all x, y in X. The characteristic properties of an inner product are as follows: For all x, y, z in X and all c in $\mathbb{R}$,

(10) $\qquad x \cdot y$ is a real number,

(11) $\qquad x \cdot (y + z) = x \cdot y + x \cdot z,$

(12) $\qquad x \cdot y = y \cdot x,$

(13) $\qquad x \cdot (cy) = c(x \cdot y),$

(14) $\qquad x \cdot x \geq 0$ with equality only for $x = 0$.

($x \cdot x$ is often expressed conveniently as x^2.)

Any inner product satisfying (10),$\cdots$,(14) on a linear space X induces a norm on X,

(15) $\qquad \|x\| = (x \cdot x)^{1/2}$ for all x in X

for which (8) and (9) hold. Conversely, (8) and (9) imply (15).

In any Hilbert space X the Cauchy-Schwarz inequality

(16) $\qquad |x \cdot y| \leq \|x\|\|y\|$

holds for all x, y in X with equality only if either $x = cy$ or $y = cx$ for some c in $\mathbb{R}$.

The finite-dimensional Hilbert spaces are just the Euclidean spaces $\mathbb{R}^n$ with norm

(17) $\qquad \|x\| = (x_1^2 + \cdots + x_n^2)^{1/2},$

inner product

(18) $\qquad x \cdot y = x_1 y_1 + \cdots + x_n y_n,$

and metric

(19) $\delta(x, y) = [(x_1 - y_1)^2 + \cdots + (x_n - y_n)^2]^{1/2}$

for $x = (x_1, \cdots, x_n), y = (y_1, \cdots, y_n)$.

On any finite-dimensional linear space X all norms are equivalent in the sense that given norms $\| \cdot \|$ and $\| \cdot \|'$ on X there exists $c \geq 1$ in $\mathbb{R}$ such that $1/c\|x\| \leq \|x\|' \leq c\|x\|$.

In a linear space X let $\widehat{x, y}$ be the line segment with endpoints x, y consisting of all $(1 - t)x + ty$ with $0 \leq t \leq 1$. (In the degenerate case $x = y$ the segment is reduced to a single point.) A subset C of X is **convex** if it contains $\widehat{x, y}$ for all x, y in C. A point z in a convex set C is an **extreme point** of C if every segment in C which contains z has z as an endpoint. The **convex hull** of a subset A of X is the union of all segments whose endpoints belong to A. It is the smallest convex subset of X containing A.

In a normed linear space X the **convex closure** $\widehat{A}$ of a subset A of X is the intersection of all closed, convex subsets of X containing A. $\widehat{A}$ is a closed, convex set and is the smallest such set containing A.

Exercises (§11.3).

1. Given a metric δ satisfying (5) and (6) on a linear space X show that (7) yields a norm satisfying (1) — (4).

2. If $\langle x_n \rangle$ is a sequence in a normed linear space X such that $\sum_{n=1}^{\infty} \|x_n\| < \infty$ then the sequence $\langle y_n \rangle$ of partial sums $y_n = x_1 + \cdots + x_n$ is Cauchy.

3. If $\langle y_n \rangle$ is a Cauchy sequence in a normed linear space X there exists a sequence $\langle x_i \rangle$ in X such that $\sum_{i=1}^{\infty} \|x_i\| < \infty$ and $y_{n_i} = x_1 + \cdots + x_i$ for some subsequence $\langle y_{n_i} \rangle$ of $\langle y_n \rangle$.

4. Show that the convex hull of a subset A of a linear space X consists of all $t_1 x_1 + \cdots + t_k x_k$ where $\{x_1, \cdots, x_k\}$ is a finite subset of A, $t_i \geq 0$ for $i = 1, \cdots, k$, and $t_1 + \cdots + t_k = 1$.

5. Show that in a normed linear space $\widehat{x,y}$ is the convex closure of the set $\{x, y\}$.

6. In a normed linear space X let B be the set of all x in X such that $\|x\| < 1$, S the set such that $\|x\| = 1$. Show that $\bar{B} = \widehat{B} = \widehat{S} = B \cup S$.

7. Show that every nonempty, convex subset C of $\mathbb{R}$ is some type of interval, $(p, q) \subseteq C \subseteq [p, q]$ for $p = \inf C, q = \sup C$.

§11.4 Topological Spaces.

A **topological space** is a set X with a relation x **clings to** A between points x and subsets A of X such that the following four conditions hold:

(i) No point x clings to $\emptyset$,

(ii) Every point x in a subset A of X clings to A,

(iii) x clings to $A \cup B$ if and only if x clings to either A or B,

(iv) If x clings to A and every point in A clings to B then x clings to B.

A metric space (X, δ) is a topological space with x clings to A defined to be $\delta(x, A) = 0$. (See Ex. 2, §11.2)

In any topological space X a point x is **exterior to** a subset A if x does not cling to A; x is **interior to** A if x is exterior to the complement $A^\sim$ of A in X; x is a **boundary point** of A if x clings to both A and $A^\sim$. The **closure** $\overline{A}$ (or A^-) of A is the set of all points in X which cling to A. The **exterior** $A^\perp$ of A is the set of all exterior points of A. That is, $A^\perp = \overline{A}^\sim$. The **interior** A° of A is the set of all points interior to A. The **boundary** $A^\bullet$ is the set of all boundary points of A. In general $A^\circ \subseteq A \subseteq \overline{A}$. A is **open** if $A = A^\circ$, **closed** if $A = \overline{A}$. Since $A^\bullet = \overline{A} - A^\circ$, A is open if it contains none of its boundary points, closed if it contains all of them. A is a **neighborhood** of x if $x \in A^\circ$.

A is **connected** if given $B \cup C = A$ with B, C nonempty
there is some point x in A that clings to both B and C. A
component of a nonempty topological space X is a maximal
connected subset D of X. The components decompose X into
a disjoint union of nonempty, closed, connected subsets.

Each subset E of X is a topological subspace of X with the
relation "x clings to A" restricted to points and subsets of E.

A mapping $f : X \to Y$ between topological spaces X, Y
is **continuous at a point** x in X if $f(x)$ clings to $f(A)$ in
Y for all subsets A of X such that x clings to A in X. f is
continuous on X if it is continuous at every point in X, that
is, if $f\overline{A} \subseteq \overline{fA}$ for all subsets A of X. This is consistent with
the special case of metric spaces in §11.2.

A point x in X is a **cluster point** of a filterbase $\mathcal{F}$ in X if x
clings to every member of $\mathcal{F}$, that is, if every neighborhood of
x is cofinal for $\mathcal{F}$. $\mathcal{F}$ **converges** to x if every neighborhood of
x is terminal for $\mathcal{F}$. In general $\mathcal{F}$ may converge to more than
one point. But this cannot happen if X satisfies the Haus-
dorff separation axiom: Given $x \neq y$ in X there exist disjoint
neighborhoods A of x and B of y. Metric spaces are Hausdorff
spaces since they have even stronger separation properties.

X is **compact** if every filterbase in X has a cluster point in
X. A continuous image fX of a compact space X is compact.
The same holds with "compact" replaced by "connected". A
metric space is compact if and only if it is complete and totally
bounded. In $\mathbb{R}^n$ (under any norm) the totally bounded subsets
are just the bounded subsets. Thus, since a closed subspace
of a complete metric space is complete, a subspace of $\mathbb{R}^n$ is
compact if and only if it is closed in $\mathbb{R}^n$ and bounded.

A subset of $\mathbb{R}$ is connected if and only if it is an interval (of
some type).

For each λ in an indexing set Λ let $f_\lambda : X \to X_\lambda$ where
X_λ is a topological space. Then there is a largest relation "x
clings to A" in X satisfying (i) – (iv) for which each f_λ is

continuous. Explicitly, let x cling to A in X whenever each finite covering $\{A_1, \cdots, A_k\}$ of A in X has some member A_i such that $f_\lambda(x)$ clings to $f_\lambda(A_i)$ for all λ in Λ. For a cartesian product $X = \Pi_{\lambda \in \Lambda} X_\lambda$ of topological spaces X_λ the **product topology** in X is induced by the projections $f_\lambda(x) = x_\lambda$ for $x = \Pi_{\lambda \in \Lambda} x_\lambda$. A product of compact spaces is compact; a product of connected spaces is connected. In $\mathbb{R}^n$ the norm topology is the same as the product topology; convergence is coordinatewise convergence.

The closure operator in a topological space X has the characteristic properties:

$(\bar{i})$ $\overline{\emptyset} = \emptyset$,

$(\overline{ii})$ $A \subseteq \overline{A}$,

$(\overline{iii})$ $\overline{A \cup B} = \overline{A} \cup \overline{B}$,

$(\overline{iv})$ $\overline{\overline{B}} = \overline{B}$

for all subsets A, B of X. $(\bar{i}), (\overline{ii}), (\overline{iii})$ are reformulations of $(i), (ii), (iii)$ respectively. $(\overline{iv})$ comes from (iv) and (iii). $((iv)$ just says that $A \subseteq \overline{B}$ implies $\overline{A} \subseteq \overline{B}$.)

Since $A^\perp$ is the complement of $\overline{A}$ the characteristic properties of the exterior operator $\perp$ can be derived from those of the closure operator:

$(i^\perp)$ $\emptyset^\perp = X$,

$(ii^\perp)$ $A^\perp \cap A = \emptyset$,

$(iii^\perp)$ $(A \cup B)^\perp = A^\perp \cap B^\perp$,

$(iv^\perp)$ $A^\perp \subseteq B^\sim$ implies $A^\perp \subseteq B^\perp$
for all subsets A, B of X.

Since $A^\circ = A^{\sim\perp}$ the characteristic properties of the interior operator are:

(i°) $X^\circ = X$,

$(ii°)$ $A° \subseteq A$,

$(iii°)$ $(A \cap B)° = A° \cap B°$,

$(iv°)$ $A°° = A°$

for all subsets A, B of X.

Order is preserved by the closure and interior operators, reversed by the exterior operator: $A \subseteq B$ implies $\overline{A} \subseteq \overline{B}$, $A° \subseteq B°$, and $B^{\perp} \subseteq A^{\perp}$.

For any subsets D, E of X $D \subseteq (E^{\perp} \cap D) \cup \overline{E} \subseteq \overline{E^{\perp} \cap D} \cup \overline{E}$ since $E^{\perp}$ and $\overline{E}$ are complementary subsets of X. Hence $\overline{D} \subseteq \overline{E^{\perp} \cap D} \cup \overline{E}$ which implies

(1) $$E^{\perp} \cap \overline{D} \subseteq \overline{E^{\perp} \cap D}.$$

For $E = A^{\sim}$ (1) takes the form

(2) $$A° \cap \overline{D} \subseteq \overline{A° \cap D}.$$

A subset A of X is **nowhere dense** in X if $A^{\perp\perp} = \emptyset$. The Baire Catagory Theorem states that no complete metric space is a countable union of nowhere dense subsets.

Exercises (§11.4).

For subsets A, B of a topological space X prove:

1. $A \cap B$ is open if A and B are open.

2. Every union of open sets is open.

3. $A \cup B$ is closed if A and B are closed.

4. Every intersection of closed sets is closed.

5. $A^{\perp \sim \perp} = A^{\perp} \subseteq A^{\perp\perp\perp}$.

6. $\overline{A}^{\perp} = A^{\perp} = A^{\perp°}$.

7. $A^{\perp\perp\perp} \cap B^{\perp\perp\perp} \cap (A \cup B)^{\perp\perp} = \emptyset$.

8. $A^{\perp\perp\perp\perp} = A^{\perp\perp}$.

9. $A^{\perp\perp} = \emptyset$ if and only if A lies in the boundary of some open set.

10. $A^{\perp\perp} = \overline{A}^{\circ}$.

§11.5 Regular Closed Sets.

A subset A of a topological space X is **regular open** if it is the interior of a closed set. Such a set must in particular be the interior of its closure,

$$(1) \qquad A = \overline{A}^{\circ} = A^{\perp\perp}.$$

A subset B of X is **regular closed** if it is the closure of an open set. Such a set must in particular be the closure of its interior,

$$(2.) \qquad B = \overline{B^{\circ}}$$

So B is regular closed if and only if its complement $A = B^{\sim}$ is regular open.

The **regular closure** $A^{\star}$ of a subset A of X is the closed set defined by

$$(3) \qquad A^{\star} = \overline{A^{\circ}}.$$

So A is regular closed if and only if $A = A^{\star}$. Although regular closure is not a topological closure the two closures have some common properties:

$$(4) \qquad \emptyset^{\star} = \emptyset \text{ and } X^{\star} = X,$$

$$(5) \qquad \text{If } A \subseteq B \text{ then } A^{\star} \subseteq B^{\star},$$

and

(6) $$A^{\star\star} = A^{\star}.$$

The closure property $A \subseteq A^{\star}$ may fail if A is not open. But in general we do have

(7) $$A^{\circ\star} = A^{\star} = A^{\star-}$$

and

(8) $$A^{\circ} \subseteq A^{\star} \subseteq \overline{A}.$$

By (6) $A^{\star}$ is regular closed for every subset A of X.

For all subsets A, B of X

(9) $$(A \cap B)^{\star} = \overline{(A \cap B)^{\circ}} = \overline{A^{\circ} \cap B^{\circ}}.$$

Now if $B = B^{\star}$ then $A^{\circ} \cap B = A^{\circ} \cap \overline{B^{\circ}} \subseteq \overline{A^{\circ} \cap B^{\circ}}$ by (2) in §11.4. So $\overline{A^{\circ} \cap B} \subseteq \overline{A^{\circ} \cap B^{\circ}}$. To reverse this we have $\overline{A^{\circ} \cap B^{\circ}} \subseteq \overline{A^{\circ} \cap B}$ since both the interior and closure operators preserve inclusion. Therefore,

(10) $$(A \cap B)^{\star} = \overline{A^{\circ} \cap B} \text{ if } B = B^{\star}.$$

If both A, B are regular closed then $(A \cap B)^{\star} = \overline{A^{\circ} \cap B^{\circ}} \subseteq \overline{A^{\circ}} = A$ and similarly $(A \cap B)^{\star} \subseteq B$. If C is any regular closed set such that $C \subseteq A \cap B$ then $C = C^{\star} \subseteq (A \cap B)^{\star}$ by (5). In summary

(11) $(A \cap B)^{\star}$ is the largest regular closed set contained in the regular closed sets A and B.

For all closed A, B in X (8) gives $(A \cup B)^{\star} \subseteq \overline{A \cup B} = \overline{A} \cup \overline{B} = A \cup B$. By (5) $A^{\star} \cup B^{\star} \subseteq (A \cup B)^{\star}$. Therefore

(12) $$(A \cup B)^{\star} = A \cup B \text{ if } A = A^{\star} \text{ and } B = B^{\star}.$$

For regular closed A define

(13) $$A' = \overline{A^{\sim}} \text{ for } A = A^{\star}.$$

A' is regular closed since $A^{\sim}$ is open. Since $A' \supseteq A^{\sim}$

(14) $$A \cup A' = X.$$

For the intersection of A with $A', (A \cap A')^{\circ} = A^{\circ} \cap A'^{\circ} \subseteq A^{\circ} \cap A' = A^{\circ} \cap \overline{A^{\sim}} \subseteq \overline{A^{\circ} \cap A^{\sim}} = \overline{\emptyset} = \emptyset$. Taking closures we get

(15) $$(A \cap A')^{\star} = \emptyset.$$

For A, B, C regular closed (10) and (12) give

$$[A \cap (B \cup C)]^{\star} = \overline{A^{\circ} \cap (B \cup C)} = \overline{(A^{\circ} \cap B) \cup (A^{\circ} \cap C)} =$$

$$\overline{A^{\circ} \cap B} \cup \overline{A^{\circ} \cap C} = (A \cap B)^{\star} \cup (A \cap C)^{\star}.$$

That is, for regular closed A, B, C

(16) $$[A \cap (B \cup C)]^{\star} = (A \cap B)^{\star} \cup (A \cap C)^{\star}.$$

We now have the following theorem.

THEOREM 1. *In any topological space $X \neq \emptyset$ the regular closed subsets ordered by containment form a Boolean algebra $\mathcal{R}$ with $\emptyset$ the zero and X the unit. The greatest lower bound $A \wedge B$ of A and B is given by*

(17) $$A \wedge B = (A \cap B)^{\star} = \overline{A^{\circ} \cap B^{\circ}} = \overline{A^{\circ} \cap B} = \overline{A \cap B^{\circ}}.$$

The least upper bound $A \vee B$ is the union

(18) $$A \vee B = A \cup B.$$

Complementation in $\mathcal{R}$ is given by

(19) $$A' = \overline{A^\sim} = A^{\sim\star}$$

which has the characteristic properties

(20) $$A' \wedge A = \emptyset \text{ and } A' \vee A = X.$$

PROOF. (17) is given by (3), (10), (11). (18) follows from (12). (19) and (20) follow from (13), (14), (15), (17). So $\mathcal{R}$ is a complemented lattice. To show that $\mathcal{R}$ is a Boolean algebra we need only verify the distributive law

(21) $$A \wedge (B \vee C) = (A \wedge B) \vee (A \wedge C).$$

Indeed, (21) follows from (16), (17), (18). □

(The regular open subsets of X form a Boolean algebra $\mathcal{R}^\circ$ with $A^\perp$ the complement of A, $A \cap B$ the meet of A and B, and $(A \cup B)^{\perp\perp}$ their join [9]. $\mathcal{R}^\circ$ is isomorphic to $\mathcal{R}$. Each of these Boolean algebras represents the homomorph of the Boolean algebra of all subsets of X modulo the nowhere dense subsets of X.)

Exercises (§11.5).

For A, B, C subsets of a topological space X prove:

1. $(A^\star \cap B^\star) \cup (A^{\star\sim} \cap B)^\star = B^\star$.

2. $[A \cap (B^\star \cup C^\star)]^\star = (A \cap B)^\star \cup (A \cap C)^\star$.

3. $(A \cap B^\star)^\star = (A \cap B)^\star$.

4. $(A \cap A^\star)^\star = A^\star$.

5. (Regular Open Sets.) The following are equivalent:

   (i) A is the interior of its closure.

   (ii) A is the interior of a closed set.

(iii) A is the interior of a regular closed set.

(iv) A is the complement of a regular closed set.

6. (Regular Closed Sets.) The following are equivalent:

(i) B is the closure of its interior.

(ii) B is the closure of an open set.

(iii) B is the closure of a regular open set.

(iv) B is the complement of a regular open set.

7. If $f : X \to Y$ is a continuous open mapping of X into a topological space Y then $f(A^\star) \subseteq (fA)^\star$. ($f$ open means $f(B)$ is open in Y for every open subset B of X.)

§11.6 Riesz Spaces.

A **Riesz space** $\mathbb{Y}$ is a real linear space satisfying (i), (ii), and (iii):

(i) $\mathbb{Y}$ is partially ordered. That is, there is a binary relation $x \leq y$ (equivalently, $y \geq x$) on $\mathbb{Y}$ which is reflexive ($x \leq x$), transitive ($x \leq y, y \leq z$ imply $x \leq z$), and antisymmetric ($x \leq y, y \leq x$ imply $x = y$).

(ii) For all x, y, z in $\mathbb{Y}, x \leq y$ implies $x + z \leq y + z, cx \leq cy$ for all $c \geq 0$ in $\mathbb{R}$, and $-y \leq -x$.

(iii) $\mathbb{Y}$ is a lattice. That is, given x, y in $\mathbb{Y}$ there exist a greatest lower bound $x \wedge y$ and least upper bound $x \vee y$ in $\mathbb{Y}$. Specifically, $x \wedge y \leq x, x \wedge y \leq y$, and $z \leq x \wedge y$ for all z in $\mathbb{Y}$ such that $z \leq x, z \leq y$. Similarly $x \vee y \geq x, x \vee y \geq y$, and $z \geq x \vee y$ for all z in $\mathbb{Y}$ such that $z \geq x, z \geq y$,
Some consequences of (i), (ii), (iii) are the identities

(1) $x \wedge (y \vee z) = (x \wedge y) \vee (x \wedge z)$,

(2) $x \vee (y \wedge z) = (x \vee y) \wedge (x \vee z)$,
and

(3) $x + y = x \vee y + x \wedge y$.

For all x in $\mathbb{Y}$ we define

(4) $|x| = x \vee (-x), x^+ = x \vee 0, x^- = -(x \wedge 0) = (-x)^+$.

Then $|x| \geq 0, x^+ \geq 0$, and $x^- \geq 0$. For $y = 0$ (3) gives

(5) $x = x^+ - x^-$

with

(6) $x^+ \wedge x^- = 0$ and $x^+ + x^- = x^+ \vee x^- = |x|$.

We also have the identities (7) $(x-y)^+ = x \vee y - y = x - x \wedge y$,

(8) $(x - y)^- = x \vee y - x = y - x \wedge y$,

and

(9) $x^+ = \frac{1}{2}(|x| + x), x^- = \frac{1}{2}(|x| - x)$.

A **Riesz ideal** in a Riesz space $\mathbb{Y}$ is a linear subspace $\mathbb{Z}$ of $\mathbb{Y}$ that is **solid**: If $y \in \mathbb{Y}, z \in \mathbb{Z}$ and $|y| \leq |z|$ then $y \in \mathbb{Z}$. Having these properties $\mathbb{Z}$ induces an equivalence relation $x \sim y$ on $\mathbb{Y}$ defined by $x - y \in \mathbb{Z}$. For each y in $\mathbb{Y}$ let $[y]$ be the set of all x in $\mathbb{Y}$ such that $x \sim y$. The set $\mathbb{Y}/\mathbb{Z}$ of these equivalence classes is a Riesz space under the operations defined by $[x] + [y] = [x + y], c[x] = [cx]$ for all real c, $[x] \leq [y]$ if $x \wedge y \sim x$. So $[x] \wedge [y] = [x \wedge y], [x] \vee [y] = [x \vee y], |[x]| = [|x|], [x]^+ = [x^+]$ and $[x]^- = [x^-]$. These relations are effective, that is, do not depend on the choice of particular members of the equivalence classes. For example, $[x]+[y]$ is well defined since $x' \sim x, y' \sim y$ imply $x' + y' \sim x + y$. That is, $[x'] = [x]$ and $[y'] = [y]$ imply $[x' + y'] = [x + y]$. The zero element in $\mathbb{Y}/\mathbb{Z}$ is $[0] = \mathbb{Z}$. The mapping $h : \mathbb{Y} \to \mathbb{Y}/\mathbb{Z}$ defined by $h(x) = [x]$ is a **Riesz homomorphism**. That is, $h(x) + h(y) = h(x + y), h(cx) = ch(x), h(x) \wedge h(y) = h(x \wedge y)$. So $h(x) \vee h(y) = h(x \vee y), |h(x)| = h(|x|), (h(x))^+ = h(x^+)$, and $(h(x))^- = h(x^-)$.

Let ν be a function on a Riesz space $\mathbb{Y}$ such that for all x, y in $\mathbb{Y}$ the following conditions hold:

(10) $0 \leq \nu(x) \leq \infty$,

(11) $\nu(0) = 0$,

(12) $\nu(cx) = |c|\nu(x)$ for all c in $\mathbb{R}$,

(13) $\nu(x + y) \le \nu(x) + \nu(y)$,

(14) $\nu(x) \le \nu(y)$ if $|x| \le |y|$.

By (14) $\nu(x) = \nu(y)$ if $|x| = |y|$. For $y = |x|$ this gives $\nu(x) = \nu(|x|)$. The function ν defines sequential convergence $x_n \to x$ in $\mathbb{Y}$ by $\nu(x_n - x) \to 0$ in $[0, \infty]$. So $\mathbb{Y}$ is a topological space in which x clings to A whenever $x_n \to x$ for some sequence $\langle x_n \rangle$ in A.

If (10) is strengthened to

(15) $0 < \nu(x) < \infty$ for all $x \ne 0$

then ν is a norm on $\mathbb{Y}$ and (14) makes ν a **Riesz norm** on $\mathbb{Y}$. If $\mathbb{Y}$ is complete under a Riesz norm ν it is called a **Banach lattice**.

Under (10),$\cdots$,(14) $\mathbb{Z} = \nu^{-1}(0)$ is a Riesz ideal in $\mathbb{Y}$. $0\varepsilon\mathbb{Z}$ by (11). (Without (11) we could have $\nu(x) = \infty$ for all x in $\mathbb{Y}$, thereby making $\mathbb{Z}$ empty.) $\mathbb{Z}$ is closed under addition by (13) and (10), under scalar multiplication by (12). So $\mathbb{Z}$ is a linear subspace of $\mathbb{Y}$. Solidity follows from (14) and (10). So $\mathbb{Y}/\mathbb{Z}$ is a Riesz space with its operations transferred homomorphically from $\mathbb{Y}$. Moreover we can transfer ν to $\mathbb{Y}/\mathbb{Z}$ by defining $\nu[x] = \nu(x)$ for all x in $\mathbb{Y}$, hence for all $[x]$ in $\mathbb{Y}/\mathbb{Z}$. This is effective since $[x] = [y]$ means $\nu(x - y) = 0$, so $\nu(x) \le \nu(y) + \nu(x - y) = \nu(y)$, hence $\nu(x) = \nu(y)$ by the symmetric role of x, y. On $\mathbb{Y}/\mathbb{Z}$ ν retains all of the properties (10), $\cdots$, (14). Moreover, $\nu[x] = 0$ if and only if $[x] = [0]$. Since the set of all x in $\mathbb{Y}$ such that $\nu(x) < \infty$ is a Riesz subspace of $\mathbb{Y}$, the set of all $[x]$ in $\mathbb{Y}/\mathbb{Z}$ such that $\nu[x] < \infty$ is a Riesz subspace of $\mathbb{Y}/\mathbb{Z}$ with ν as a Riesz norm.

Exercises (§11.6).

1. Show that for all x, y, z in a Riesz space $\mathbb{Y}$:

(i) $|x \vee z - y \vee z| \leq |x - y|$,

(ii) $|x \wedge z - y \wedge z| \leq |x - y|$,

(iii) $||x| - |y|| \leq |x - y|$,

(iv) $|x^+ - y^+| \leq |x - y|$,

(v) $|x^- - y^-| \leq |x - y|$.

2. Let ν be a function on a Riesz space $\mathbb{Y}$ satisfying (10),$\cdots$,(14). Let $x_n \to x$ in $\mathbb{Y}$ under the convergence $\nu(x_n - x) \to 0$ induced by ν. Prove:

(i) $|x_n| \to |x|$,

(ii) $x_n^+ \to x^+$,

(iii) $x_n^- \to x^-$,

(iv) If $x_n \leq x_{n+1}$ for all n then $x = \bigvee_{n=1}^{\infty} x_n$.

§11.7 The Inclusion-Exclusion Formula.

A function S on a boolean algebra $\mathcal{B} = \{a, b, \cdots\}$ is called **additive** if

(1) $\qquad S(a \vee b) = S(a) + S(b)$ for $a \wedge b = 0$.

(1) implies

(2) $\qquad S(a \vee b) = S(a) + S(b) - S(a \wedge b)$

for all a, b in $\mathcal{B}$. (2) is the case $n = 2$ of the **inclusion-exclusion** formula: Given S additive on a boolean algebra $\mathcal{B}$ and a finite sequence $a_1, \cdots, a_n$ in $\mathcal{B}$,

(3) $\qquad S(a_1 \vee \cdots \vee a_n) = \sum_{m=1}^{n} (-1)^{m-1} S_m(a_1, \cdots, a_n)$

where

$$(4) \quad S_m(a_1, \cdots, a_n) = \sum_{1 \leq j_1 < \cdots < j_m \leq n} S(a_{j_1} \wedge \cdots \wedge a_{j_m}).$$

(If $m > n$ then the sum in (4) has no summands, hence equals zero.)

We prove (3) by induction. For $n = 1$ (3) is just the trivial identity $S(a_1) = S(a_1)$.

Now let (3) hold for some $n \geq 1$. We must prove

$$(5) \quad S(a_1 \vee \cdots \vee a_{n+1}) = \sum_{m=0}^{n} (-1)^m S_{m+1}(a_1, \cdots, a_{n+1}).$$

For the summand in (5) with $m = 0$ we get

$$(6) \qquad S_1(a_1, \cdots, a_{n+1}) = S_1(a_1, \cdots, a_n) + S(a_{n+1})$$

from (4). For $m > 0$ we have

$$(7) \quad S_{m+1}(a_1, \cdots, a_{n+1}) = S_{m+1}(a_1, \cdots, a_n) + S_m(b_1, \cdots, b_n)$$

where $b_i = a_i \wedge a_{n+1}$ for $i = 1, \cdots, n$. Applying (2) with $a = a_1 \vee \cdots \vee a_n$ and $b = a_{n+1}$ we get

$$(8) \qquad \begin{aligned} S(a_1 \vee \cdots \vee a_{n+1}) = \\ S(a_1 \vee \cdots \vee a_n) + S(a_{n+1}) - S(b_1 \vee \cdots \vee b_n). \end{aligned}$$

Applying the induction hypothesis to the first and last terms on the right-hand side of (8) we get

(9)
$$S(a_1 \vee \cdots \vee a_{n+1}) = S(a_{n+1})+$$

$$\sum_{m=1}^{n}(-1)^{m-1}[S_m(a_1,\cdots,a_n) - S_m(b_1,\cdots,b_n)]$$

$$= S_1(a_1,\cdots,a_{n+1})+$$

$$\sum_{m=1}^{n}(-1)^m[S_{m+1}(a_1,\cdots,a_n) + S_m(b_1,\cdots,b_n)]$$

$$= S_1(a_1,\cdots,a_{n+1}) + \sum_{m=1}^{n}(-1)^m S_{m+1}(a_1,\cdots,a_{n+1})$$

by (6), (7). Finally, (9) gives (5).

Exercises (§11.7).

1. For S any function on a boolean algebra $\mathcal{B}$ prove that (1) holds if and only if $S(0) = 0$ and (2) holds for all, a, b in $\mathcal{B}$.

2. Verify that (6) and (7) follow from definition (4), and that (8) and (9) are valid.

3. Let S be additive on a boolean algebra $\mathcal{B}$. Let $a_1,\cdots,a_n$ be members of $\mathcal{B}$ such that $S(a_{j_1} \wedge \cdots \wedge a_{j_m}) = r^m$ for all $1 \le j_1 < \cdots < j_m \le n$ and some real number r. Prove:

   (i) $S_m(a_1,\cdots,a_n) = \binom{n}{m} r^m$ in (4)

   and

   (ii) $S(a_1 \vee \cdots \vee a_n) = 1 - (1-r)^n$ in (3).

4. Let S be additive on a boolean algebra $\mathcal{B}$ such that S is idempotent, $S(a) = 0$ or 1 for all a in $\mathcal{B}$. Given $a_1,\cdots,a_n$ in $\mathcal{B}$ let k be the number of indices j such that $S(a_j) = 1$. Prove:

(i) $S_m(a_1, \cdots, a_n) = \begin{pmatrix} k \\ m \end{pmatrix}$ by (4)

and

(ii) $S(a_1 \vee \cdots \vee a_n) = sgn\, k$ by (3).

(*Note*: The nonzero, additive idempotents S on $\mathcal{B}$ form the **Stone Space** [9] of $\mathcal{B}$.)

REFERENCES

[1]   S. Banach, *Sur les lignes rectifiables et les surfaces dont l'aire est finie*, Fund. Math. **7** (1925), 225–236.

[2]   R. P. Boas, *Indeterminate forms revisited*, Math. Mag. no. 3 **63** (1990), 155–159.

[3]   F. S. Cater, *Upper and lower generalized Riemann integrals*, Real Analysis Exchange **16** (1990–91), 215–237.

[4]   P. Cousin, *Sur les fonctions de n variables complexes*, (Jahrbuch **26**, 456), Acta Math. **19** (1895), 1–62.

[5]   J. Dieudonné, *Foundations of Modern Analysis*, Academic Press, New York, 1960.

[6]   R. J. Fleissner, *On the product of derivatives*, Fund. Math. **88** (1975), 173–178.

[7]   R. A. Gordon, *The Integrals of Lebesgue, Denjoy, Perron, and Henstock*, Graduate Studies in Math., Amer. Math. Soc. (1994).

[8]   R. A. Gordon, *The use of tagged partitions in elementary real analysis*, Amer. Math. Monthly, no. 2 **105** (1998), 107–117.

[9]   P. R. Halmos, *Lectures on Boolean Algebras*, Van Nostrand, Princeton, 1963.

[10]   D. Hartig, *L'Hôpital's rule via integration*, Amer. Math. Monthly, no. 2, **98** (1991), 156–157.

[11]   R. Henstock, *Definitions of Riemann type of the variational integrals*, Proc. London Math. Soc., no. 3 **11** (1961), 402–418.

[12]   R. Henstock, *A short history of integration theory*, Southeast Asian Bull. Math., no. 2, **12** (1988), 75–95.

[13]   R. Henstock, *Lectures on the Theory of Integration*, World Scientific, Singapore (1988).

347

[14]   R. Henstock, *The General Theory of Integration*, Clarendon Press, Oxford, U.K., 1991.

[15]   X.-C. Hwang, *A discrete L'Hôpital's rule*, College Math. J. **19** (1988), 321–329.

[16]   J. Jarník and J. Kurzweil, *A nonabsolutely convergent integral which admits transformation and can be used for integration on manifolds*, Czech. Math. J. **35** (1985), 116–139.

[17]   W. B. Jurkat and D.J.F. Nonenmacher, *An axiomatic theory of nonabsolutely convergent integrals in* $\mathbb{R}^n$, Fund. Math. **145** (1994), 221–242.

[18]   A. Kolmogorov, *Untersuchen über den Integralbegriff*, Math. Ann. **103** (1930), 654–696.

[19]   J. Kurzweil, *Generalized ordinary differential equations and continuous dependence on a parameter*, Czech. Math. J., no. 82, **7** (1957), 418–446.

[20]   J. Kurzweil, *Nichtabsolut konvergente Integrale*, Teubner, Leipzig, 1980.

[21]   S. Leader, *What is a differential? A new answer from the generalized Riemann integral*, Amer. Math. Monthly **93** (1986), 348–356.

[22]   S. Leader, *A concept of differential based on variational equivalence under generalized Riemann integration*, Real Analysis Exchange **12** (1986-87), 144–175.

[23]   S. Leader, *1-differentials on 1-cells: A further study*, New Integrals, Lecture Notes in Math., vol. 1419, Springer, 1990, pp. 82–96.

[24]   S. Leader, *Variation of f on E and Lebesgue outer measure of fE*, Real Analysis Exchange, no. 2, **16** (1990-91), 508–515.

[25]   S. Leader, *Basic convergence principles for the Kurzweil-Henstock integral*, Real Analysis Exchange **18** (1992-93), 95–114.

[26]   S. Leader, *Uniform Kurzweil-Henstock integrability*, Real Analysis Exchange **19** (1993-94), 173–193.

[27]   S. Leader, *Conversion formulas for the Lebesgue-Stieltjes integral*, Real Analysis Exchange, no. 2, **20** (1994-95), 527–535.

[28]   S. Leader, *Transforming Lebesgue-Stieltjes integrals into Lebesgue integrals*, Real Analysis Exchange, no. 2, **20** (1994-95), 603–616.

[29]   P. Y. Lee, *Lanzhou Lectures on Henstock Integration*, World Scientific, Singapore (1989).

[30]   P. Y. Lee and R. Vyborny, *The Integral - An Easy Approach after Kurzweil and Henstock*, Austral. Math. Soc. Lect. Ser. 14, Cambridge Univ. Press, New York (2000).

[31]   W. A. J. Luxemburg and A. C. Zaanen, *Riesz spaces 1*, North Holland, Amsterdam, London (1971).

[32]   J. Mawhin, *Generalized multiple Perron integrals and the Green-Goursat theorem for differentiable vector fields*, Czech. Math. J. **31** (1981), 614–632.

[33]   R. M. McLeod, *The Generalized Riemann Integral*, Carus Math. Monographs, no. 20, Math. Assoc. Amer., 1980.

[34]   E. J. McShane, *Unified Integration*, Academic Press, New York, 1983.

[35]   W. F. Pfeffer, *The divergence theorem*, Trans. Amer. Math. Soc., no. 2, **295** (1986), 665–685.

[36]   W. F. Pfeffer, *The Gauss-Green theorem*, Real Analysis Exchange **14** (1988-89), 523–527.

[37]   W. F. Pfeffer, *The Riemann Approach to Integration*, Cambridge Univ. Press, London, 1993.

[38]   S. Saks, *Theory of the Integral*, 2-nd revised edition, Dover, New York, 1964.

[39]   W. L. C. Sargent, *On the integrability of products*, J. London Math. Soc. **23** (1948), 28–34.

[40]   T. J. Stieltjes, *Recherches sur les fractions continues*, Annales de la Faculté des Sciences de Toulouse **8** (1894), J.1 - J.122.

[41]   E. C. Titchmarsh, *The Theory of Functions*, (2nd Ed.), Oxford Univ. Press, London, 1939.

INDEX

Abutting cells 9
additive 21, 342
approximating sums 16

Banach indicatrix 267, 272
— lattice 341
— space 328
Boolean algebras/fields 10
Borel sets 130
boundary 326, 331
bounded sets 323

Cauchy extension 6
— integral 2
— - Schwarz inequality 329
— sequence 324
cell 9
— interior 9, 285
— summant 15, 21
  tagged — 11
closed, closure 326, 331
cluster point 332
cofinal sets 319
complete metric space 324
component, connected 332
Cousin's lemma 13
continuity 96, 324, 332
  absolute — 249, 250, 251, 258
  left/right — 181
  uniform — 324
  uniform absolute — 263, 265

convergence 332
convex 330

Damper 92, 99
δ-division 12
δ-fine 12
δ-neighborhood 12
δ-sum (lower/upper) 16, 17
differentiable functions 187
differential 3, 6, 53
  absolutely integrable — 60
  archimedean — 99
  conditionally integrable — 60
  — coefficient 6, 73, 187
  continuous — 96
  — convergence 57
  dampable — 99
  damper-summable — 99
  — equivalence 53
  idempotent — 92
  — norm 57
  — of a function 66
  Stieltjes — 77
  summable 58
  tag-finite — 87
  unit — ω 6, 91
diameter 323
Dini derivatives 61
directed set 321
division 11

Essential
  — boundedness 235
  — continuity 238
  — infimum/supremum 235

— lower/upper bounds 235
— variation 241
exterior 326, 331
extreme point 330

Figure 9
filter 319
  — base 319
  ultra — 319
frequently 319
gauge 11
gradient 187

Hahn decomposition 133
Hilbert space 328

Indicator 10
inner product 187, 328, 329
integrable 18, 54
  uniformly — 33, 36
integral 1-6, 18, 53, 54
interior 9, 285, 326, 331
invariants 324

Lebesgue space $\mathcal{L}_1$ 78
limits (lower/upper) 320

Measurable 129-131, 137, 139
measure 145
  Lebesgue — 132
  outer — 169
metric 322

n-cell 285
  — lower/upper faces 291

— vertex 285
n-differential 125, 286
n-summant 125
neigborhood 331
net 321
nonexpansive mapping 327
norm 327, 328
nowhere dense set 334

Open set 331
orientation summant Q 16, 25, 196
overlapping cells/figures 9

Path 303, 304
  initial/terminal point of — 304
  length of — 304
  parametrized — 303
  polygonal — 308
  rectifiable — 304
  weakly archimedean — 307
product topology 333

Q (orientation summant) 16, 25, 196

Region 307
regular closed set 9, 335
— closure 335
— open set 335
regulated function 145
Riesz homomorphism 340
— ideal 340
— norm 341
— space 339

Sigma algebra 129

σ-essentials limits 223
σ-everywhere 73
σ-null 73, 126
simple arc 305
simple closed curve 306
solid set 55, 340
step function 145
Stone space 345
summant 15
  additive — 21
  cell — 15, 21
  sub/superadditive — 22
  Stieltjes — 16
  tame — 99

Terminal set 319
topological space 339
  — equivalence 325
totally bounded set 325

Ultimate lower/upper bound 320
ultimately 319
uniform equivalence 324
uniformly discrete set 325

Variation
  bounded/unbounded — 68
  total — 68
vertex of n-cell 285